AF675994

Sensors for Fire and Smoke Monitoring

Sensors for Fire and Smoke Monitoring

Editors

Robert S. Allison
Joshua Johnston
Martin Wooster

MDPI • Basel • Beijing • Wuhan • Barcelona • Belgrade • Manchester • Tokyo • Cluj • Tianjin

Editors

Robert S. Allison
Electrical Engineering &
Computer Science
York University
Toronto
Canada

Joshua Johnston
Canadian Forest Service
Sault Ste Marie
Canada

Martin Wooster
Earth Observation Science
King's College London
London
UK

Editorial Office
MDPI
St. Alban-Anlage 66
4052 Basel, Switzerland

This is a reprint of articles from the Special Issue published online in the open access journal *Sensors* (ISSN 1424-8220) (available at: www.mdpi.com/journal/sensors/special_issues/fire_smoke).

For citation purposes, cite each article independently as indicated on the article page online and as indicated below:

LastName, A.A.; LastName, B.B.; LastName, C.C. Article Title. *Journal Name* **Year**, *Volume Number*, Page Range.

ISBN 978-3-0365-1985-2 (Hbk)
ISBN 978-3-0365-1984-5 (PDF)

© 2021 by the authors. Articles in this book are Open Access and distributed under the Creative Commons Attribution (CC BY) license, which allows users to download, copy and build upon published articles, as long as the author and publisher are properly credited, which ensures maximum dissemination and a wider impact of our publications.
The book as a whole is distributed by MDPI under the terms and conditions of the Creative Commons license CC BY-NC-ND.

Contents

sensors

MDPI

Editorial

Sensors for Fire and Smoke Monitoring

Robert S. Allison [1,*], Joshua M. Johnston [2] and Martin J. Wooster [3]

1 Centre for Vision Research, Department of Electrical Engineering and Computer Science, York University, 4700 Keele St., Toronto, ON M3J 1P3, Canada
2 Canadian Forest Service, Great Lakes Forestry Centre, 1219 Queen St. E., Sault Ste. Marie, ON P6A 2E5, Canada; Joshua.Johnston@canada.ca
3 Leverhulme Centre for Wildfires, Environment and Society, Department of Geography, King's College London, Aldwych, London WC2B 4BG, UK; martin.wooster@kcl.ac.uk
* Correspondence: allison@eecs.yorku.ca

Mastery of fire is intimately linked to advances in human civilization, culture and technology. It is also a key component of renewal and regeneration of natural ecosystems. At the same time fire continues to be a threat to lives, property and the environment. Our complex relationship with fire has been the impetus for many innovations in direct and remote sensing of key properties of fires including flame, heat and smoke. This Special Issue on fire and smoke detection and monitoring has curated and collected the latest research on sensors and systems to detect and quantify wildland, structural, and industrial combustion, and the emissions of smoke produced.

Public interest surrounds detecting and monitoring fire and smoke behavior for fire protection and suppression in residential and industrial settings, or for the design and test of safer materials and structures. The time lag required to detect smoke in a building fire is a key factor in the likelihood of harm to occupants. Jang and Hwang [1] carefully controlled smoke velocity and concentration to measure obscuration thresholds for different detectors with different fuels. They found that sensitivity of photo-electric and ionization sensors depended on the type of fuel. The results will be useful to predict detector activation times and determine the required safe egress time.

check for updates

Citation: Allison, R.S.; Johnston, J.M.; Wooster, M.J. Sensors for Fire and Smoke Monitoring. *Sensors* **2021**, *21*, 5402. https://doi.org/10.3390/s21165402

Received: 4 August 2021
Accepted: 5 August 2021
Published: 10 August 2021

Publisher's Note: MDPI stays neutral with regard to jurisdictional claims in published maps and institutional affiliations.

Copyright: © 2021 by the authors. Licensee MDPI, Basel, Switzerland. This article is an open access article distributed under the terms and conditions of the Creative Commons Attribution (CC BY) license (https://creativecommons.org/licenses/by/4.0/).

The majority of the papers concentrated on wildland fires. This is likely due to a combination of a large number of concerted research programs and recent advances, combined with increased public interest. The devastation from the recent wildfires in California, Australia, and elsewhere has captured the attention of the public, government officials and funding agencies and highlighted the need to detect and monitor wildfires to protect and preserve life and property. Image-based remote sensing from the air and space across different wavelength regions has revolutionized the monitoring and management of wildfires (for review see [2]). Barmpoutis and colleagues [3] review the current state-of-the-art in optical remote fire detection systems from terrestrial, airborne or space platforms and present a bibliometric analysis of the literature in addition to their review.

Recent advances in low resource detector technology and the rise in the availability of data from small satellites has seen a surge in innovative detector technologies and in the applications to which such data are put. Dufour and colleagues [4] present an innovative sensor system for measurements of biomass burning. The system is based on a bi-spectral microbolometer that can make radiometric measurements at long- and mid-wavelength IR. Such a system would enable low-resource sensor platforms such as nanosats or small UAVs for wildfire monitoring. Johnston et al. [5] detail the system features and performance requirements for a wildfire monitoring satellite in the context of the WildFireSat mission. The user-centric systems approach illustrates the complex interplay of observational, measurement and precision requirements for a successful and useful system.

Progress depends on the ability to extract meaningful information from the sensor systems, and many papers focused on processing of the sensor data to extract meaningful fire characteristics. At the experimental level, Dickson et al. [6] provided new insights into fire dynamics through careful analysis of direct measurements to better characterize sensible heat flux within a spreading flame-front. Fisher et al. [7] were able to conduct regional analysis of particulate emissions from El Niño exacerbated wildfires in Indonesia using geostationary satellite observations, expanding our understanding of the contribution of different types of fires to these types of harmful emission during severe fire events.

Processing of sensor data and classification of such data with limited resources was considered by several papers. Ifimov and colleagues [8] were confronted with a mid-wave IR airborne image dataset without reliable geolocation data. They were able to use a two-step semi-automatic geocorrection process to generate informational products such as fire radiative power density (FRPD). This technique could be used by other researchers to process airborne infrared imaging data without the need for highly accurate positioning data. Sousa et al. [9] focus on mobile platforms and the integration of off-the-shelf thermal imaging cameras into mobile robot image processing pipelines. Tlig et al. [10] adapt PCA-based image processing techniques to extract relevant fire management characteristics from wildfire imagery. Pan et al. [11] apply convolutional neural networks for forest fire detection and prune the network based on Fourier domain similarity to reduce the resource requirements for edge computing implementations of the technique.

This Special Issue presents a unique cross section of current research priorities related to fire and smoke sensing in the scientific community. Although far from a complete survey, the articles presented demonstrate a growing interest in the use of remote sensing for wildfire research and monitoring, and the pursuit of new technologies to broaden the potential of this field. Notably, there is a strong interest in processing techniques for extracting increasingly complex products from existing data types. However, the use of direct measurement sensors and the careful analysis of the data they produce continues to provide new insights into wildfire dynamics and human safety in domestic settings. It is the view of the guest editors that this is an accurate reflection of the present focus of our field, and that this Special Issue is a suitable introduction to the topic of Sensors for Fire and Smoke Monitoring.

Acknowledgments: The guest editors thank the authors and reviewers for their contributions that made this special issue possible. We would also like to express our gratitude for the constant support, diligent work and attention to detail provided by the Sensors editorial staff.

Conflicts of Interest: The authors declare no conflict of interest.

References

1. Jang, H.Y.; Hwang, C.H. Obscuration Threshold Database Construction of Smoke Detectors for Various Combustibles. *Sensors* **2020**, *20*, 6272. [CrossRef]
2. Allison, R.S.; Johnston, J.M.; Craig, G.; Jennings, S. Airborne optical and thermal remote sensing for wildfire detection and monitoring. *Sensors* **2016**, *16*, 1310. [CrossRef] [PubMed]
3. Barmpoutis, P.; Papaioannou, P.; Dimitropoulos, K.; Grammalidis, N. A Review on Early Forest Fire Detection Systems Using Optical Remote Sensing. *Sensors* **2020**, *20*, 6442. [CrossRef] [PubMed]
4. Dufour, D.; Le Noc, L.; Tremblay, B.; Tremblay, M.N.; Généreux, F.; Terroux, M.; Vachon, C.; Wheatley, M.J.; Johnston, J.M.; Wotton, B.M.; et al. A Bi-Spectral Microbolometer Sensor for Wildfire Measurement. *Sensors* **2021**, *21*, 3690. [CrossRef] [PubMed]
5. Johnston, J.M.; Jackson, N.; McFayden, C.; Ngo Phong, L.; Lawrence, B.; Davignon, D.; Wooster, M.J.; van Mierlo, H.; Thompson, D.K.; Cantin, A.S.; et al. Development of the User Requirements for the Canadian WildFireSat Satellite Mission. *Sensors* **2020**, *20*, 5081. [CrossRef] [PubMed]
6. Dickinson, M.B.; Wold, C.E.; Butler, B.W.; Kremens, R.L.; Jimenez, D.; Sopko, P.; O'Brien, J.J. The Wildland Fire Heat Budget—Using Bi-Directional Probes to Measure Sensible Heat Flux and Energy in Surface Fires. *Sensors* **2021**, *21*, 2135. [CrossRef] [PubMed]
7. Fisher, D.; Wooster, M.J.; Xu, W.; Thomas, G.; Lestari, P. Top-Down Estimation of Particulate Matter Emissions from Extreme Tropical Peatland Fires Using Geostationary Satellite Fire Radiative Power Observations. *Sensors* **2020**, *20*, 7075. [CrossRef] [PubMed]
8. Ifimov, G.; Naprstek, T.; Johnston, J.M.; Arroyo-Mora, J.P.; Leblanc, G.; Lee, M.D. Geocorrection of Airborne Mid-Wave Infrared Imagery for Mapping Wildfires without GPS or IMU. *Sensors* **2021**, *21*, 3047. [CrossRef] [PubMed]

9. Sousa, M.J.; Moutinho, A.; Almeida, M. Thermal Infrared Sensing for Near Real-Time Data-Driven Fire Detection and Monitoring Systems. *Sensors* **2020**, *20*, 6803. [CrossRef] [PubMed]
10. Tlig, L.; Bouchouicha, M.; Tlig, M.; Sayadi, M.; Moreau, E. A Fast Segmentation Method for Fire Forest Images Based on Multiscale Transform and PCA. *Sensors* **2020**, *20*, 6429. [CrossRef] [PubMed]
11. Pan, H.; Badawi, D.; Cetin, A.E. Computationally Efficient Wildfire Detection Method Using a Deep Convolutional Network Pruned via Fourier Analysis. *Sensors* **2020**, *20*, 2891. [CrossRef]

sensors

Article

A Bi-Spectral Microbolometer Sensor for Wildfire Measurement

Denis Dufour [1,*], Loïc Le Noc [1], Bruno Tremblay [1], Mathieu N. Tremblay [1], Francis Généreux [1], Marc Terroux [1], Carl Vachon [1], Melanie J. Wheatley [2], Joshua M. Johnston [3], Mike Wotton [3] and Patrice Topart [1]

1 INO (Institut National d'Optique), 2740 Einstein Street, Québec, QC G1P 4S4, Canada; Loic.Le.Noc@ino.ca (L.L.N.); Bruno.Tremblay@ino.ca (B.T.); MathieuN.Tremblay@ino.ca (M.N.T.); Francis.Genereux@ino.ca (F.G.); Marc.Terroux@ino.ca (M.T.); Carl.Vachon@ino.ca (C.V.); Patrice.Topart@ino.ca (P.T.)

2 Faculty of Forestry, University of Toronto, 33 Willcocks Street, Toronto, ON M5S 3B3, Canada; melanie.wheatley@mail.utoronto.ca

3 Canadian Forest Service, Great Lakes Forestry Centre, 1219 Queen St. E., Sault Ste. Marie, ON P6A 2E5, Canada; Joshua.johnston@canada.ca (J.M.J.); mike.wotton@canada.ca (M.W.)

* Correspondence: Denis.Dufour@ino.ca

Abstract: This study describes the development of a prototype bi-spectral microbolometer sensor system designed explicitly for radiometric measurement and characterization of wildfire mid- and long-wave infrared radiances. The system is tested experimentally over moderate-scale experimental burns coincident with FLIR reference imagery. Statistical comparison of the fire radiative power (FRP; W) retrievals suggest that this novel system is highly reliable for use in collecting radiometric measurements of biomass burning. As such, this study provides clear experimental evidence that mid-wave infrared microbolometers are capable of collecting FRP measurements. Furthermore, given the low resource nature of this detector type, it presents a suitable option for monitoring wildfire behaviour from low resource platforms such as unmanned aerial vehicles (UAVs) or nanosats.

Keywords: wildfire; microbolometer; FRP; radiometric; UAV; satellite

check for updates

Citation: Dufour, D.; Le Noc, L.; Tremblay, B.; Tremblay, M.N.; Généreux, F.; Terroux, M.; Vachon, C.; Wheatley, M.J.; Johnston, J.M.; Wotton, M.; et al. A Bi-Spectral Microbolometer Sensor for Wildfire Measurement. *Sensors* **2021**, *21*, 3690. https://doi.org/10.3390/s21113690

Academic Editor: Gamani Karunasiri

Received: 15 April 2021
Accepted: 19 May 2021
Published: 26 May 2021

Publisher's Note: MDPI stays neutral with regard to jurisdictional claims in published maps and institutional affiliations.

Copyright: © 2021 by the authors. Licensee MDPI, Basel, Switzerland. This article is an open access article distributed under the terms and conditions of the Creative Commons Attribution (CC BY) license (https://creativecommons.org/licenses/by/4.0/).

1. Introduction

Globally, wildfires burn nearly 4.3 million km^2 annually [1], with climate driving and controlling fire regimes [2]. However, anthropogenic burning (regardless of intent) is the dominant cause of wildfire ignition the world over [3–5]. Global wildfire activity is increasing under climate change [6,7]. This trend is worsened by continued population growth and the expansion of wildland urban interface across many regions of the world [8,9]. Not surprisingly, there has also been an increased frequency of catastrophic wildfire events in recent years (e.g., Australia, 2009, 2019/20; California, 2018; South America, 2019; the Mediterranean/Greece, 2018; the Arctic, 2019; Canada, 2016, 2017, 2018).

Emerging technologies may provide a means of supporting increasingly complex operations resulting from the growing threat of wildfires. Of particular interest has been the use of new remote sensing technologies, tools and information systems in wildfire operations [10]. Regionally, some countries implement Earth observation (EO) monitoring systems (e.g., Brazil, Canada, USA, South Africa), while the Global Wildfire Information System (GWIS) provides global EO fire monitoring services. Near real-time smoke monitoring derived from these EO products is available both regionally [11,12] and globally [13].

In an effort to expand the scientific and operational value of infrared (IR) imaging, there is a growing body of evidence suggesting that it can be used to characterize wildfire behaviour under experimental conditions. Specifically, through the automation of fire structure detection [14,15] and temporal analysis, it is possible to map and measure the rate and direction of spread [16–18], arguably with higher precision than traditional methods [19]. Additionally, the fire radiative power (FRP; W) [20] emitted from actively spreading flame fronts has been shown to be a strong predictor of fire line intensity (kW m^{-1}) [21,22].

The potential utility of infrared remote sensing in the operational management of wildfires has spurred numerous technology developments. The government of Canada recently announced its intention to build the world's first dedicated operational wildfire monitoring satellite, WildFireSat [23]. The technology behind this system has been proposed for smaller-scale applications [24]. In the airborne domain, there has also been substantial interest in the expanded use of unmanned aerial vehicles (UAV) in wildfire operations [25–28]. However, UAV applications are highly sensitive to payload mass and power requirements, which presents varying challenges in terms of integrating the type of IR systems necessary for comprehensive wildfire monitoring.

An ideal means to achieve infrared (IR) band radiometric imaging measurement in a small, lightweight and low-power system suitable for resource-constrained platforms such as UAVs is to use uncooled microbolometer sensors. Although initially primarily of interest for military applications, today, microbolometer sensors have become much less expensive to fabricate than cooled infrared detectors and are finding more uses in commercial applications. Notably, Institut national d'optique (INO) built the microbolometer sensors for the new infrared sensor technology (NIRST) instrument, which was launched on NASA's SACD-Aquarius satellite in 2011 [29] and was a foundational step in the evolution of the WildFireSat mission. A dual-camera system operating in both the mid-wave infrared (MWIR) and long-wave infrared (LWIR) bands, NIRST was a pioneering demonstration of the use of microbolometer technology for wildfire monitoring from space, using a 512×3 format focal plane array. Since then, INO has been developing other microbolometer array formats for custom applications, such as the INO-384 detector, which has 384×288 pixels of 35 μm pitch [30]. The INO-384 focal plane array (FPA), in combination with the development of novel broadband absorber technologies such as gold black and resonant structures, has been used for nearly 10 years in a variety of leading-edge systems developed at INO, such as see-through Terahertz systems and hyperspectral imagers [31,32].

Notably, the wide infrared wavelength band sensitivity made possible by INO's gold black absorber technology, combined with the recent availability of wide infrared band optics, led INO to develop the Microxcam-384-MLWIR camera, which offers uniform sensitivity in the 3 to 14 μm spectral range in a compact, lightweight and low power format. By fitting this camera with appropriate bandpass filters and a calibration system, it is possible to adapt it for wildfire measurement. Its low weight and power demand makes it ideal for UAV platforms.

In this study we describe the development of a novel bi-spectral microbolometer-based thermal imaging system test bench, based on the Microxcam-384-MLWIR camera, known as Bomberos. The suitability of Bomberos for wildfire monitoring applications is evaluated through a statistical comparison of its MWIR and LWIR bands to reference imagery collected simultaneously with a commercial photovoltaic detector-based MWIR radiometric imager (FLIR SC6703 camera) over a series of small-scale experimental biomass fires. Specifically, we compare the similarity of FRP measurement between the Bomberos MWIR and FLIR MWIR bands, and between the Bomberos MWIR and LWIR bands. In doing so, we demonstrate that Bomberos is an effective means of accurately characterizing wildfire energy, and present the very first field-based, observational evidence that MWIR microbolometers are capable of measuring FRP.

2. Materials and Methods

2.1. Bomberos

2.1.1. System Description

Bomberos was devised as a test bench to evaluate the performance of INO's broadband microbolometer technology in FRP measurement. It consists of: (1) two infrared cameras mounted on a motorized pan-tilt actuator, (2) two blackbody calibration targets, (3) a thermal control system, (4) a visible-band camera for reference, (5) a tripod, and (6) two laptop computers for data acquisition. Photos of Bomberos are shown in Figure 1.

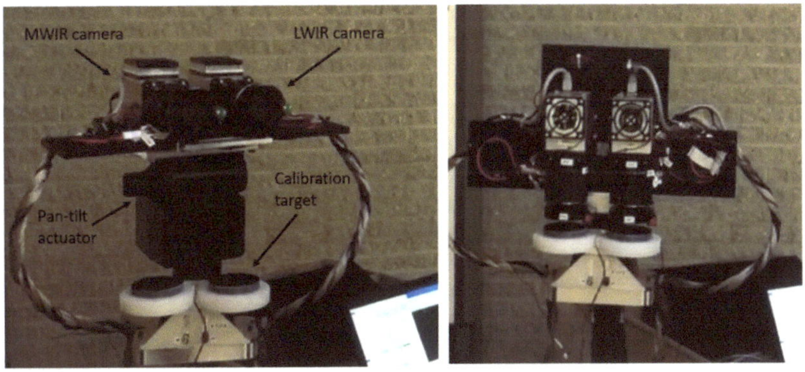

Figure 1. The Bomberos bi-spectral imaging system, shown in observation (left) and calibration (right) modes.

It should be noted that a potential future implementation for UAVs would be far lighter and more compact. Indeed, the very wide IR sensitivity of the INO gold black bolometers could make it possible to have a single camera, with butcher-block filters near the focal plane, to cover both the MWIR and LWIR bands. Such a camera would consume far less power than a cooled infrared camera. Indeed, the INO Microxcam-384-MLWIR camera core typically draws less than three Watts of power, excluding the detector's thermo-electric cooler (TEC) that is used for temperature stabilization and which typically consumes one additional Watt. A single embedded computer would suffice to acquire imagery from such a bi-spectral camera on a UAV. Moreover, the motorized pan-tilt actuator plus blackbody plate assembly could be replaced by a single lightweight mechanical shutter with a high emissivity surface of known, measured temperature for calibration.

On the Bomberos system, the MWIR and LWIR cameras are modified INO Microxcam-384-MLWIR cameras that are identical aside from the bandpass filters and aperture masks. They both have a compact (61 × 61 × 65 mm, 360 g) Microxcam camera core, containing 3 to 14 μm gold-black coated 384 × 288 microbolometer detectors, and are configured with identical commercial broadband IR optics (Ophir SupIR radiometric lens model #65119, with F/1 aperture and 35 mm focal length). The Microxcam core contains a TEC element under the detector to maintain its temperature to a stable value (typically 25 °C). Each Microxcam core interfaces to a data acquisition and control computer via a Gig-E communication link and is powered by a 12 VDC supply. The two IR cameras are synchronized to measure the same scene at the same time.

Wavelength selectivity is provided by bandpass MWIR (3.4 to 4.2 μm) and LWIR (10.4 to 12.3 μm) one-inch diameter filter elements mounted on custom-designed filter holders between the sensor and the imaging lens. The wideband detectors can be quite sensitive to small temperature variations of these filters, particularly the MWIR filter, which is emissive in the LWIR range. Thereby, the filter and detector environment is temperature-controlled to 27 °C using TEC elements and temperature monitoring feedback. As described in more detail later, custom-designed aperture-limiting masks are added to the ends of the lenses to prevent excessive irradiance from reaching the detectors when exposed to high-temperature fires. Finally, mechanical shutters are installed just outside the lenses, to protect the detectors from excessive thermal signals (remanence effects, as described in the next section) when not in image acquisition mode.

The two cameras are mounted on pan-tilt actuators that can be used to scan different viewing angles, and to provide occasional downward-looking views of the blackbody calibration targets. These targets serve as a known temperature reference that enable converting detector signal intensity to scene radiance and apparent temperature. The targets consist of aluminum plates covered with high-emissivity paint (black velvet). The target temperatures are measured since these temperature values are required to apply radiometric correction, as described in the next sub-section. The targets are thermally passive, and therefore their temperature can vary depending on the ambient temperature.

The cameras, pan-tilt actuators and blackbody targets are all installed on a ruggedized tripod mount. A commercial small visible-band video camera was also installed next to the two infrared cameras, to provide contextual information for data analysis.

2.1.2. Calibration and Measurement Procedure

Wildfire measurement requires the infrared cameras to be configured appropriately, in particular because of the large amount of power incident on the pixels when viewing a high temperature event. This can cause two problems: detector saturation and remanence. For the purposes of the Bomberos test campaign described later in this article, we assumed the maximum apparent fire temperature to be 900 °C. This is based on the previously measured intensity of similar experimental fires and the ground-projected pixel size. Under normal operation settings, the Bomberos camera signals (in particular the MWIR band one) would saturate well below this point. A simple way to avoid saturation is to reduce the bolometer integration time to reduce the responsivity. However, reducing the responsivity has the effect of reducing the pixel sensitivity, which is defined either as noise-equivalent spectral radiance (NESR) or noise-equivalent temperature difference (NETD). To counter this, the Bomberos cameras were configured to acquire images in a high dynamic range (HDR) mode with two alternating integration times, one to optimize the sensitivity for low temperature scenery, and the other to allow unsaturated fire temperature measurements up to 900 °C. The image frames with the two integration times are taken in rapid succession, 20 ms apart. The change in the scene radiance is assumed to be minimal within this time interval. Similar HDR sampling approaches have been used for FRP measurements elsewhere (e.g., [22]) and are used in the FLIR reference system used in this study (Section 2.2.2). Reconstructed images are then obtained by taking the high sensitivity/low dynamic range image and replacing saturated portions with image data from the low sensitivity/high dynamic range image.

As for the remanence effect, this phenomenon occurs when the VOx material in the bolometer is heated beyond a critical temperature point, for example, as a result of excessive incident irradiance on the pixels. The result is an extra offset signal on the bolometer that gradually goes away with time (on the order of several hours, but can be up to several days in severe cases), plus a loss of responsivity that also gradually returns to normal over time. The remanence effect was characterized in laboratory measurements by exposing the Bomberos cameras to various calibrated hot blackbody temperatures for different lengths of time. These measurements allowed the determination of the incident irradiance threshold on the pixels at which the remanence effect starts to occur. Based on these results, aperture-limiting circular masks were placed in front of the two cameras to restrict the amount of incident irradiance when the detector is exposed to the highest foreseen scene radiance (corresponding to 900 °C temperature exposure for 1 s every minute) to just below the remanence irradiance threshold. For the MWIR camera, a mask with a 6.5-mm hole was used, and for the LWIR camera, a 20-mm hole. Given the 35-mm diameter of the lenses, this reduces the effective lens aperture from F/1 to F/5.38 (MWIR) and F/1.75 (LWIR). Since the masks are in the field of view of the lenses, their thermal emission will contribute to the detector signal and must therefore be compensated for. It should be noted that these masks will reduce the sensitivity to low temperature scenery measurement, by a factor proportional to the square of the F-number.

To enable accurate radiometric measurements with a thermal imaging camera, all the factors that contribute to the signal measured on the detector must be considered. These can be expressed as in the following equation, where S is the digitized signal measured by a detector pixel (counts), R is the pixel responsivity (in counts/W), and O is the pixel offset signal (in counts):

$$S(T_{scene}) = R \cdot \left\{ \int_{\lambda=0}^{\infty} (G_{mask}(\lambda) \cdot L(\lambda, T_{mask}) + G_{env}(\lambda) \cdot L(\lambda, T_{env}) + G_{scene}(\lambda) \cdot L(\lambda, T_{scene})\ )\ d\lambda \right\} + O \quad (1)$$

The integral terms express that the power incident on a pixel is the sum of the thermal spectral radiances ($L(\lambda,T)$, representing Planck's blackbody function as a function of wavelength and temperature from: (1) the scene being imaged (T_{scene}), (2) the environment around the detector (T_{env}), and (3) the aperture-limiting mask (T_{mask}). The wavelength-dependent factors, $G(\lambda)$ in the equation, are thermal-radiance-to-incident-power weighting factors for the different thermal contributors, defined by the camera geometry and optical properties (e.g., spectral emissivity and reflectance of the different components). To achieve accurate measurement of the scene radiance, all the other contributors to the signal should be removed. This can be achieved by performing an offset measurement close in time to the scene measurement, such that the detector environment and aperture mask temperatures are unchanged and can thus be eliminated by subtraction. The offset measurement consists of viewing the reference passive temperature targets, which are at a known temperature T_{offset}. The resulting signal is as follows:

$$S\left(T_{offset}\right) = R \cdot \left\{\int_{\lambda=0}^{\infty} (G_{mask}(\lambda) \cdot L(\lambda, T_{mask}) + G_{env}(\lambda) \cdot L(\lambda, T_{env}) + G_{scene}(\lambda) \cdot L\left(\lambda, T_{offset}\right) \,)\, d\lambda \right\} + O \tag{2}$$

Subtracting the offset signal from the scene signal and rearranging the terms, we obtain the following expression for the scene radiance:

$$L_{scene,band} = \frac{S(T_{scene}) - S\left(T_{offset}\right)}{R \cdot G_{scene,band}} + L_{offset,band} \tag{3}$$

where it was approximated that the $G(\lambda)$ transfer function is uniform across the narrow wavelength range of the MWIR or LWIR filter through which the scene is viewed, such that the integration can be approximated as multiplication by a constant, i.e.,

$$\int_{\lambda=0}^{\infty} G_{scene}(\lambda) \cdot L(\lambda, T_{scene})\, d\lambda \approx G_{scene,band} \cdot L_{scene,band} \tag{4}$$

This approximation was confirmed by measurements that showed that the detector signal S is indeed a linear function of the scene radiance L.

To summarize, the scene radiance can be determined by: (1) measuring the scene and offset detector signals, (2) calculating the in-band radiance $L_{offset,band}$ of the passive blackbody target of known temperature used for the offset measurement, and (3) determining the value of $R \cdot G_{scene,band}$ for the given detector pixel.

The $R \cdot G_{scene,band}$ values for each pixel and HDR setting were determined by laboratory measurement of calibrated laboratory blackbody sources by the Bomberos cameras as the blackbodies were set to a range of temperatures from 20 °C to 900 °C. Caution was taken to ensure the hot blackbody was not heating the cameras and mask during the measurements, which would cause an additional signal drift, inducing calibration errors. Accordingly, a small (one-inch diameter) high-temperature cavity blackbody, which provides a limited field of view to the camera and thereby limits heat transfer to the cameras, was used for the higher temperature measurements (above 300 °C). The calibration measurements were obtained at a range of pan and tilt settings to scan the entire field of view.

The Bomberos camera image acquisition and pan-tilt actuator control procedure was fully automated to provide the most accurate and repeatable measurements possible. A measurement cycle was set to last 60 s, and has the following steps:

t = 0 s: tilt the cameras downwards towards the calibration targets, then open the shutters every 3 s for a duration of 1 s each and acquire offset images.
t = 25 s: close the shutters and orient the cameras towards the scene of interest.
t = 29 s: open the shutters, acquire 1 s of scene images, then close the shutters.
t = 35 s: tilt the cameras downwards towards the calibration targets, then open the shutters every 3 s for a duration of 1 s each and acquire offset images.
t = 60 s: repeat the cycle.

As can be seen, the cameras spend most of their time with the shutters closed and looking at the calibration targets to minimize remanence effects. However, in future developments of its microbolometer-based fire sensors, INO plans to use their new high resolution 1024 × 768 array, which is expected to exhibit a lesser remanence effect, and therefore can be operated for longer periods with the shutter open. Additionally, the remanence effect would be significantly less in a push-broom type scanning configuration where the hot fire image would pass rapidly across the pixels, in contrast to this experiment where the cameras were staring at a hot fire at very close range.

2.2. Experimental Design and Protocol

2.2.1. Layout

A series of experimental burns was conducted for this study at the Canadian Forest Service's Rose Experimental Burn Station [19,22] over three days in September of 2019. The Bomberos system was mounted ~21.6 m up a scaffold tower 4.8 m east of the burn platform. A commercial MWIR camera was mounted on the same level of the tower to serve as reference imagery for comparison. The view zenith angle for the imagers was ~21 degrees, as shown in Figure 2.

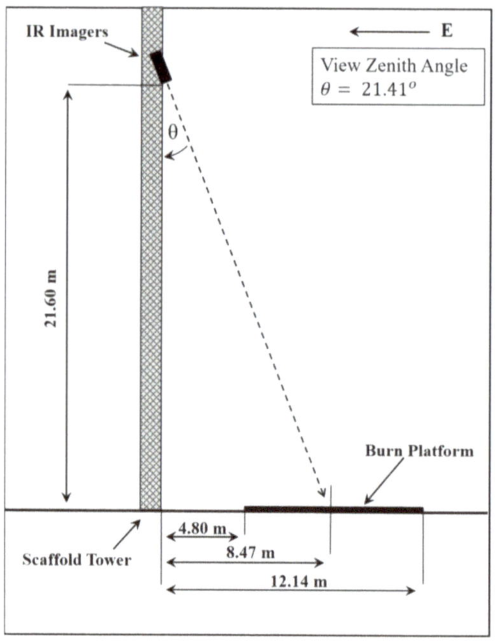

Figure 2. Visualization of the position of the Bomberos and FLIR SC6703 systems on the scaffold tower relative to the burn platform (not to scale).

2.2.2. Reference Imagery

A FLIR SC6703 with a narrow pass 3.9 µm spectral filter was used as the baseline MWIR reference point. The narrow spectral band and wide dynamic range (~423 to 1123 K) of this camera (achieved through HDR sampling of three integration times) has been proven to provide highly reliable FRP measurements [22]. In this study, it was configured to collect imagery at a frame rate of 30 Hz.

2.2.3. Data Collection Protocol

Prior to ignition, the FLIR camera collected several frames in ambient imaging mode. This was to provide spatial reference points prior to collecting data in the higher temperature settings, which omit ambient temperatures. Once this was complete, the reference imager was placed in high-temperature mode, and image collection began at a rate of 30 Hz. At the same time, the Bomberos cameras were set to acquire continuously, as described in Section 2.1.2. The Bomberos cameras measured low-temperature background and high-temperature fires simultaneously due to the HDR acquisition mode.

Once imaging had begun, a series of experimental fires were lit on the burn platform using recently harvested hay (primarily Timothy grass; *Phleum pratense*) from nearby agricultural activities. Although not a forest fuel itself, there is precedence for its use in FRP studies [21]. The experimental fires themselves ranged in size from 0.25 to 4 m in diameter and, when smaller diameters were burning, multiple fires would be spaced throughout the platform in varying spatial configurations (Figure 3). Throughout the burning process, fires were allowed to smolder down and be stoked up by the periodic addition of more fuel, resulting in longer durations for sampling and a broad range of intensities.

Figure 3. Example of experimental burn trials conducted 18 September 2019 during this experiment. Pictured here, researchers add fuel to a ~2 m diameter fire with a pitchfork while it is sampled by Bomberos and the reference FLIR imager on the tower adjacent to the burn platform.

2.3. Data Processing and Analysis

2.3.1. Temporal and Spatial Characterization

Temporal alignment between the Bomberos and FLIR images was performed prior to calculating FRP for each image. The Bomberos images were recorded in one-second intervals, whereas the FLIR images were recorded continuously throughout the burns, and, therefore, the FLIR images had to be segmented to match the Bomberos sequences. Time synchronization was done by visually examining the images of the Bomberos and FLIR data to look at commonalities between the images (e.g., people appearing in the frames, more fuel being added to the fire). Once the images were matched visually, they were checked by creating a time series of the mean FRP for frame and ensuring that the images were comparable.

Spatial characterization between the Bomberos and FLIR images was determined by using ambient images from both cameras and known measurements of burn pad panels to calculate the approximate pixel size for each of the Bomberos and FLIR image outputs. Measurements were taken from four different locations across the burn pad to produce four different pixel sizes for each image. These four pixel sizes were then averaged to produce one pixel size for each of the Bomberos and FLIR images. New measurements were taken every time the camera angle changed to ensure that the new position of the camera angle did not affect the pixel size. On average, the FLIR pixel size was 0.025 m in diameter and the Bomberos was 0.022 m. These approximate pixel sizes for each of the Bomberos and FLIR images were used in the FRP calculation.

2.3.2. Fire Radiative Power Calculations

Fire radiative power was calculated separately for both the MWIR and LWIR bands. For the Bomberos MWIR and FLIR MWIR FRP calculations, the MWIR radiance method of Wooster et al. [20] was used. The FRP calculation coefficient was tailored to the specific spectral response functions of the Bomberos and FLIR cameras. Before calculating FRP, all pixels with a temperature of less than 500 K were removed from the images. This was to remove pixels that were detected in the Bomberos imagery at a lower threshold than the FLIR imagery, ensuring that the datasets were comparable. For the Bomberos LWIR camera, FRP was calculated from the brightness temperature with an assumption that the fire behaved as a blackbody, filling the entirety of each pixel area, allowing FRP to be calculated using the Stefan–Boltzmann law. For both the MWIR and LWIR FRP calculations, FRP was calculated for each pixel in the image, and then summed to produce total FRP per frame for the fire in kilowatts (kW).

2.3.3. Bomberos MWIR—FLIR MWIR Intercomparison

The Bomberos MWIR and FLIR MWIR intercomparison for FRP was carried out by comparing the mean FRP for each sequence. The sum of FRP per frame was taken for both the Bomberos MWIR and FLIR MWIR data, and then the mean was calculated for each one-second sequence. The field of view for the Bomberos and FLIR cameras were not the same, so in instances where there were multiple fires on the burn pad, each fire was isolated in both the Bomberos and FLIR images so that the FRP comparison would only be for a single fire. In cases where there was only one fire on the burn pad, the entire field of view for each camera was compared in the FRP comparison. The mean FRP values for each sequence of the Bomberos MWIR and FLIR MWR data were then plotted against each other and compared using linear regression with the intercept constrained at 0. In this analysis, the slope from the linear regression estimated the deviance of the FRP values from the line of perfect agreement. A separate *t*-test was also conducted to determine if the slope of the relationship between the FLIR MWIR and Bomberos MWIR FRP data was significantly different from the line of perfect agreement. There were 67 one-second fire sequences, resulting in one mean FRP estimation each, which were compared in this analysis.

2.3.4. Bomberos MWIR—LWIR Intercomparison

The Bomberos MWIR and LWIR intercomparison was analyzed using the same methods described above. A linear regression model with the intercept constrained at zero was used and results were analyzed to determine the deviance of the MWIR and LWIR FRP values from the line of perfect agreement. Similarly to the Bomberos and FLIR MWIR intercomparison, a *t*-test was conducted to determine if the slope of the relationship between Bomberos MWIR and LWIR was significantly different from the line of perfect agreement. There were 67 fire sequences compared in this analysis.

3. Results

3.1. Bomberos MWIR and FLIR MWIR FRP Comparison

The linear regression between the Bomberos MWIR and FLIR MWIR FRP shows a significant relationship ($R^2 = 0.998$, $p < 0.0001$; see Figure 4). Most data points follow the line of perfect agreement. Overall, there was more data collected at lower FRP values than higher FRP values. The agreement between FRP values from Bomberos and FLIR is extremely good for lower FRP values; however, as FRP increases, there is deviation from the line of perfect agreement, with Bomberos estimating slightly higher FRP values than FLIR. The slope of the relationship between Bomberos MWIR and FLIR MWIR is 1.06, meaning that there is a 6% deviance between the FLIR MWIR and Bomberos MWIR relationship and the line of perfect agreement. This difference between the slope and the line of perfect agreement is significant, $t(66) = 9.31$, $p < 0.001$. Figure 5a,c show the FRP values per pixel for the Bomberos MWIR and FLIR MWIR images, respectively, for an example frame in a sequence, with the two images being very comparable in pixel FRP values and overall mean FRP.

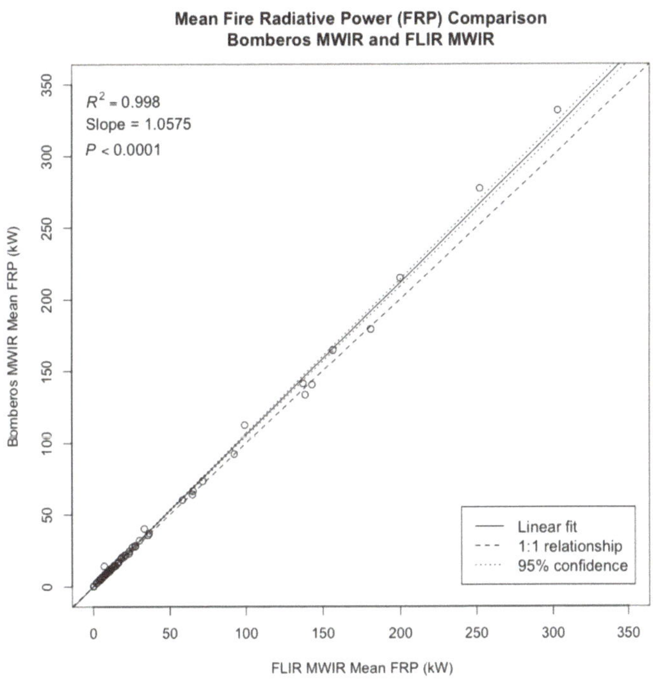

Figure 4. Linear regression between the FLIR MWIR and Bomberos MWIR fire radiative power (FRP). Data points consist of 67 different one-second sequences recorded from experimental fires conducted between 16 and 18 September 2019.

3.2. Bomberos MWIR and LWIR FRP Comparison

The linear regression between the Bomberos MWIR and LWIR FRP shows a significant relationship ($R^2 = 0.999$, $p < 0.0001$; see Figure 6). The mean FRP for the Bomberos MWIR and LWIR align closely with the line of perfect agreement, with deviation from the 1:1 line increasing as FRP increases. At higher values of FRP, the Bomberos LWIR estimates higher FRP values than the Bomberos MWIR. The slope of the relationship between the Bomberos MWIR and LWIR mean FRP, 0.94, is significantly different from the line of perfect agreement, $t(66) = -14.21$, $p < 0.01$.

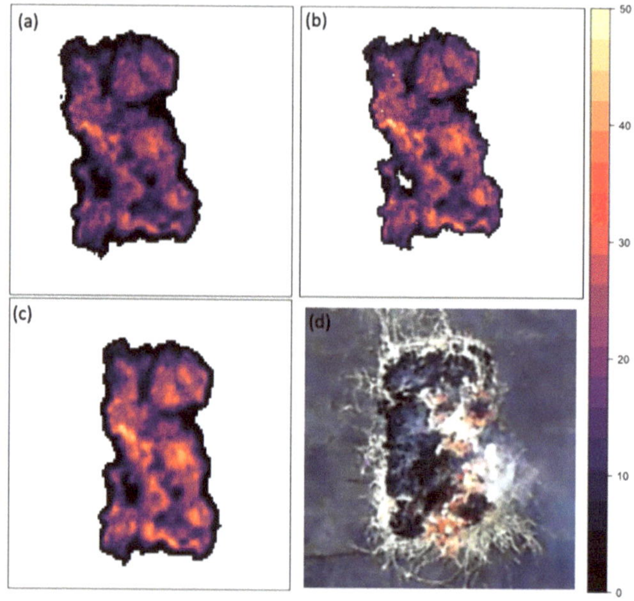

Figure 5. Images of a 1.5 × 3 m fire on 16 September 2019. (**a**) MWIR Bomberos image of FRP per pixel in Watts. The total FRP for the fire in the frame is 64,236.2 W (64.3 kW); (**b**) LWIR Bomberos image of FRP per pixel in Watts. The total FRP for the fire in the frame is 72,651.0 W (72.7 kW); (**c**) MWIR FLIR image of FRP per pixel in Watts. The total FRP for the fire in the frame is 63,642.9 W (63.6 kW); (**d**) Visual imagery of the fire from a camera fixed to a tower 21.6 metres above the fire.

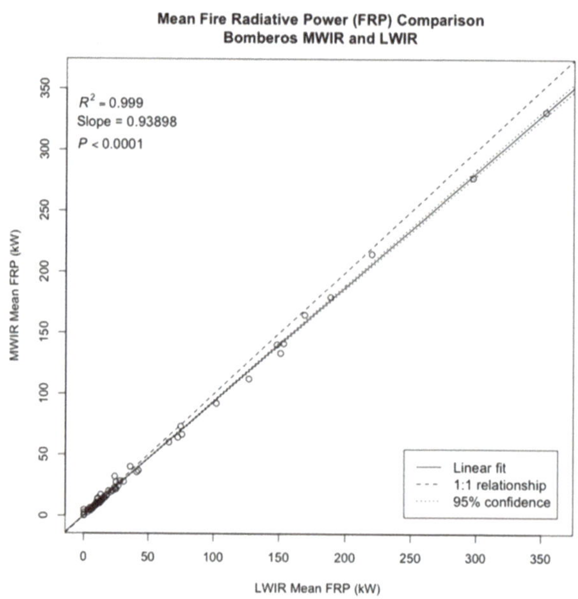

Figure 6. Linear regression between the Bomberos MWIR and LWIR fire radiative power (FRP). Data points consist of 67 different one-second sequences recorded from experimental fires conducted between 16 and 18 September 2019.

4. Discussion

This study examines the use of microbolometer-based mid- and long-wave IR detectors for collecting radiometric measurements from biomass combustion. Through direct comparison between measurements made with the Bomberos microbolometer test bench and coincident FLIR imagery of 67 experimental burning samples, it was determined that the microbolometer detectors were highly reliable in retrieving FRP in both the mid- and long-wave IR. That said, in both spectral bands, the relationship between Bomberos and FLIR FRP did show statistically significant deviation from the line of perfect agreement.

Regression analysis of the Bomberos and FLIR MWIR FRP retrievals demonstrate exceptional correlation ($R^2 = 0.998$, $p < 0.0001$), though the slope from this comparison, 1.06, is a small but statistically significant deviation from the line of perfect agreement ($t(66) = 9.31$, $p < 0.001$). This trend towards over-estimation (of ~6%) by Bomberos is more pronounced at higher FRP and may be explained by a number of potential factors. Notably, the FLIR and Bomberos spatial resolutions differed (~0.022 m and ~0.025 m diameter for Bomberos and FLIR, respectively) due to their detector dimensions and optical configurations. Given that pixel area corrections were estimated manually in the analysis, it is possible that minor errors in these calculations could have impacted the accuracy of either the FLIR or Bomberos FRP calculations. Additionally, the temporal co-registration of the imagery is not exact, which could have influenced the comparison; however, the comparison of mean total FRP over the sample periods was designed to minimize these errors. Though the two cameras have different saturation points, in both data sets, saturation was not observed. Aside from the fundamental difference in detector technology, the most significant difference between the Bomberos and FLIR MWIR detectors is their spectral response functions. While the FLIR system was equipped with a narrow pass (~0.15 μm) filter that stops short of 4.0 μm, the Bomberos MWIR bandpass filter is much wider, extending from 3.4 to 4.2 μm and potentially intersecting the CO_2 band around 4.2 μm. Given that biomass combustion releases large amounts of hot CO_2 (which varies substantially throughout the vertical profile of the smoke plume), and since the absorption of ambient CO_2 is a strong factor even in short-range measurements, even a small deviation within this spectral range could introduce substantial differences in the FRP observations. It should also be noted that the lower temperature threshold for the data segmentation was determined based on the brightness temperature ranges, and not modified to account for the varying spectral response functions of the two cameras; however, the net effect of this discrepancy is minimal on overall FRP calculations. Notably, despite these potential causes, the agreement between the FRP observations remains strong.

Regression analysis of the Bomberos MWIR and LWIR FRP retrievals demonstrate similarly strong correlation ($R^2 = 0.999$, $p < 0.0001$) but with a slope of 0.94, which is again a statistically significant deviation from the line of perfect agreement ($t(66) = -14.21$, $p < 0.01$). In this case, the trend is towards over-estimation (of ~6%) in the LWIR. Though the bias is small, here, the over-estimation is more consistently observed across the full range of observations. It should be noted that not only are the MWIR and LWIR observations being recorded in completely different spectral bands, but they are also calculating FRP in very different ways. Based on the variation in the calculation method (MWIR radiance method vs. Stefan–Boltzmann law), disagreement in the results is to be expected. With this in mind, the agreement between the MWIR and LWIR FRP retrievals is considered very successful.

5. Conclusions

Early efforts to image biomass burning events using MWIR microbolometer technology demonstrated that it was possible to detect these phenomena from space-borne platforms (e.g., [29,33]). However, this demonstration failed to include complete radiometric calibration capabilities, resulting in the inability to derive FRP from the data. As a result, some debate has persisted within the community as to the applicability of MWIR microbolometer technology in providing comprehensive fire monitoring datasets. In this study, we have demonstrated that it is possible to perform high-accuracy radiometric observations of biomass burning events in the MWIR and LWIR using microbolometer technology. To our knowledge, this study represents the first scientific evidence that MWIR microbolometers are highly reliable sources of FRP measurements. In doing so, these findings suggest that these types of low resource detectors are suitable for deployment for wildfire monitoring applications, particularly where onboard resources are limited (e.g., UAV and nanosat type scenarios).

The encouraging results from this study are expected to stimulate further developments towards a compact system specifically designed for radiometric airborne and space-based wildfire measurement. This could include combining MWIR and LWIR bands in a single camera—for example, by using butcher block filters near the focal plane array, or with a filter wheel. Moreover, the development of a less resource-intensive thermal offset compensation method would likely be required. This could potentially be achieved by using a lightweight mechanical shutter, or by using temperature measurements at strategic points in the camera and correcting the images accordingly.

Author Contributions: Bomberos instrument design and manufacture, L.L.N., B.T., M.T. and F.G.; test planning and execution, J.M.J., D.D. and M.N.T.; data analysis, M.J.W. and L.L.N.; software, C.V.; project administration P.T., J.M.J. and B.M.W.; writing, D.D., M.J.W. and J.M.J. All authors have read and agreed to the published version of the manuscript.

Funding: This research was funded in part by the Government of Canada's Mitacs program.

Data Availability Statement: The data presented in this study are available on request from the corresponding author.

Acknowledgments: We gratefully acknowledge the support of the Canadian Space Agency and Natural Resources Canada for allowing us to benefit from the experimental burn campaign to collect the data for this study.

Conflicts of Interest: The authors declare no conflict of interest. The funders had no role in the design of the study; in the collection, analyses, or interpretation of data; in the writing of the manuscript, or in the decision to publish the results.

References

1. Giglio, L.; Boschetti, L.; Roy, D.P.; Humber, M.L.; Justice, C.O. The Collection 6 MODIS burned area mapping algorithm and product. *Remote Sens. Environ.* **2018**, *217*, 72–85. [CrossRef] [PubMed]
2. Bowman, D.M.J.S.; Balch, J.K.; Artaxo, P.; Bond, W.J.; Carlson, J.M.; Cochrane, M.A.; D'Antonio, C.M.; DeFries, R.S.; Doyle, J.C.; Harrison, S.P.; et al. Fire in the Earth System. *Science* **2009**, *324*, 481–484. [CrossRef] [PubMed]
3. Archibald, S.; Staver, A.C.; Levin, S.A. Evolution of human-driven fire regimes in Africa. *Proc. Natl. Acad. Sci. USA* **2012**, *109*, 847. [CrossRef] [PubMed]
4. Taylor, A.H.; Trouet, V.; Skinner, C.N.; Stephens, S. Socioecological transitions trigger fire regime shifts and modulate fire–climate interactions in the Sierra Nevada, USA, 1600–2015 CE. *Proc. Natl. Acad. Sci. USA* **2016**, *113*, 13684. [CrossRef]
5. Balch, J.K.; Bradley, B.A.; Abatzoglou, J.T.; Nagy, R.C.; Fusco, E.J.; Mahood, A.L. Human-started wildfires expand the fire niche across the United States. *Proc. Natl. Acad. Sci. USA* **2017**, *114*, 2946. [CrossRef]
6. Flannigan, M.D.; Krawchuk, M.A.; de Groot, W.J.; Wotton, B.M.; Gowman, L.M. Impacts of changing climate for global wildland fire. *Int. J. Wildland Fire* **2009**, *18*, 483–507. [CrossRef]
7. Jolly, W.M.; Cochrane, M.A.; Freeborn, P.H.; Holden, Z.A.; Brown, T.J.; Williamson, G.J.; Bowman, D.M.J.S. Climate-induced variations in global wildfire danger from 1979 to 2013. *Nat. Commun.* **2015**, *6*, 7537. [CrossRef]
8. Theobald, D.M.; Romme, W.H. Expansion of the US wildland–urban interface. *Landsc. Urban Plan.* **2007**, *83*, 340–354. [CrossRef]
9. Johnston, L.M.; Wang, X.; Erni, S.; Taylor, S.W.; McFayden, C.B.; Oliver, J.A.; Stockdale, C.; Christianson, A.; Boulanger, Y.; Gauthier, S.; et al. Wildland fire risk research in Canada. *Environ. Rev.* **2020**, 1–23. [CrossRef]
10. Allison, R.S.; Johnston, J.M.; Craig, G.; Jennings, S. Airborne optical and thermal remote sensing for wildfire detection and monitoring. *Sensors* **2016**, *16*, 1310. [CrossRef]
11. Larkin, N.K.; O'Neill, S.M.; Solomon, R.; Raffuse, S.; Strand, T.; Sullivan, D.C.; Krull, C.; Rorig, M.; Peterson, J.; Ferguson, S.A. The BlueSky smoke modeling framework. *Int. J. Wildland Fire* **2009**, *18*, 906–920. [CrossRef]
12. Chen, J.; Anderson, K.; Pavlovic, R.; Moran, M.D.; Englefield, P.; Thompson, D.K.; Munoz-Alpizar, R.; Landry, H. The FireWork v2.0 air quality forecast system with biomass burning emissions from the Canadian Forest Fire Emissions Prediction System v2.03. *Geosci. Model Dev.* **2019**, *12*, 3283–3310. [CrossRef]

13. Kaiser, J.W.; Heil, A.; Andreae, M.O.; Benedetti, A.; Chubarova, N.; Jones, L.; Morcrette, J.-J.; Razinger, M.; Schultz, M.G.; Suttie, M.; et al. Biomass burning emissions estimated with a global fire assimilation system based on observed fire radiative power. *Biogeosciences* **2012**, *9*, 527–554. [CrossRef]
14. Valero, M.M.; Verstockt, S.; Rios, O.; Pastor, E.; Vandecasteele, F.; Planas, E. Flame filtering and perimeter localization of wildfires using aerial thermal imagery. *SPIE 10214-Thermosense Therm. Infrared Appl.* **2017**, *10214*, 1–8.
15. Valero, M.M.; Rios, O.; Pastor, E.; Planas, E. Automated location of active fire perimeters in aerial infrared imaging using unsupervised edge detectors. *Int. J. Wildland Fire* **2018**, *27*, 241–256. [CrossRef]
16. McRae, D.J.; Jin, J.-Z.; Conard, S.G.; Sukhinin, A.I.; Ivanova, G.A.; Blake, T.W. Infrared characterization of fine-scale variability in behaviour of boreal forest fires. *Can. J. For. Res.* **2005**, *35*, 2194–2206. [CrossRef]
17. Pastor, E.; Àgueda, A.; Andrade-Cetto, J.; Muñoz, M.; Pérez, Y.; Planas, E. Computing the rate of spread of linear flame fronts by thermal image processing. *Fire Saf. J.* **2006**, *41*, 569–579. [CrossRef]
18. Paugam, R.; Wooster, M.J.; Roberts, G. Use of Handheld Thermal Imager Data for Airborne Mapping of Fire Radiative Power and Energy and Flame Front Rate of Spread. *IEEE Trans. Geosci. Remote Sens.* **2013**, *99*, 1–15. [CrossRef]
19. Johnston, J.M.; Wheatley, M.J.; Wooster, M.J.; Paugam, R.; Davies, G.M.; DeBoer, K.A. Flame-Front Rate of Spread Estimates for Moderate Scale Experimental Fires Are Strongly Influenced by Measurement Approach. *Fire* **2018**, *1*, 16. [CrossRef]
20. Wooster, M.J.; Roberts, G.; Perry, G.L.W.; Kaufman, Y.J. Retrieval of biomass combustion rates and totals from fire radiative power observations: FRP derivation and calibration relationships between biomass consumption and fire radiative energy release. *J. Geophys. Res.* **2005**, *110*, D24311. [CrossRef]
21. Smith, A.M.S.; Wooster, M.J. Remote classification of head and backfire types from MODIS fire radiative power and smoke plume observations. *Int. J. Wildland Fire* **2005**, *14*, 249–254. [CrossRef]
22. Johnston, J.M.; Wooster, M.J.; Paugam, R.; Wang, X.; Lynham, T.J.; Johnston, L.M. Direct Estimation of Byram's fire intensity from infrared remote sensing imagery. *Int. J. Wildland Fire* **2017**, *26*, 668–684. [CrossRef]
23. Johnston, J.M.; Jackson, N.; McFayden, C.; Phong, L.N.; Lawrence, B.; Davignon, D.; Wooster, M.J.; van Mierlo, H.; Thompson, D.K.; Cantin, A.S.; et al. Development of the User Requirements for the Canadian WildFireSat Satellite Mission. *Sensors* **2020**, *20*, 5081. [CrossRef] [PubMed]
24. Ngo Phong, L.; Dufour, D.; Johnston, J.; Chevalier, C.; Côté, P.; Fisette, B.; Wang, M.; Châteauneuf, F. A low resource imaging radiometer for nanosatellite based fire diagnosis. In *Infrared Remote Sensing and Instrumentation XXVI*; Proc. SPIE: San Francisco, CA, USA, 2018; Volume 10765.
25. Maza, I.; Caballero, F.; Capitán, J.; Martínez-de-Dios, J.R.; Ollero, A. Experimental Results in Multi-UAV Coordination for Disaster Management and Civil Security Applications. *J. Intell. Robot. Syst.* **2011**, *61*, 563–585. [CrossRef]
26. Yuan, C.; Liu, Z.; Zhang, Y. UAV-based forest fire detection and tracking using image processing techniques. In Proceedings of the 2015 International Conference on Unmanned Aircraft Systems (ICUAS), Denver, CO, USA, 9–12 June 2015.
27. Fraser, R.; van der Sluijs, J.; Hall, R. Calibrating Satellite-Based Indices of Burn Severity from UAV-Derived Metrics of a Burned Boreal Forest in NWT, Canada. *Remote Sens.* **2017**, *9*, 279. [CrossRef]
28. Sherstjuk, V.; Zharikova, M.; Sokol, I. Forest Fire-Fighting Monitoring System Based on UAV Team and Remote Sensing. In Proceedings of the 2018 IEEE 38th International Conference on Electronics and Nanotechnology (ELNANO), Kyiv, Ukraine, 24–26 April 2018.
29. Marraco, H.; Ngo Phong, L. NIRST: A satellite-based IR instrument for fire and sea surface temperature measurement. In *Non-Intrusive Inspection Technologies*; Proc. SPIE: San Francisco, CA, USA, 2006; Volume 6213.
30. Fisette, B.; Tremblay, M.; Oulachgar, H.; Généreux, F.; Béland, D.; Beaupré, P.; Julien, C.; Gay, D.; Deshaies, S.; Tremblay, B.; et al. Novel vacuum packaged 384 × 288 broadband bolometer FPA with enhanced absorption in the 3–14 μm wavelength range. In *Infrared Technology and Applications XLIII*; Proc. SPIE: San Francisco, CA, USA, 2017; Volume 10177.
31. Marchese, L.; Terroux, M.; Généreux, F.; Tremblay, B.; Bolduc, M.; Bergeron, A. Review of the characteristics of 384 × 288 pixel THz camera for seethrough imaging. In *Millimetre Wave and Terahertz Sensors and Technology VI*; Proc. SPIE: San Francisco, CA, USA, 2013; Volume 8900.
32. Béland, D.; Spisser, H.; Dufour, D.; Le Noc, L.; Picard, F.; Topart, P. Portable LWIR hyperspectral imager based on MEMS Fabry-Perot interferometer and broadband microbolometric detector array. In *MOEMS and Miniaturized Systems XVII*; Proc. SPIE: San Francisco, CA, USA, 2018; Volume 10545.
33. Ngo Phong, L.; Pancrati, O.; Marchese, M.; Châteauneuf, F. Spaceborne linear arrays of 512 × 3 microbolometers. In *Reliability, Packaging, Testing, and Characterization of MOEMS/MEMS and Nanodevices XII*; Proc. SPIE: San Francisco, CA, USA, 2013; Volume 8614, p. 86140N-1.

sensors

MDPI

Article

Geocorrection of Airborne Mid-Wave Infrared Imagery for Mapping Wildfires without GPS or IMU

Gabriela Ifimov [1,*], Tomas Naprstek [1], Joshua M. Johnston [2], Juan Pablo Arroyo-Mora [1], George Leblanc [1] and Madeline D. Lee [3]

1 Flight Research Laboratory, Aerospace Research Centre, National Research Council Canada, 1920 Research Private, Ottawa, ON K1V 1J8, Canada; Tomas.Naprstek@nrc-cnrc.gc.ca (T.N.); JuanPablo.Arroyo-Mora@nrc-cnrc.gc.ca (J.P.A.-M.); George.Leblanc@nrc-cnrc.gc.ca (G.L.)
2 Canadian Forest Service, Great Lakes Forestry Centre, 1219 Queen St. E., Sault Ste. Marie, ON P6A 2E5, Canada; Joshua.Johnston@canada.ca
3 Department of Geoscience and Petroleum, Faculty of Engineering, University of Science and Technology, S. P. Andersens veg 15a, 7031 Trondheim, Norway; Madeline.Lee@ntnu.no
* Correspondence: Gabriela.Ifimov@nrc-cnrc.gc.ca

Abstract: The increase in annual wildfires in many areas of the world has triggered international efforts to deploy sensors on airborne and space platforms to map these events and understand their behaviour. During the summer of 2017, an airborne flight campaign acquired mid-wave infrared imagery over active wildfires in Northern Ontario, Canada. However, it suffered multiple position-based equipment issues, thus requiring a non-standard geocorrection methodology. This study presents the approach, which utilizes a two-step semi-automatic geocorrection process that outputs image mosaics from airborne infrared video input. The first step extracts individual video frames that are combined into orthoimages using an automatic image registration method. The second step involves the georeferencing of the imagery using pseudo-ground control points to a fixed coordinate systems. The output geocorrected datasets in units of radiance can then be used to derive fire products such as fire radiative power density (FRPD). Prior to the georeferencing process, the Root Mean Square Error (RMSE) associated with the imagery was greater than 200 m. After the georeferencing process was applied, an RMSE below 30 m was reported, and the computed FRPD estimations are within expected values across the literature. As such, this alternative geocorrection methodology successfully salvages an otherwise unusable dataset and can be adapted by other researchers that do not have access to accurate positional information for airborne infrared flight campaigns over wildfires.

Keywords: wildfire; fire monitoring; mid-wave infrared; airborne sensor; geocorrection

check for updates

Citation: Ifimov, G.; Naprstek, T.; Johnston, J.M.; Arroyo-Mora, J.P.; Leblanc, G.; Lee, M.D. Geocorrection of Airborne Mid-Wave Infrared Imagery for Mapping Wildfires without GPS or IMU. *Sensors* **2021**, *21*, 3047. https://doi.org/10.3390/s21093047

Academic Editor: Stefano Sfarra

Received: 31 March 2021
Accepted: 23 April 2021
Published: 27 April 2021

Publisher's Note: MDPI stays neutral with regard to jurisdictional claims in published maps and institutional affiliations.

cc BY

Copyright: © 2021 by the authors. Licensee MDPI, Basel, Switzerland. This article is an open access article distributed under the terms and conditions of the Creative Commons Attribution (CC BY) license (https://creativecommons.org/licenses/by/4.0/).

1. Introduction

As wildfire events have increased in many areas of the world with catastrophic consequences [1–4], the ability to accurately monitor these events at different spatial and temporal scales is becoming a key concern and focus for governments around the globe [5–7]. Active wildfire monitoring e.g., [8], rapid turn-around of data processing, and computation of fire mapping products are all fundamental to understanding the life cycle of fires, detecting spatial patterns, mitigating their potential damage, and implementing response and recovery activities. Efforts to deploy different observation platforms [8–13] is an essential task aiming to better understand, for instance, how climate change influences wildfire behaviour [14–16]. Remotely sensed data from infrared sensors (e.g., mid-wave infrared and long-wave infrared) are able to capture information from both naturally-occurring and controlled-fires regardless of time of day [17–19].

The importance of accurate detection of wildfires is that it allows the users of this technology to better address aspects such as fire behaviour [20], build fuel consumption models [21], and learn how they impact their surrounding environments [17,22]. To date, spaceborne and manned airborne platforms are commonly used to study wildfires [17,23,24], with the inclusion of unmanned aerial vehicles (UAV) into the airspace still being developed [19,25,26]. However, each platform provides unique data over top of the fire. For instance, satellite based infrared sensors provide landscape level information at a moderate (e.g., ~0.4–1.0 km) spatial resolution e.g., [8,27], however, they have very defined orbital paths and revisit periods which are often inadequate for operational purposes. Moreover, spaceborne sensors require independent data for the calibration and validation of fire satellite products [28,29]. Given the highly dynamic behaviour of wildfires [30], the use of airborne platforms for wildfire mapping [17] are ideal for near real-time characterization at operational scales (e.g., 10 km × 10 km) and fine spatial resolution (~1–5 m). Thus, airborne infrared imagery acquired over wildfires is adequate to generate end-products such as fire-front locations [30], rate of spread [24], fire radiative power and fire-line intensity [31]. These estimations can be further compared and/or integrated with satellite imagery, such as Sentinel-3 Sea and Land Surface Temperature Radiometer (SLSTR) [12], the Visible Infrared Imaging Radiometer Suite (VIIRS) [8] and the future Canadian WildFireSat [13], to establish a wildfire management system at local and national scales.

Considering airborne platforms, a key aspect of wildfire mapping is the process of geocorrection and geo-registration [9,24]. The objective of the geocorrection process is to compensate platform and/or sensor motion so that the final image is represented in a grid format and usually resampled to a specific cell size [32]. Geo-registration or georeferencing refers to the process where the goal is to align the corrected image to a local coordinate system with a given ellipsoid and datum [32]. Processing thermal infrared data to detect, for example, fires' heat signatures and registering them to a coordinate system is ideal for most applications [17]. However, the geocorrection and registration process can be challenging within wildfire mapping, particularly at the airborne scale when Inertial Measurement Unit (IMU) and precise Global Navigation Satellite System (GNSS) data are not available [9,33]. For an effective georeferenced dataset, a user will, in general, require a quality GNSS unit capable of differential post processing [34]. As well, a high-frequency inertial measurement unit (IMU) that records the sensor's attitude (i.e., roll, pitch, yaw) [32] is often required. While GNSS positioning is a standard feature on most survey aircraft's navigation system, ideally the thermal sensor would also have dedicated high accuracy GNSS and IMU devices. However, this might not be the case due to on-board space limitations, or budgetary constraints [17]. As such, a geocorrection method, that can achieve high accuracy in the absence (or malfunction) of GNSS/IMU, can be valuable for maximizing the utility of the thermal imagery. While not necessarily common, equipment malfunction can occur during a field campaign, particularly in the context of wildfires where activities are carried out under both environmentally challenging and high operator stress conditions. As airborne data is expensive and labour intensive, if a malfunction in the GNSS or IMU does occur then solutions to recover the geolocation information are needed.

Here we present a novel data loss minimization methodology based upon the malfunction of both GNSS and IMU instruments in an airborne thermal infrared field campaign over an active wildfire. While specific to the issues encountered in this field campaign, the method can be easily applied to any scenario where infrared data does not have associated GPS or IMU information, whether due to equipment issues or imagery that has been acquired without the associated positional instruments. The method is a two-step semi-automatic geocorrection process that outputs image mosaics from airborne infrared video input. The first step involves taking the infrared video as individual frames and using an automatic image registration method to develop the initial mosaic. This step was followed by georeferencing the imagery using high-resolution GeoEye (1.84 m) satellite imagery acquired over the same area. While the images produced by this first step can be used for some basic, non-georeferenced analysis, the second step is highly beneficial as it aligns the orthoimage with distinct geographical features, as well as, assists in accounting for larger warping effects that occur in the imagery due to significant aircraft motion. The final output is a set of accurate orthoimages that can be compared with other GPS-referenced data and analyzed for purposes such as fire end-products (e.g., fire radiative power).

2. Data Collection

2.1. Field Campaign

The airborne field campaign was a collaborative effort between the Flight Research Lab of the National Research Council Canada (NRC) and the Canadian Forestry Service of Natural Resources Canada. The two primary goals of the campaign were: (1) quantify wildland fires intensity and progression, and, (2) compare the fire's radiative emissions of the airborne campaign to those determined from a coincidental overpass of Sentinel-3a Sea and Land Surface Temperature Radiometer (SLSTR). The project employed NRC's Twin Otter survey aircraft that collected Forward Looking InfraRed (FLIR) mid-wave infrared (MWIR) imagery (Section 2.2) over the 37th Being Observed (BOB) wildfire in the Sioux Lookout district (SLK-37), near Pickle Lake, Northern Ontario, Canada (Figure 1A,B).

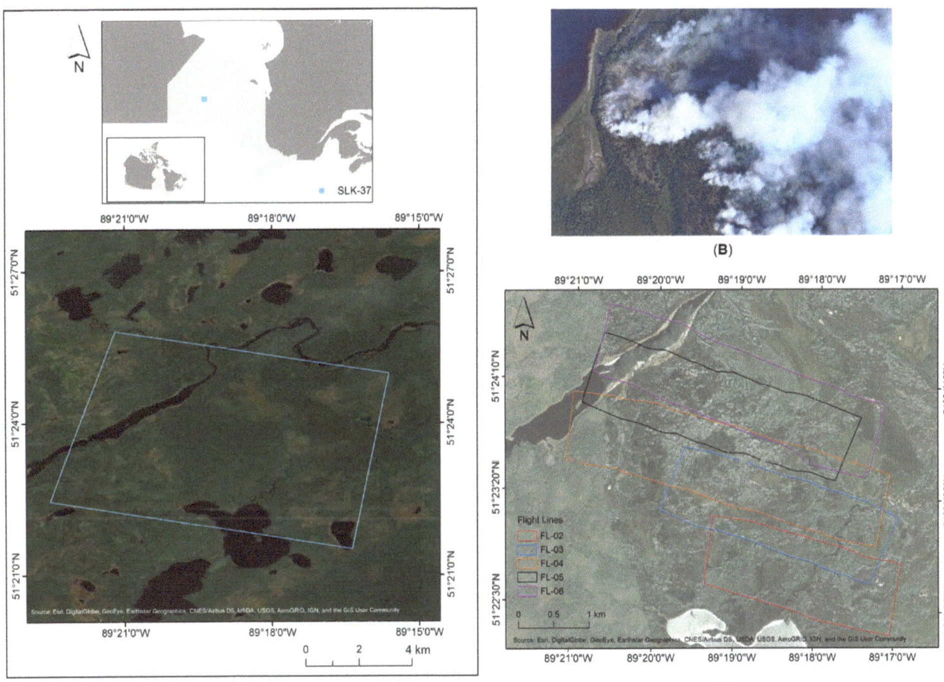

Figure 1. (**A**) Sioux Lookout (SLK-37) site located near Pickle Lake, Northern Ontario, Canada. FLIR MWIR data was recorded twice in successive alternating directions over eight flight lines between 2 August and 3 August 2017. (**B**) Photograph taken over the study area acquired on 3 August 2017, at 11:27 (EST). (**C**) Boundaries of the five selected flight lines (i.e., FL-02 to FL-06) for data analysis.

A total of six flights (F) were conducted, however, due to weather and technical issues, only two, F-03 and F-04, produced reliable datasets. The details of these two flights are presented in Table 1. The flight patterns were flown in successive alternating directions, designed to provide overlapping data (i.e., 20–30% overlap) at the edge of each flight line. The overlap ensured total coverage of the area and allowed for the development of good quality mosaicked images. Taking into account the dynamic nature of a fire, the flight lines were planned according to the information provided by the local fire management headquarters. To cover the full extent of SLK-37, eight flight lines were preplanned and collected twice during both nighttime and daytime flight campaigns. A total of 32 flight lines (FL) were flown during both daytime (16 flight lines) and nighttime (16 flight lines) operations between 2 August and 3 August 2017. Active fire was present on five of these eight flight lines, which were selected for data analysis. The overall area (~6.5 km by 7.5 km), covered by these five flight lines (i.e., FL-02 to FL-06), is shown in Figure 1C.

Table 1. Summary of presented FLIR data acquired over the Sioux Lookout (SLK-37) site, located near Pickle Lake, Northern Ontario, Canada, during the 2017 campaign. Average altitude and ground speed were extracted from the aircraft's avionics GPS system.

Flt No.	Date	Time Period	No. of Lines	Start Time (EST)	End Time (EST)	Altitude (m MSL)	Ground Speed (Knots)
F-03	August 2 [a]	Night	16 [b]	22:20	23:37	2888.2	92.5
F-04	August 3	Day	16 [b]	10:26	11:41	2915.1	93.9

[a] UTC = Local (EST) +5. [b] Eight flight lines acquired twice.

2.2. FLIR Sensor

The thermal system used in this study was the FLIR SC8303 sensor (FLIR Systems Inc., Wilsonville, OR, USA) that covers the 3.0 µm to 5.0 µm MWIR region (Table 2). For this project, the infrared sensor was equipped with a 3.74 µm narrow bandpass flame filter. This flame filter has the same spectral band center location as the S7 and F1 channels of the Sentinel-3a SLSTR instrument [12].

Table 2. Summary of key characteristics of the SC8303 FLIR MWIR sensor.

Characteristic	Value
Spectral Range (µm)	3.0–5.0
Detector Pitch (µm)	14
Frame Rate (Hz)	Up to 125
Resolution (pixels)	1344 × 784
Temperature Accuracy	±2 or 2%
Standard Temperature Range (°C)	−20 to +350
Operating Temperature Range (°C)	−40 to 50
Weight without lens (kg)	4.5

The FLIR instrument can acquire data at different integration times in succession. An integration time is defined as the length of time (ms) that the sensor's shutter is open for a single frame. Because wildfires are unique situations that have a broad array of temperatures that can range from ambient up to nearly 1000 °C [35], the integration time has a great impact on the imagery collected by the sensor. Short integration times only capture high temperature values, while long integration times are required to measure low temperature values. As such, the ability to collect multiple images at different integration times allows for a high dynamic range image and minimizes saturation by extreme values. In any digital imager the measured energy causes the detector to produce a signal voltage that is transferred to an A/D converter and results in a specific digital count related to the voltage amplitude. The FLIR sensor's 14-bit dynamic range creates count values from 0 to 16,383 proportional to the incident energy at that integration time. These digital counts can then be transformed into temperature and radiance values via calibration relationships that are specific to the individual detector. Different integration times are selected so that the dynamic range of the signal counts can be maximised to ensure proper confidence in the measurement and visualization of the fire.

For the FLIR instrument in this work, pre-flight ground testing and laboratory measurements were performed in July 2017 prior to its installation aboard the aircraft to ensure accuracy of measurements and identify optimal integration times [36]. The full range of the blackbody source (0–9999) was captured under four integration times: 1.4 ms, 0.3 ms,

0.04 ms, and 0.0021 ms. Data was acquired using all four integration times during the 2017 flight campaign.

2.3. Twin Otter Survey Aircraft Hardware

While in the aircraft environment, there are three separate sensor systems used to capture the wildfire data. The first is the FLIR sensor that collects imagery at a rate of 30 Hz (i.e., 4 × 7.5 Hz given the four integration times) and is GPS time-tagged from an Inter-Range Instrumentation Group (IRIG) Time B, a generalized time measuring device for electronics, whose category "B" outputs 100 Hz data. The second system, the Databoss, records a spatial subset of the FLIR data at 30 Hz, as well as real-time aircraft system GPS information (i.e., GPS time, latitude, longitude, and heading) at 1 Hz. This system is in place for the on-board survey team to see the FLIR data as it is recorded and that the ground target area is properly acquired. This is necessary for real-time quality assurance to ensure no frames are dropped due to lack of available CPU resources, as the FLIR recording software pauses the live IR camera feed while recording. Lastly, the aircraft's onboard Data Acquisition System (DAS) collects raw IMU data at 100 Hz and raw GPS pseudo-range data at 1 Hz. Typically, the Twin Otter raw GPS is processed using a nearby GPS base station to arrive at differential GPS data, and this is post-processed with the raw IMU data in a Kalman filter [37]. This process outputs a combined position/attitude solution at 100 Hz which is then associated with the GPS-time tagged FLIR data.

There were two separate points of failure in this process during the 2017 campaign: (1) the FLIR sensor never received proper timing information from the IRIG Time B, and (2) the aircraft's IMU failed to record at the correct frequency. As a result, there was no GPS time to associate the FLIR imagery with the rest of the GPS information, and no high-resolution Kalman filter combined solution that could be calculated.

3. Methods

3.1. Geocorrection Process

To minimize the impact of the hardware issues encountered during the 2017 field campaign, an alternative geocorrection method was developed (Figure 2). The methodology can be divided into the following steps: Databoss alignment, initial geolocation, frame registration, and gridding. This methodology was developed and performed using MATLAB Version 2016a, and the Image Processing Toolbox (The MathWorks, Inc., Natick, MA, USA). The specific code for this process can be found in [38].

3.1.1. Initial Imagery Setup and Databoss Alignment

The first step in the process was to export each FLIR video frame for all integration times into separate mat files using the FLIR ResearchIR software Version 4.90.9.30 (FLIR Systems Inc.). Each integration time was exported for each flight line and processed as a complete dataset. The following step was the alignment to the Databoss GPS information. By comparing the recorded Databoss video, used as a secondary viewing system in-flight, with the raw FLIR imagery, a linear transformation from the raw FLIR frame number to the Databoss GPS could be found. This gave us access to real-time processed GPS position, altitude, and heading at 1 Hz.

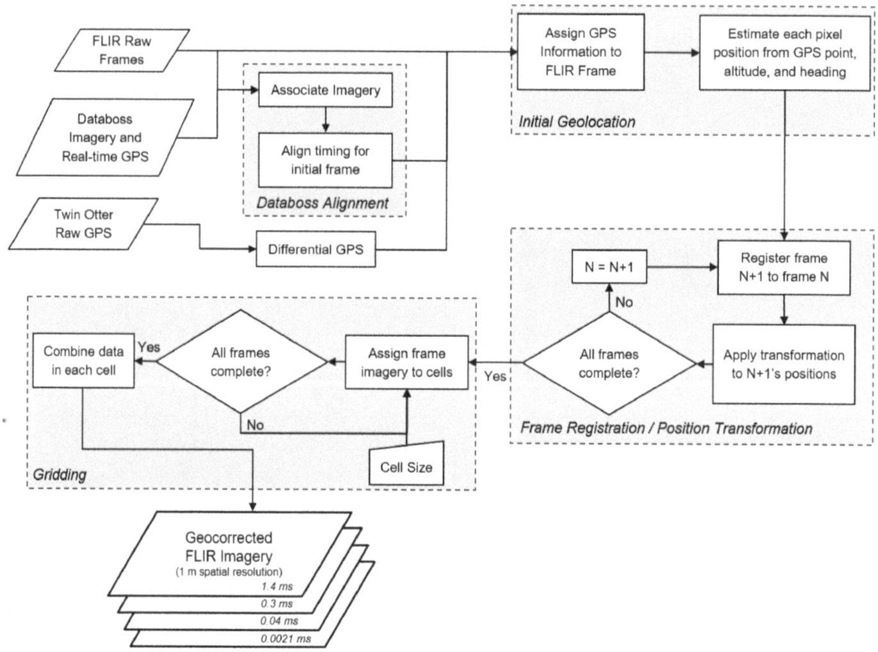

Figure 2. Workflow representation of the adjusted data and processing steps to compute GPS-tagged orthoimages of FLIR airborne data.

3.1.2. Initial Geolocation

Using the associated 1 Hz GPS data found in the previous step, an initial geolocation could be determined. Due to lack of IMU information, these readings were assumed to be recorded during flat and level flight (i.e., the sensor is always pointed nadir). Taking into account the lever arm distance of the GPS antenna to the FLIR sensor, we could use the GPS information and FLIR lens properties to calculate the horizontal and vertical fields of view, H_{FoV} and V_{FoV}, respectively:

$$H_{FoV} = z_{center}\left(\frac{S_H}{EFL}\right), \quad (1)$$

$$V_{FoV} = z_{center}\left(\frac{S_V}{EFL}\right), \quad (2)$$

where, S_H and S_V are the sensor horizontal and vertical dimensions, respectively, EFL is the effective focal length, and z_{center} is the aircraft-ground separation below the central pixel (Figure 3). All variables are measured in meters. First assuming a heading of 0° (i.e., North), each pixel was assigned a GPS location based on the frame size (1344 × 784 pixels):

$$x_{pos,0} = \left(x_{center} - \frac{H_{FoV}}{2} : \frac{H_{FoV}}{1344} - 1 : x_{center} + \frac{H_{FoV}}{2}\right), \quad (3)$$

$$y_{pos,0} = \left(y_{center} - \frac{V_{FoV}}{2} : \frac{V_{FoV}}{784} - 1 : y_{center} + \frac{V_{FoV}}{2}\right), \quad (4)$$

where x_{center} and y_{center} are the GPS coordinates at the central pixel. Note that this required the center pixel's coordinates to be in the Universal Transverse Mercator (UTM) system, as H_{FoV} and V_{FoV} were measured in metres.

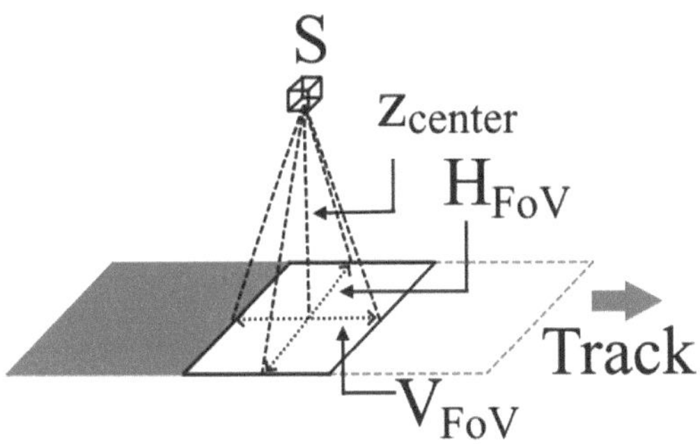

Figure 3. Calculation for the horizontal (H_{FoV}) and vertical (V_{FoV}) field of view (FoV) with respect to the sensor (S) and the aircraft-ground separation below the central pixel (Z_{center}).

Next, a rotation matrix was applied to the coordinates using the GPS heading, θ, with the center pixel as the origin:

$$\begin{bmatrix} x_{pos} & y_{pos} \end{bmatrix} = \begin{bmatrix} \cos\theta & -\sin\theta \\ \sin\theta & \cos\theta \end{bmatrix} \cdot \begin{bmatrix} -x_{pos,0}+x_{center} \\ y_{pos,0}-y_{center} \end{bmatrix} + \begin{bmatrix} x_{center} & y_{center} \end{bmatrix} \quad (5)$$

This gave a final matrix where each pixel (whose value was in raw counts) was assigned an initial estimated Easting (x_{pos}) and Northing (y_{pos}) location (Figure 4).

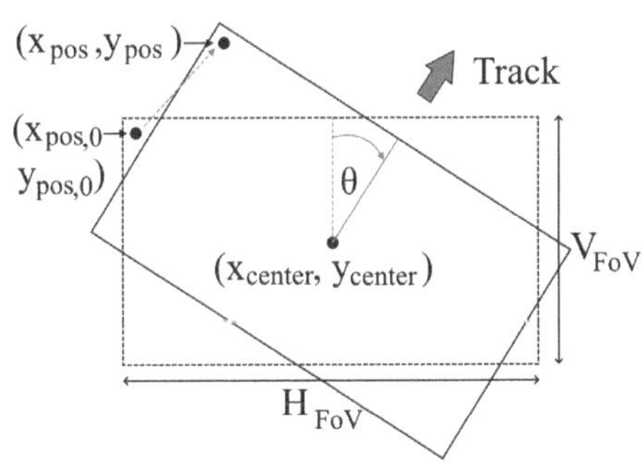

Figure 4. The initial pixel position assignment ($x_{pos,0}$, $y_{pos,0}$), shown in the dashed-line box. The post-rotation position (x_{pos}, y_{pos}) is shown in the solid-line box. H_{FoV} and V_{FoV} are the horizontal and vertical dimensions.

3.1.3. Frame Registration

Once the data had initial geolocation estimates assigned, the process of building a geocorrected image using frame registration could begin. This was the most time-consuming process in the methodology presented here, as it is an iterative calculation, checking each pair of chronological frames in the IR video. Image registration is a process of automated data alignment, based on finding similarity in two images, and attempting to fit them together. This process is completed by centering a single image (the "fixed" image) and transforming a second image (the "moving" image), such that similarities between the two line up. The transform, T, is defined as:

$$T = \begin{bmatrix} a & b & 0 \\ c & d & 0 \\ e & f & 1 \end{bmatrix}, \quad (6)$$

where the values of a to f are variable, depending on the type of transformation that is being allowed. Given the expected small variability in space and amplitude from one frame to the next, and the previous assumption of a reasonably flat, level, and straight flight line, the simplest transform, translation ($T_{translation}$), was used for the processing:

$$T_{translation} = \begin{bmatrix} 1 & 0 & 0 \\ 0 & 1 & 0 \\ t_x & t_y & 1 \end{bmatrix}, \quad (7)$$

where t_x is the translation in the x direction and t_y is the translation in the y direction. Once $T_{transform}$ was developed for a pair of frames, it was applied to the moving frame to arrive at the new, "registered", coordinates:

$$f_r(x, y) = f_f(x, y) + f_f\left(t_x x_x + t_y y_x,\ t_x x_y + t_y y_y\right), \quad (8)$$

where $f_r(x,y)$ is the moving frame's registered coordinates, $f_f(x,y)$ is the fixed frame's coordinates, x_x is the change in Eastings (in meters) of one x-value shift, x_y is the change in Northings (in meters) of one x-value shift, y_x is the change in Eastings (in meters) of one y-value shift, and y_y is the change in Northings (in meters) of one y-value shift. The last four of these variables are calculated from:

$$x_x = y_y = d_{pixel} \cos\theta, \quad (9)$$

$$x_y = y_x = d_{pixel} \sin\theta, \quad (10)$$

where d_{pixel} is the length of the (square) pixel.

Applying this transformation, new coordinates for the moving frame were created, based on the fixed image's coordinates. By iteratively moving through each frame, updating the moving and fixed frames in sequence, this process was repeated until all frames were registered (Figure 5).

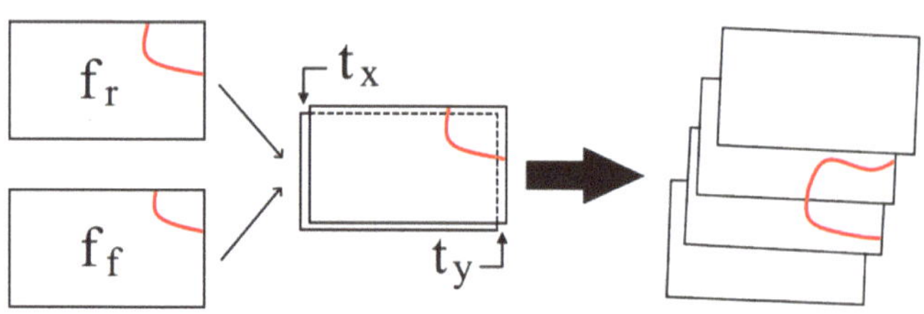

Figure 5. An outline of the general image registration process on an idealized feature (red line). Note that the spacing between the frames in the far right set have been exaggerated for clarity, and the middle pair of frames are more indicative of the usual spatial shift from frame to frame. f_r is the moving frame's registered coordinates, f_f is the fixed frame's coordinates, t_x is the translation in the x direction, and t_y is the translation in the y direction.

3.1.4. Gridding

With all frames aligned, a method to combine them into a single mosaic was required. To do this, a grid system, $G_{(x, y)}$, was developed and data points from each frame were assigned to the grid in their respective grid cell locations. The cell size was chosen to be 1 m by 1 m, based on the average spacing between pixels, which was generally between 0.7 m and 0.8 m. This size ensured that no "holes" in the data would occur where no raw pixels existed. With many overlapping frames, most grid cells contained multiple data points, not all of which were the same value, and therefore a process to combine these was developed. A simple approach to this problem would be to take the mean of all data that falls within a cell. However, this causes an undesirable smoothing of the data. Due to the sharp edges of the fire features, this smoothing can mean a significant change in grid cell value. Instead, a modified minimum distance approach was used where the pixel that came from the FLIR frame whose center was closest to the grid cell of interest was found and its value used in the grid cell (Figure 6).

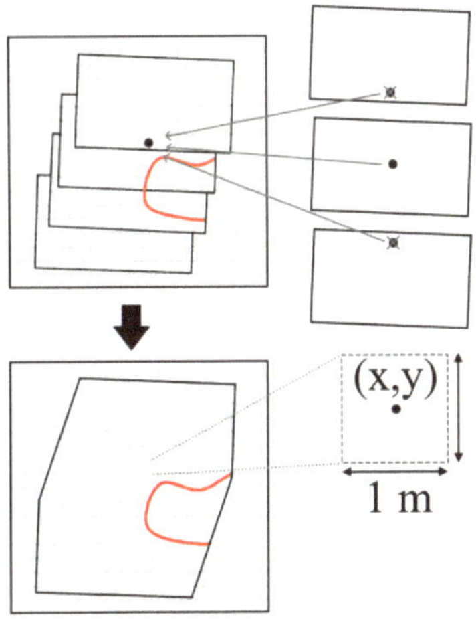

Figure 6. An exaggerated example of the gridding process on an idealized feature (red line). The dashed gray lines indicate the grid of cells. The left grids show the before and after representation of imagery during the gridding process. The three frames on the top right exemplify the process of determining what value should be in the final cell. Example shown for a 1 m pixel size.

3.2. Georeferencing Process

The geocorrection solution required further improvements to reduce the geolocation error (≤30 m) to a more acceptable threshold for comparison to mid-resolution satellite sensors (≤30–300 m). Figure 7 shows the georeferencing processing flow. The geocorrected FLIR data was georeferenced to GeoEye satellite imagery (2 m geolocation error) available in ArcGIS 10.6 (Environmental Systems Research Institute, Redlands, CA, USA). To reduce registration errors, the image with the longest integration time, 1.4 ms, was used as the baseline as it contained ambient temperature features. Firstly, the longitude (X dimension), latitude (Y dimension) information and the FLIR data in counts units (Z dimension) were exported from MATLAB and a geocorrected FLIR image was built in ENVI 5.5 (Exelis Visual Information Solutions, Boulder, CO, USA). Next, pseudo-ground control points (PGCPs) were registered throughout the FLIR image based on the GeoEye imagery in ENVI. To facilitate the selection of PGCPs, a point grid (50 m by 50 m) was created with GeoEye imagery over-layered in ArcMap. The first PGCP was selected over a recognizable area (e.g., riverbed), while the rest of the PGCPs were selected manually using the 50 m point grid at different distances from the initial point. Priority was given to PGCPs over recognizable areas (i.e., riverbed located around the image edges) and to points that did not cause significant image wrapping and showed less than 30 m error. Once a range of PGCPs was chosen, the FLIR imagery was georeferenced to the UTM Zone 16 N, WGS84 coordinate system at a pixel size of 1 m. The other integration times (i.e., 0.3 ms, 0.04 ms and 0.0021 ms) were subsequently georeferenced to the 1.4 ms integration time image for each of the flight lines.

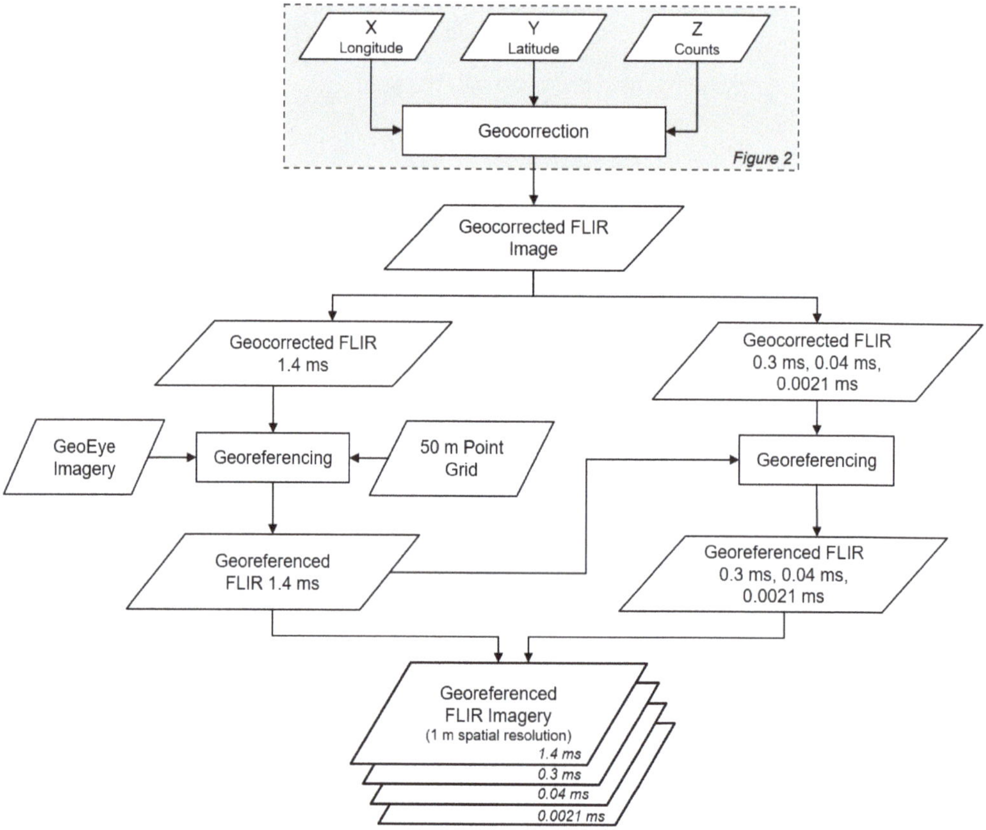

Figure 7. Workflow displaying the georeferencing methodology developed to reduce the geolocation error of the acquired FLIR data.

3.3. Data Products

As mentioned in Section 2.2., calibration relationships between counts and temperatures were derived from the laboratory measurements. As temperatures are derived directly from the measured counts, calculation of the radiance produced at a specific wavelength from a specific temperature can be obtained by using the following version of the Planck function equation:

$$L(\lambda, T) = \frac{c_1}{\lambda^5\left(e^{\frac{c_2}{\lambda T}} - 1\right)} \tag{11}$$

where λ is wavelength (μm) and in this case equal to the flame filter, 3.74 μm, T is brightness temperature (temperature when the emissivity = 1), L(λ,T) is spectral radiance in Wm^{-2} sr^{-1} μm^{-1}, and c_1 is the first radiation constant (value $1.191044024 \times 10^{-16}$ Wm^2), c_2 is the second radiation constant (value 1.4387687×10^{-2} mK). As shown in Figure 8, linear calibration relationships were found between recorded counts and calculated radiance [36]. High confidence R^2 values are reported at 0.9995 for 1.4 ms, 0.9972 for 0.3 ms, 0.9993 for 0.04 ms, and 0.9996 for the 0.0021 ms integration time. The equations yielded by the linear best fit were used to calculate radiance across each frame recorded by the sensor for each integration time.

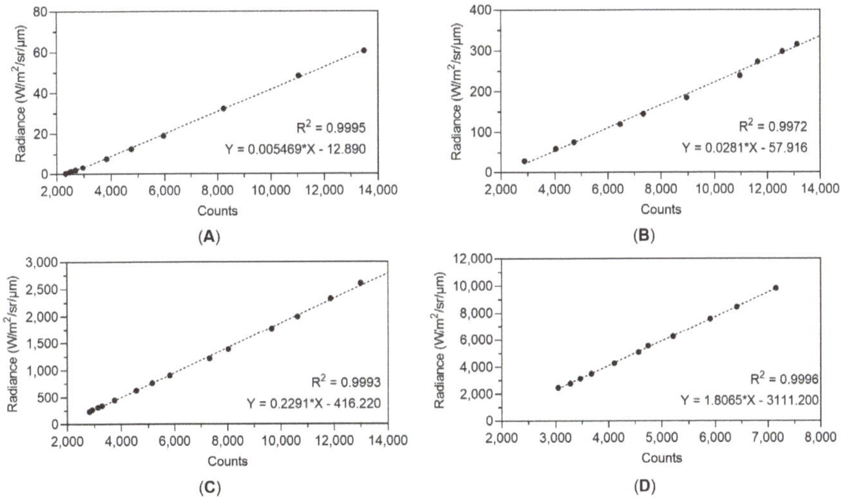

Figure 8. Relationship between counts recorded by the FLIR sensor and calculated radiance using the Planck function and a wavelength of 3.74 µm at integration times of (**A**) 1.4 ms, (**B**) 0.3 ms, (**C**) 0.04 ms, and (**D**) 0.0021 ms modified from [36]. The equations yielded by the trend line at each integration time were used to calculate radiance across each frame recorded by the sensor.

Atmospheric correction was applied to the radiance values in the imagery in order to remove the influences of the atmosphere in accordance with convention for modelling the spectral radiance of a sub-pixel wildfire event [11]. The PcModWin 6.0.0.5 (Ontar Corporation, North Andover, MA, USA) software, a MODTRAN® 6 [39] interface, was used to retrieve atmospheric variables for the correction process [40]. In this study, the standard MODTRAN parameters for sub-arctic summer atmospheric and rural aerosol models were used for the atmospheric compensation calculations.

To test the validity of the data, Fire Radiative Power Density (FRPD) was calculated using the MWIR method [11]. A threshold was applied to the data to remove the background and saturated pixel values. These background thresholds were based on data distribution at 10 Wm^{-2} FRPD for 1.4 ms, 60 Wm^{-2} FRPD for 0.3 ms, and 470 Wm^{-2} FRPD for the 0.04 ms integration time. Saturated pixels were defined as pixels with a count value above 13,500 because the detector's behaviour beyond this value is highly non-linear and may either induce noise or provide false measurements. In this case, the saturation FRPD thresholds were 1098 Wm^{-2} for 1.4 ms, 5825 Wm^{-2} for 0.3 ms, and 48,500 Wm^{-2} for the 0.04 ms integration time. No saturated pixels were recorded at the shortest integration time 0.0021 ms. No fire pixels were recorded for FL-02 and FL-03 at the 0.0021 ms integration time during both nighttime (F-03) and daytime (F-04) flights.

4. Results

4.1. Geocorrection Assessment

To confirm the validity of the geocorrection process, we compared the distribution of data before and after it was applied to the daytime FL-04 dataset, acquired on F-04 with an integration time of 1.4 ms. In Figure 9 it can be seen that, while dramatically reducing the amount of data, the general distribution of radiance values remains highly similar after the geocorrection process. This is supported by the statistics of each dataset (Table 3), which indicate that the mean, median, minimum value, and maximum value have changed by less than 2%, while the standard deviation has changed by 7.14%. The skewness has only changed by 6.23%, however the kurtosis has deviated more, with an almost 13% change from the initial distribution to the geocorrected distribution. As the primary difference between the two distributions is in the higher radiance values (tail-end of the distribution), the larger kurtosis change compared to the other statistics is expected. Overall, the statistical changes between the two distributions are low, confirming that the geocorrection process has maintained the integrity of the raw FLIR data.

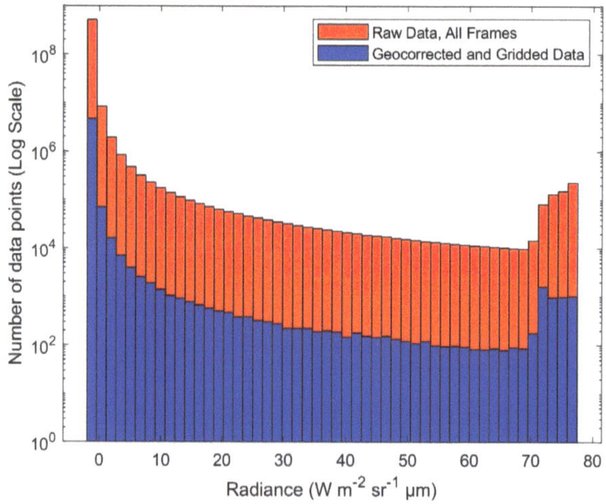

Figure 9. A histogram of the FLIR data in Radiance (Wm^{-2} sr^{-1} µm^{-1}). The distribution of count values remains very similar through the geocorrection process, even though the amount of data has been greatly reduced.

Table 3. Statistics of all frames of raw data, and the geocorrected, gridded data. Despite the reduction of data points to arrive at the gridded dataset, the data has not been greatly altered, indicated by the similar statistics of the two datasets. Note that all values, other than the total number of data points, skewness, and kurtosis, are in radiance (Wm^{-2} sr^{-1} µm).

Variable	Raw, All Frames	Geocorrected, Gridded	Change (%)
Total Data Points	527,901,696	4,701,690	99.11
Mean	−1.18	−1.20	1.70
Median	−1.44	−1.44	0.00
Standard Deviation	3.08	2.86	7.14
Minimum	−1.66	−1.64	1.21
Maximum	77.39	77.05	0.44
Skewness	21.18	22.50	6.23
Kurtosis	489.75	553.33	12.98

An example of the raw imagery during daytime FL-04, acquired along F-04 at an integration of 1.4 ms, before georeferencing is shown in Figure 10. Misalignment issues can be observed at the edges of the imagery over the water bodies (Figure 10B,C). Based on 19 selected ground control points, a Root Mean Squared Error (RMSE) between 112.12 m and 312.73 m is reported. The mean RMSE found for FL-04, F-04, before the georeferencing step is at 233.99 m ± 73.28 m (based on 20 PGCPs). The northwest corner (Figure 10B) shows lower RMSE between 112.12 m and 178.01 m, while the southeast one (Figure 10C) shows higher RMSE between 178.01 m and 312.73 m.

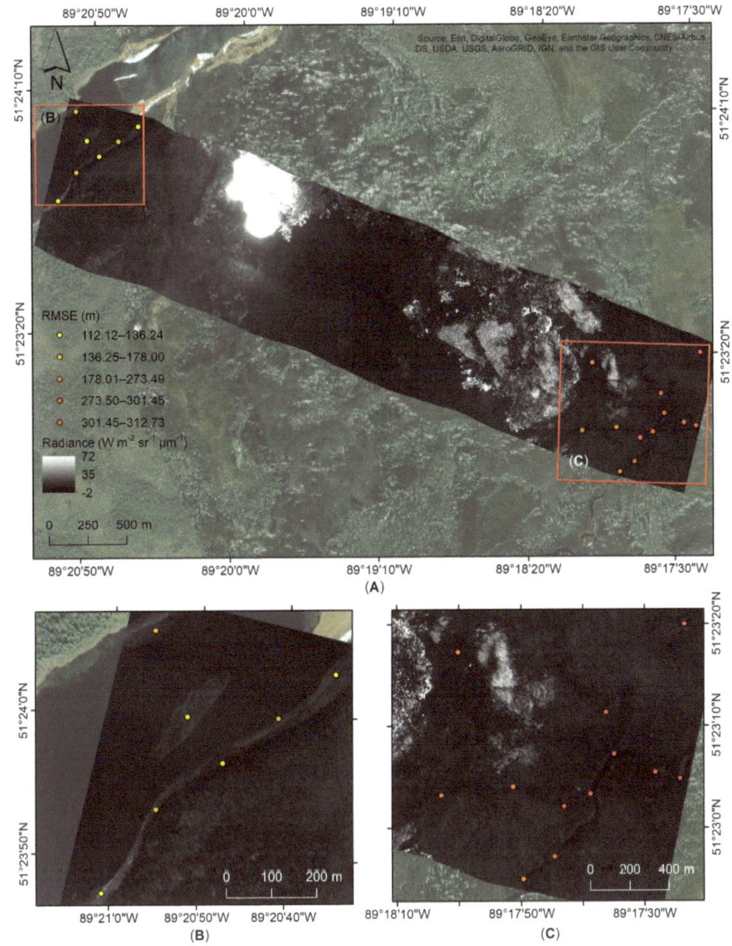

Figure 10. (**A**) An example of reported RMSE of the orthoimage (before the georeferencing process) in units of radiance (Wm^{-2} sr^{-1} μm^{-1}) of FL-04, at an integration time of 1.4 ms, from F-04 acquired on 3 August 2017. Areas showing the misalignments and RMSE found following the geocorrection process for (**B**) the northwest and (**C**) southeast corners of the flight line.

4.2. Georeferencing Assessment

Table 4 shows the geolocation errors of the 1.4 ms integration time that was used as the baseline for the georeferencing process for the five processed flight lines. The total mean RMSE reported is 11.90 m ± 7.26 m with 6.96 m ± 5.21 m for Easting and 8.66 m ± 6.65 m for Northing. Absolute Easting error ranges between 0.30 m and 24.13 m, while Northing between 0.25 m and 25.43 m. The total minimum and maximum RMSE error are 0.87 m and 29.25 m, respectively. The Northing absolute median error is reported to be higher at 7.46 m, in comparison with Easting at 5.30 m, and a total RMSE error of 10.75 m.

Figure 11 shows the reported RMSE, ranging from 0.87 m to 29.25 m, following the georeferencing process along Radiance (Wm^{-2} sr^{-1} μm^{-1}) at an integration of 1.4 ms for the 3 August (F-04) flight lines. While, the northwest river shows variable RMSE with seven PGCPs showing above 14 m error (Figure 11B), a RSME below 14.60 m is reported along the southeast riverbed (Figure 11C).

Table 4. Reported absolute (Abs.) georeferencing errors for the FLIR imagery acquired on 3 August 2017, at the longest integration time of 1.4 ms of the five processed flight lines. Results reported over 51 pseudo-ground control points (PGCPs).

Variable	Abs. Easting (m)	Abs. Northing (m)	Total RMSE (m)
Mean	6.96	8.66	11.90
Minimum	0.30	0.25	0.87
Maximum	24.13	25.43	29.25
Median	5.30	7.46	10.75
Standard deviation	5.21	6.65	7.26

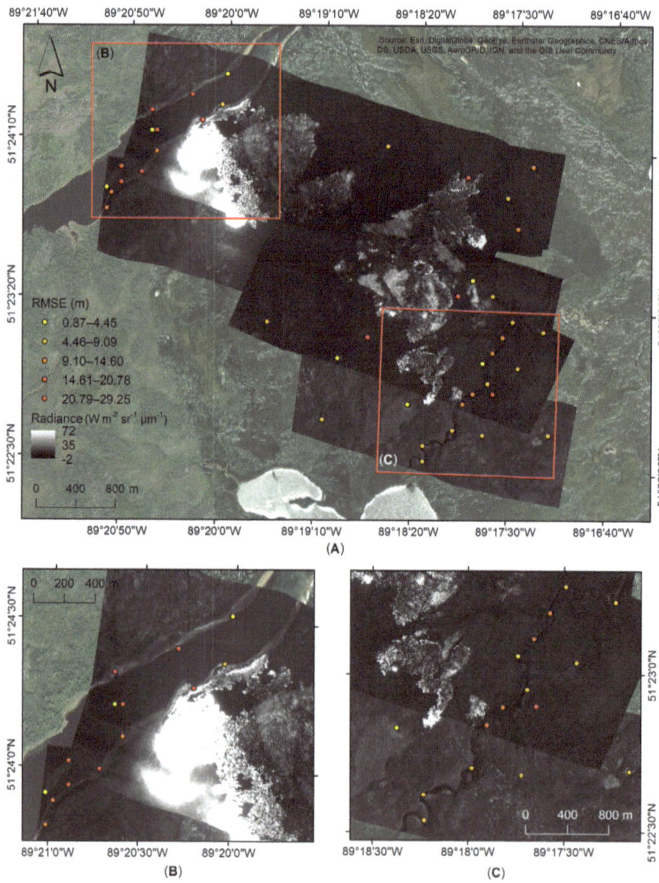

Figure 11. (**A**) Reported RMSE following the georeferencing process of the FLIR imagery in units of Radiance (Wm^{-2} sr^{-1} μm^{-1}) acquired on 3 August 2017, F-04, at the 1.4 ms integration time. Areas showing RMSE following the georeferencing process for (**B**) the northwest and (**C**) southeast corners of the flight lines.

To minimize geolocation errors between the 2 and 3 August datasets, the 2 August flight lines at 1.4 ms was georeferenced to the 3 August ones (Table 5). The mean total RMSE between the two datasets is 3.86 m, with a mean absolute Easting error of 2.38 m and 2.53 m for Northing. The total RMSE errors ranges between 0.17 m and 14.28 m with values between 0.01 m and 9.42 m for Easting and 0.04 m and 13.17 m for Northing. The total median RSME is 3.21 m at a standard deviation of 2.92 m.

Table 5. Reported absolute (Abs.) georeferencing errors for the data acquired on 2 August (F-03) at the longest integration time of 1.4 ms to the baseline imagery collected at an integration of 1.4 ms on 3 August (F-04) during the 2017 campaign over the SLK-37 site.

Variable	Abs. Easting (m)	Abs. Northing (m)	Total RMSE (m)
Mean	2.38	2.53	3.86
Minimum	0.01	0.04	0.17
Maximum	9.42	13.17	14.28
Median	2.02	1.55	3.21
Standard deviation	1.96	2.76	2.92

Following the georeferencing process, the average RMSE errors is 15.11 m $\pm$ 3.54 m for the 2 August dataset and 13.53 m $\pm$ 2.60 m for 3 August (Table 6). Between 9 and 13 PGCPS were used in the georeferencing processing depending on the flight line. An RMSE between 8.73 m for FL-02 at 0.3 ms integration time, and 24.5 m for FL-05 for an integration time of 1.4 ms, can be observed. The 3 August dataset overall show lower RMSE errors between 8.83 m for FL-02 at an integration time of 1.4 ms, and 17.44 m for FL-05 at 0.04 ms. This is expected at the daytime flight show more landscape features that were used in the georeferencing process.

Table 6. Reported total RMSE (m) of pseudo-ground control points (PGCPs) for data acquired during night flight (F-03, 2 August) and day flight (F-04, 3 August) during the 2017 campaign over the SLK-37 site.

Flight Line		Night Flight (August 2)		Day Flight (August 3)	
	IT (ms)	PGCPs	RMSE (m)	PGCPs	RMSE (m)
FL-02	1.4	9	10.85	10	8.83
	0.3	9	8.73	10	8.83
	0.04	9	9.67	9	9.71
FL-03	1.4	9	16.22	9	14.28
	0.3	10	14.54	9	14.28
	0.04	10	14.54	9	14.93
FL-04	1.4	9	17.32	13	11.76
	0.3	10	12.46	13	11.76
	0.04	9	13.36	12	12.58
	0.0021	9	13.36	12	12.39
FL-05	1.4	9	24.50	10	16.65
	0.3	10	16.99	10	16.65
	0.04	9	18.21	12	17.44
	0.0021	9	18.21	12	17.33
FL-06	1.4	10	17.33	10	13.84
	0.3	10	14.57	10	13.84
	0.04	9	15.58	12	14.64
	0.0021	9	15.58	12	14.62

4.3. Data Products

After thresholding (Section 3.3), all integration times' datasets were superimposed upon each other creating a new combined dataset. Pixels with radiance below 1 Wm^{-2} sr^{-1} μm^{-1} were assigned as background and were not taken into consideration for the purpose of this analysis. Figure 12 shows the radiance (Wm^{-2} sr^{-1} μm^{-1}) captured during the nighttime flight, F-03, on 2 August, where the presence of fire is detected over a few locations. In one particular area (Figure 12A), pixels with radiance values of up to 1765 Wm^{-2} sr^{-1} μm^{-1} are observed. A lower intensity radiance is shown in Figure 12C where fire pixels are present in different areas with a few pixels showing up to 662 Wm^{-2} sr^{-1} μm^{-1} and the majority of the fire pixels being between 6 and 26 Wm^{-2} sr^{-1} μm^{-1}. In the southeast corner, more combustion pixel are observed at a radiance of up to 7 Wm^{-2} sr^{-1} μm^{-1} with a few pixels up to 13 Wm^{-2} sr^{-1} μm^{-1} (Figure 12B) and a small hotspot of up to 895 Wm^{-2} sr^{-1} μm^{-1} (Figure 12D).

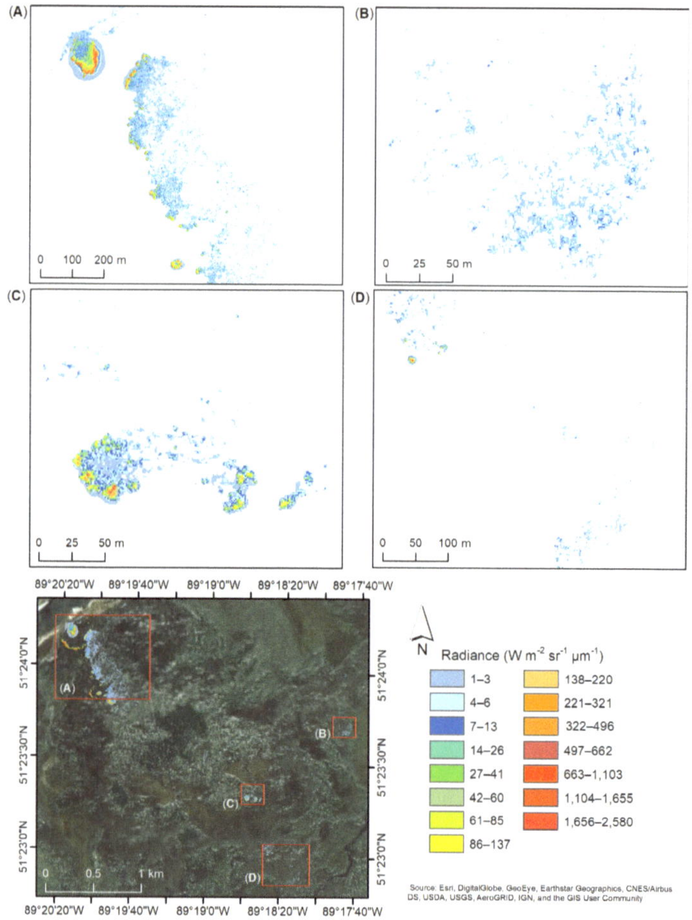

Figure 12. Radiance (Wm^{-2} sr^{-1} μm^{-1}) captured during the nighttime flight, F-03, acquired on 2 August 2017, over four different area (**A–D**) of the SLK-37 site. Radiance of the 1.4 ms, 0.3 ms, 0.04 ms, and 0.0021 ms integration times superimposed based on the set thresholds (Section 3.3). Radiance below 1 Wm^{-2} sr^{-1} μm^{-1} (white) not displayed.

Similar trends are observed for the daytime flight, F-04, collected on 3 August (Figure 13). In the northwest side of the site (Figure 13A) the presence of fire extended and increased in intensity with radiance up to 2579 Wm^{-2} sr^{-1} μm^{-1} over two areas. These areas show marginally more thermal halo pattern generated by the hot soot within the smoke plume. The same mid-center area (Figure 13C), as that shown in Figure 12C, appears to have diminished in intensity, with radiance maximum of only 321 Wm^{-2} sr^{-1} μm^{-1}, and increased within the area with a few pixels of up to 662 Wm^{-2} sr^{-1} μm^{-1}. As recorded during the nighttime flight (FL-03), the southeast corner of the site shows radiance up to 13 Wm^{-2} sr^{-1} μm^{-1} (Figure 13B) and a small hotspot with a maximum radiance of 850 Wm^{-2} sr^{-1} μm^{-1} in the north-west area (Figure 13D).

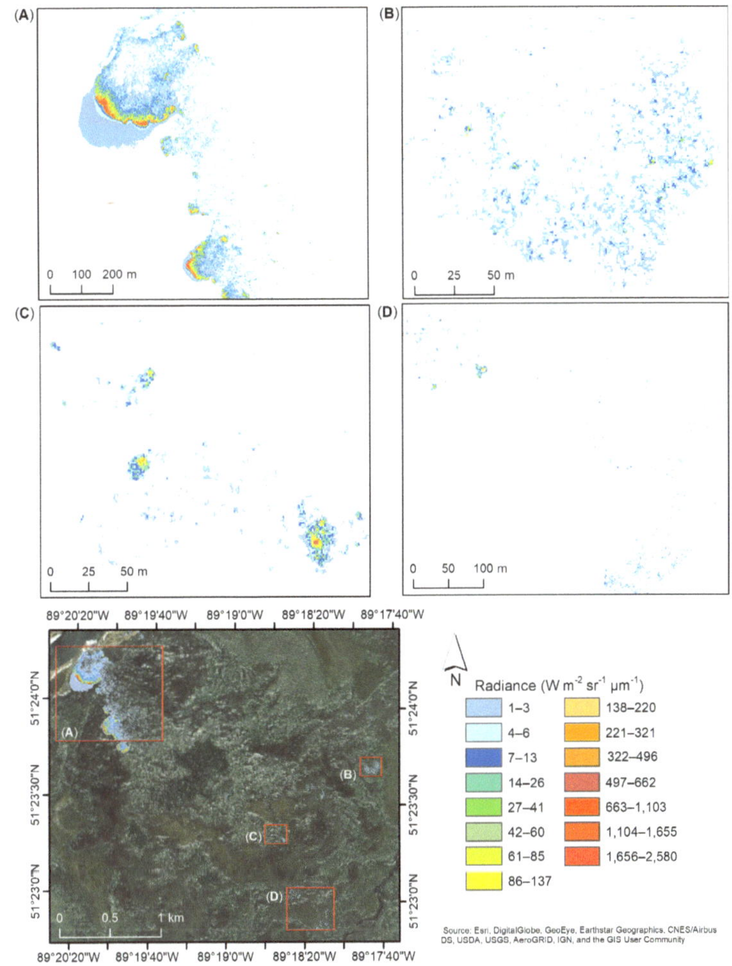

Figure 13. Radiance ($Wm^{-2}\ sr^{-1}\ \mu m^{-1}$) captured during the nighttime flight, F-04, acquired on 3 August 2017, over four different area (**A–D**) of the SLK-37 site. Radiance of the 1.4 ms, 0.3 ms, 0.04 ms, and 0.0021 ms integration times superimposed based on the set thresholds (Section 3.3). Radiance below 1 $Wm^{-2}\ sr^{-1}\ \mu m^{-1}$ (white) not displayed.

Probability density and standard statistics (minimum, maximum, mean, standard deviation and sum) of FRPD (kWm^{-2}) were computed for each flight line at the different integration times acquired during both nighttime and daytime flights (Figure 14). Given the dynamics of fire, FRPD for each flight line and integration time was calculated separately.

For FL-02 (Figure 14A), higher FRPD are observed during the nighttime flight (F-03) with a maximum of 16.98 kWm^{-2} and a mean of 4.69 kWm^{-2} for the 0.04 ms, in comparison with the daytime flight (F-04) at a maximum of 15.40 kWm^{-2} and a mean of 4.16 kWm^{-2}. Similar trends are seen for FL-03 (Figure 14B) where overall lower FRPD is recorded during the daytime. The FRPD recorded at the longer integration times decreased between nighttime flight (F-03) and daytime flight (F-04) from a sum of 1004.55 kWm^{-2} to 550.77 kWm^{-2} at 1.4 ms and from 1151.72 kWm^{-2} to 633.09 kWm^{-2} at 0.3 ms. Even though the FRPD at 0.04 ms is larger during the daytime flight (F-04), from a maximum of 16.73 kWm^{-2} to 36.10 kWm^{-2}, the total area of FRPD decreased from 816.95 kWm^{-2}

to 622.20 kWm^{-2} between night (F-03) and the day (F-04), respectively. This suggests that fire reduced in overall intensity but increased in FRPD over the area covered by FL-03. In contrast, higher FRPD were recorded over flight lines FL-04 to FL-06 (Figure 14C–E) in comparison to FL-02 and FL-03, but also during the daytime flight in comparison with the nighttime one. The highest FRPD is observed (Figure 14E) in the area covered by both flight lines FL-05 and FL-06 with a value of 47.65 kWm^{-2} for FL-06 at 0.04 ms. For FL-05 (Figure 14D), higher intensity FPRD is recorded during the day, from a maximum of 31.96 kWm^{-2} to 46.79 kWm^{-2} with a total FRPD also increasing from 25,531.18 kWm^{-2} to 54,959.26 kWm^{-2} from nighttime flight (F-03) to daytime flight (F-04).

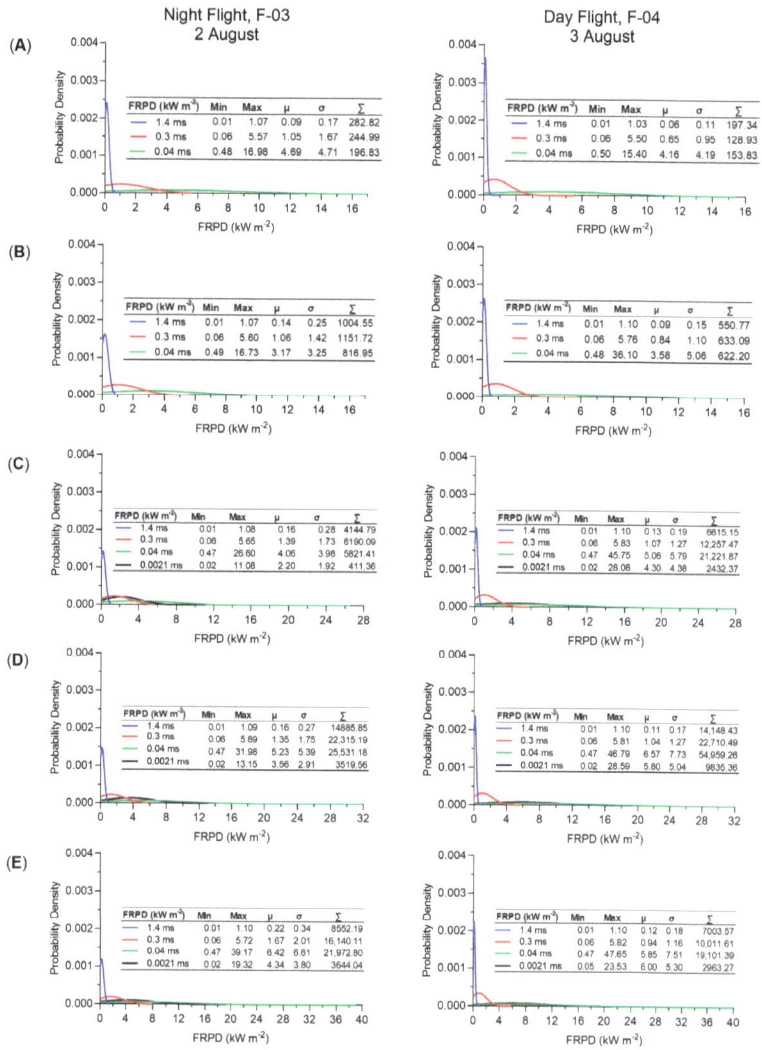

Figure 14. Probability density and statistics (minimum, maximum, mean, standard deviation and sum) of FRPD (kW^{-2}) at the different integration times (i.e., 1.4 ms, 0.3 ms, 0.04 ms, and 0.0021 ms) for (**A**) FL-02, (**B**) FL-03, (**C**) FL-04, (**D**) FL-05, and (**E**) FL-06 acquired during night flight (F-03) and day flight (F-04) for the 2017 campaign.

Figure 15 showcases an example of the highest FRPD recorded over FL-05 during both nighttime and daytime flights. The distributions of FRPD for the night (Figure 15A) and day (Figure 15B) flights clearly demonstrate the bias toward higher FRPD values during the day (i.e., increased flaming combustion) and lower FRPD during the night (i.e., increased smouldering combustion), which is consistent with diurnal trends in fire behaviour. As observed in the radiance (Figure 13A), the day flight FRPD shows marginally more halo generated by the soot in the smoke plume.

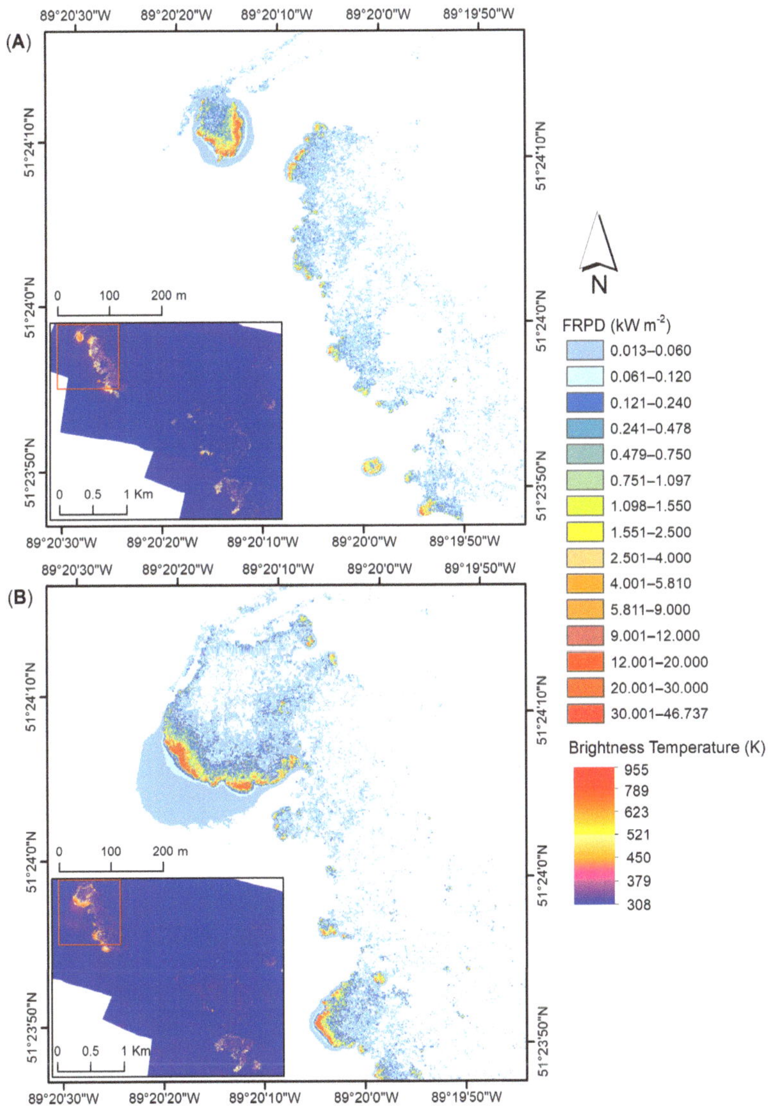

Figure 15. Example of FRPD (kWm^{-2}) captured over FL-05 during both (**A**) nighttime flight, FL-03, on 2 August, and, (**B**) daytime flight, F-04, on 3 August, during the 2017 campaign over the SLK-37 site. Calculated FRDP of the 1.4 ms, 0.3 ms, 0.04 ms, and 0.0021 ms integration times superimposed based on the set thresholds (Section 3.3). Overall inset area shown in brightness temperature (K) as a reference.

5. Discussion

Our study provides a novel geocorrection methodology for a case where GPS and IMU airborne equipment malfunctions occurred during a 2017 airborne wildfire campaign collecting FLIR MWIR images. This solution required developing an approach to assign GPS coordinates to FLIR video imagery, followed by combining the data through image registration. This process alone, while effective, required an additional georeferencing method to reduce the RMSE of the positional accuracy. Using this approach, the RMSE was reduced from a baseline error of up to 300 m (Figure 10) to an average of 15.11 m for the nighttime and 13.53 m for the daytime datasets (Figure 11, Table 6). By comparison, under optimal equipment conditions, a similar method for the geocorrection of ATIR FireMapper 2.0 airborne infrared imagery, acquired to estimate fire spread rates in southern California, shows a geocorrection RMSE of 0.33 pixels (4.13 m) [24]. In addition, Ref. [41] reports up to 2 m along-track error when using a georeferencing gridding in combination with associated GPS and IMU information for processing of airborne Specim AISA Eagle and Hawk hyperspectral data. Using a frame center matching technique, Ref. [42] was able to reduce misregistration errors of multi-temporal ADAR 5500 airborne imagery. The study found RSME errors of up to 8.30 m when using automatic ground control points during the georeferencing process in combination with introducing a DEM during the geocorrection process. Using the same frame registration method, Ref. [43] found an average 2.1 pixels (0.32 m) geolocation error, which can increase to 4.4 pixels (0.67 m) for moderate terrain (95–105 m elevation, 10–20 m slope) and up to 16.2 pixels (2.47 m) for extreme terrain (145–155 m elevation, 20–30 m slope), after orthorectification of ADS40 imagery. These methods show lower errors in comparison with our study due to the use of DEM, LIDAR and GPS/IMU data in the geocorrection process.

Our study also relied on GeoEye imagery available through the ArcGIS software, which has a reported error of 2 m [44]. Although not necessarily optimal by comparison, our approach was successful in recovering what otherwise may have been unusable data. Given the initial purpose of comparing this dataset to satellite imagery (30 m resolution) we considered the overall below 30 m RMSE error (Figure 11, Tables 4–6) to be acceptable when taking into consideration the extra cost and processing time involved in reducing further the geolocation error.

The distribution of raw FLIR data as compared to the gridded and geocorrected dataset (Figure 9) and associated statistics (Table 3) are key to showing that this method is not biasing the data. The gridding process, which effectively distills 99% of the frame data to arrive at the final, single dataset, should not be averaged. As described in Section 3.1.4., the sharp peaks of the fire, represented by the tail ends of the distributions, would be affected by an averaging process, smoothing out the results. Keeping the tail of the distribution is important to maintain, as this is where most of the information in the data products (e.g., FRPD) is held. Even without averaging, it is clear that the largest difference between the two distributions is in the tail, indicated by the change in the kurtosis (12.98%) values. This change, while still overall small, is due to the sampling process during gridding. The peaks of the fires were generally one or two pixels large, and as such, they were more likely to be affected by the cell size being slightly larger than the raw pixel size. This caused the peak values (the right side of the distribution tail) to be disproportionately lost during the gridding's subsampling process, thus leading to a slight increase in representation on the left side of the tail (the more numerous fire values surrounding the peaks). However, the shift is small (radiance of high 80 s to low 70 s) and contained to the tail of the distribution. This is supported by the statistics, particularly the change in mean (1.70%), median (0%), and standard deviation (7.14%) being low. For these reasons, we are confident that the gridding process did not bias or alter the raw FLIR data by a significant amount.

A primary limitation to this method is that the initial georeferencing process assumes the sensor was pointed nadir in the aircraft. Unfortunately, without any IMU information this will always be a necessary assumption (unless of course the camera is mounted at an angle on the airborne platform). This assumption is the key reason that the georeferencing step was developed in our methodology. Even with a level flight path, small deviations (primarily variations in the instantaneous roll and pitch angles) will add up over the course of the flight line, leading to an accumulation of positional errors. This can be clearly seen in Figure 10, where the eastern side of the image has a greater positional inaccuracy as compared to the western side where the image registration process began. The initial frame of data will also influence this "drift" of error. We use the GPS data and ground elevation to determine the coordinates for all pixels in this initial frame. However, all subsequent frames are registered against this, and as such, any error in the locations due to a small pitch or roll during this initial frame will be carried through all other frames. The georeferencing step greatly alleviates these overall errors, as can be seen in the final positional error analysis (Figure 11, Table 6). An area for future investigation would be to determine if there is a method of using any GPS information (in this case, the 1 Hz real-time processed locations) during the image registration process to develop a combined solution. This would likely assist in reducing the drift error, thus requiring less adjustment during the georeferencing process.

The decision to limit the image registration process to only the translation transform came about after experimentation and analysis was performed with the other transformation options, namely rotation, scaling, and shearing factors. Due to the 7.5 Hz collection rate for an individual integration time, combined with the straight and level flying during the survey, little would change between most frames. Therefore, even when given the additional freedom of transforms, the translation parameters would dominate.

Based on the MWIR radiance method [11], our study shows FRPD values up to 39.17 kWm^{-2} (Figures 14 and 15A) during the nighttime and up to 47.65 kWm^{-2} (Figures 14 and 15B) for data acquired during daytime. This is consistent with diurnal trends in fire behaviour [45] where increased flaming combustion is observed during the day and increased smouldering combustion during the night. Even though our study found an FRPD of up to 48 kWm^{-2}, based on the probability density results (Figure 14) most pixels show values up to 6 kWm^{-2} with a few showing the fire front (Figure 15) between 9 and 47 kWm^{-2}. Though occupying a far larger combustion zone, our FRPD and temperature range observations correspond with measurements taken in study [11] that describes the MWIR method. Over a 50-min fire, the study found a peak of 7 kW with most reported values between 5 and 2 kW. Furthermore, at temperatures of 900–950 K they report FRPD values (referred to as FRP) of up to approximately 50–60 kWm^{-2}. In comparison, our data shows temperatures up to 955 K (Figure 15) for a FRPD of up to 47 kWm^{-2}. Other studies across the literature reported similar fire intensity data ranges. Using an AGEMA 550 thermal camera, Ref. [9] found an FRP of up to 6 kW for smouldering fire and between 15 and 35 kW FRP for flaming fire for different small-scale fire scenarios (1.0–1.5 cm pixel resolution) collected at an off nadir angle. In the same study, fire intensity was investigated over an open burn plot acquired with the same AGEMA 550 thermal camera attached to a helicopter. They found an FRPD (referred to as FRF) between 12 and 14 KWm^{-2} of a 500 s controlled burn fire. Another study reports a radiative fire intensity between 50 and 250 kWm^{-1} when using the AGEMA 550 and FLIR SC6703 sensors [31]. Based on the current literature, our FRPD values are within expected/compared values in other studies.

The initial goal of this survey was to compare the airborne infrared imagery to coincident satellite imagery to complete a multiscale analysis, specifically the Sentinel-3 SLSTR, which suffered a protective shutdown during the flight campaign [46]. Based on the Mission Status report [46], the satellite experienced a computer double bit-error and no data was acquired between 31 July and 6 August 2017. This malfunction happened as it passed through the South Atlantic Anomaly, an area of increased radiation due to increased proximity to Earth's surface. The relevance of geocoding airborne wildfire imagery extends beyond its use in evaluating space-borne retrievals. Once corrected, this imagery provides significantly more value in terms of characterizing wildfire fire radiative power [9], rate of spread [47] and fire-line intensity [31] providing the unique capacity to study a naturally occurring wildfire without the need for ground sampling. Furthermore, our approach is valuable in experimental situations where the availability of precise positional data is limited (e.g., [9]) due to budget and or aircraft type (e.g., short term contract and/or UAV).

While the method and results described here are specific to the 2017 field campaign data shown, in essence, this method solves the georeferencing step in any airborne campaign where no high-frequency, high-accuracy IMU data is collected. As such, with minimal effort, most of the process can be used in other situations and contexts. To facilitate this, open-source MATLAB code of the image registration and gridding processes has been released [38]. Some effort is required by the user to prepare their data to fit this methodology, most notably setting the initial position estimations for the first frame in the dataset. However, the process itself requires basic expertise in MATLAB or image processing, and an example dataset (several frames from the wildfire campaign presented here) is given to guide a user through the requirements.

6. Conclusions

To validate satellite products, an airborne campaign was carried out to collect thermal imaging of wildland fires near Pickle Lake, Northern Ontario, in August 2017. For this campaign, multiple BOB fires were imaged using an on-board MWIR broadband FLIR SC8303 sensor equipped with a flame filter (3.74 μm) at four different integration times (i.e., 1.4 ms, 0.3 ms, 0.04 ms, and 0.002 ms). Unfortunately, there were two separate points of hardware failure during the data acquisition: the sensor never received GPS timing information and the aircraft's IMU failed to record. An alternative geolocation method was then developed to address these two major issues. The FLIR frames were aligned with a backup GPS-tagged camera and assigned geolocations, before being combined into full orthoimages through an image-registration process that did not require IMU information. The orthoimages were then further georeferenced using a developed grid system to improve the geolocation error. This second step was necessary to allow for an effective comparison with mid-resolution satellite sensors (≤30–300 m). Using our approach, the geolocation RMSE error was reduced from up to 300 m to 15.11 m for the nighttime data set and to 13.53 m for the daytime dataset. Our study also shows that the computed FRPD estimations, derived from the geocorrected FLIR data, are within expected values across the literature. Overall, the process led to an effective geocorrection of the infrared data, which has since been used for various analyses, such as fire radiative power, to inform wildfire scientists.

The methodology here can be easily adapted and applied to any wildfire campaign where no IMU or GPS information is available, whether due to a lack of hardware or hardware issues like those in the wildfire campaign presented here. It is a simple and effective basis for orthoimage creation that will keep sharp edges present at wildfire peaks, while being geographically accurate for comparison to associated satellite imagery, or other airborne data.

Author Contributions: Conceptualization, G.I., T.N. and J.P.A.-M.; methodology, G.I. and T.N.; software, T.N. and G.I.; formal analysis, G.I. and T.N.; investigation, J.P.A.-M. and J.M.J.; resources, G.L.; data curation, T.N., G.I. and M.D.L.; writing—original draft preparation, G.I., T.N. and J.P.A.-M.; writing—review and editing, G.I., T.N., J.P.A.-M., J.M.J., G.L. and M.D.L.; visualization, G.I. and T.N.; supervision, J.P.A.-M.; project administration, M.D.L. and J.P.A.-M.; funding acquisition, J.M.J. All authors have read and agreed to the published version of the manuscript.

Funding: Funding for the airborne data acquisition was provide by the Canadian Space Agency. The geocorrection methodological research received no external funding.

Institutional Review Board Statement: Not applicable.

Informed Consent Statement: Not applicable.

Data Availability Statement: Not applicable.

Acknowledgments: The authors would like to thank the Ontario Aviation Forest Fire and Emergency Services division, and in particular the Pickle Lake Fire Management Headquarters for their operational support throughout the field campaign.

Conflicts of Interest: The authors declare no conflict of interest.

References

1. Doerr, S.H.; Santín, C. Global trends in wildfire and its impacts: Perceptions versus realities in a changing world. *Philos. Trans. R. Soc. Lond. B Biol. Sci.* **2016**, *371*, 20150345. [CrossRef] [PubMed]
2. Coogan, S.C.P.; Robinne, F.-N.; Jain, P.; Flannigan, M.D. Scientists' warning on wildfire—A Canadian perspective. *Can. J. For. Res.* **2019**, *49*, 1015–1023. [CrossRef]
3. Flannigan, M.D.; Krawchuk, M.A.; de Groot, W.J.; Wotton, B.M.; Gowman, L.M. Implications of changing climate for global wildland fire. *Int. J. Wildland Fire* **2009**, *18*, 483–507. [CrossRef]
4. O'Neill, S.J.; Handmer, J. Responding to bushfire risk: The need for transformative adaptation. *Environ. Res. Lett.* **2012**, *7*, 014018. [CrossRef]
5. Johnston, L.M.; Wang, X.; Erni, S.; Taylor, S.W.; McFayden, C.B.; Oliver, J.A.; Stockdale, C.; Christianson, A.; Boulanger, Y.; Gauthier, S.; et al. Wildland fire risk research in Canada. *Environ. Rev.* **2020**, *28*, 164–186. [CrossRef]
6. Baldwin, C.; Ross, H. Beyond a tragic fire season: A window of opportunity to address climate change? *Australas. J. Environ. Manag.* **2020**, *27*, 1–5. [CrossRef]
7. Schoennagel, T.; Nelson, C.R.; Theobald, D.M.; Carnwath, G.C.; Chapman, T.B. Implementation of National Fire Plan treatments near the wildland-urban interface in the western United States. *Proc. Natl. Acad. Sci. USA* **2009**, *106*, 10706–10711. [CrossRef]
8. Schroeder, W.; Oliva, P.; Giglio, L.; Csiszar, I.A. The New VIIRS 375m active fire detection data product: Algorithm description and initial assessment. *Remote Sens. Environ.* **2014**, *143*, 85–96. [CrossRef]
9. Paugam, R.; Wooster, M.J.; Roberts, G. Use of Handheld Thermal Imager Data for Airborne Mapping of Fire Radiative Power and Energy and Flame Front Rate of Spread. *IEEE Trans. Geosci. Remote Sens.* **2013**, *51*, 3385–3399. [CrossRef]
10. Radke, L.F.; Clark, T.L.; Coen, J.L.; Walther, C.A.; Lockwood, R.N.; Riggan, P.J.; Brass, J.A.; Higgins, R.G. The WildFire Experiment (WiFE): Observations with Airborne Remote Sensors. *Can. J. Remote Sens.* **2000**, *26*, 406–417. [CrossRef]
11. Wooster, M.J.; Roberts, G.; Perry, G.L.W.; Kaufman, Y.J. Retrieval of biomass combustion rates and totals from fire radiative power observations: FRP derivation and calibration relationships between biomass consumption and fire radiative energy release. *J. Geophys. Res. Atmos.* **2005**, *110*. [CrossRef]
12. Wooster, M.J.; Xu, W.; Nightingale, T. Sentinel-3 SLSTR active fire detection and FRP product: Pre-launch algorithm development and performance evaluation using MODIS and ASTER datasets. *Remote Sens. Environ.* **2012**, *120*, 236–254. [CrossRef]
13. Johnston, J.M.; Jackson, N.; McFayden, C.; Ngo Phong, L.; Lawrence, B.; Davignon, D.; Wooster, M.J.; van Mierlo, H.; Thompson, D.K.; Cantin, A.S.; et al. Development of the User Requirements for the Canadian WildFireSat Satellite Mission. *Sensors* **2020**, *20*, 5081. [CrossRef]
14. Davis, K.T.; Dobrowski, S.Z.; Higuera, P.E.; Holden, Z.A.; Veblen, T.T.; Rother, M.T.; Parks, S.A.; Sala, A.; Maneta, M.P. Wildfires and climate change push low-elevation forests across a critical climate threshold for tree regeneration. *Proc. Natl. Acad. Sci. USA* **2019**, *116*, 6193–6198. [CrossRef]
15. Williams, A.P.; Abatzoglou, J.T.; Gershunov, A.; Guzman-Morales, J.; Bishop, D.A.; Balch, J.K.; Lettenmaier, D.P. Observed Impacts of Anthropogenic Climate Change on Wildfire in California. *Earth's Future* **2019**, *7*, 892–910. [CrossRef]
16. Goss, M.; Swain, D.L.; Abatzoglou, J.T.; Sarhadi, A.; Kolden, C.A.; Williams, A.P.; Diffenbaugh, N.S. Climate change is increasing the likelihood of extreme autumn wildfire conditions across California. *Environ. Res. Lett.* **2020**, *15*, 094016. [CrossRef]
17. Allison, R.; Johnston, J.; Craig, G.; Jennings, S. Airborne optical and thermal remote sensing for wildfire detection and monitoring. *Sensors* **2016**, *16*, 1310. [CrossRef] [PubMed]
18. Giglio, L.; Csiszar, I.; Restás, Á.; Morisette, J.T.; Schroeder, W.; Morton, D.; Justice, C.O. Active fire detection and characterization with the advanced spaceborne thermal emission and reflection radiometer (ASTER). *Remote Sens. Environ.* **2008**, *112*, 3055–3063. [CrossRef]
19. Yuan, C.; Zhang, Y.; Liu, Z. A survey on technologies for automatic forest fire monitoring, detection, and fighting using unmanned aerial vehicles and remote sensing techniques. *Can. J. For. Res.* **2015**, *45*, 783–792. [CrossRef]
20. Castillo, M.; Plaza, Á.; Garfias, R. A recent review of fire behavior and fire effects on native vegetation in Central Chile. *Glob. Ecol. Conserv.* **2020**, *24*, e01210. [CrossRef]
21. Ottmar, R.D. Wildland fire emissions, carbon, and climate: Modeling fuel consumption. *For. Ecol. Manag.* **2014**, *317*, 41–50. [CrossRef]
22. Bento-Gonçalves, A.; Vieira, A. Wildfires in the wildland-urban interface: Key concepts and evaluation methodologies. *Sci. Total Environ.* **2020**, *707*, 135592. [CrossRef] [PubMed]
23. Dickinson, M.B.; Hudak, A.T.; Zajkowski, T.; Loudermilk, E.L.; Schroeder, W.; Ellison, L.; Kremens, R.L.; Holley, W.; Martinez, O.; Paxton, A.; et al. Measuring radiant emissions from entire prescribed fires with ground, airborne and satellite sensors–RxCADRE 2012. *Int. J. Wildland Fire* **2016**, *25*, 48–61. [CrossRef]

24. Stow, D.; Riggan, P.; Schag, G.; Brewer, W.; Tissell, R.; Coen, J.; Storey, E. Assessing uncertainty and demonstrating potential for estimating fire rate of spread at landscape scales based on time sequential airborne thermal infrared imaging. *Int. J. Remote Sens.* **2019**, *40*, 4876–4897. [CrossRef]
25. He, X.; Yang, X.; Luo, Z.; Guan, T. Application of unmanned aerial vehicle (UAV) thermal infrared remote sensing to identify coal fires in the Huojitu coal mine in Shenmu city, China. *Sci. Rep.* **2020**, *10*, 13895. [CrossRef]
26. Ciullo, V.; Rossi, L.; Pieri, A. Experimental Fire Measurement with UAV Multimodal Stereovision. *Remote Sens.* **2020**, *12*, 3546. [CrossRef]
27. Freeborn, P.H.; Wooster, M.J.; Roy, D.P.; Cochrane, M.A. Quantification of MODIS fire radiative power (FRP) measurement uncertainty for use in satellite-based active fire characterization and biomass burning estimation. *Geophys. Res. Lett.* **2014**, *41*, 1988–1994. [CrossRef]
28. Wan, Z.; Zhang, Y.; Ma, X.; King, M.D.; Myers, J.S.; Li, X. Vicarious calibration of the Moderate-Resolution Imaging Spectroradiometer Airborne Simulator thermal-infrared channels. *Appl. Opt.* **1999**, *38*, 6294–6306. [CrossRef]
29. Hall, J.V.; Zhang, R.; Schroeder, W.; Huang, C.; Giglio, L. Validation of GOES-16 ABI and MSG SEVIRI active fire products. *Int. J. Appl. Earth Obs. Geoinf.* **2019**, *83*, 101928. [CrossRef]
30. Johnston, J.M.; Wheatley, M.J.; Wooster, M.J.; Paugam, R.; Davies, G.M.; DeBoer, K.A. Flame-Front Rate of Spread Estimates for Moderate Scale Experimental Fires Are Strongly Influenced by Measurement Approach. *Fire* **2018**, *1*, 16. [CrossRef]
31. Johnston, J.M.; Wooster, M.J.; Paugam, R.; Wang, X.; Lynham, T.J.; Johnston, L.M. Direct estimation of Byram's fire intensity from infrared remote sensing imagery. *Int. J. Wildland Fire* **2017**, *26*, 668–684. [CrossRef]
32. Müller, R.; Lehner, M.; Müller, R.; Reinartz, P.; Schroeder, M.; Vollmer, B. A program for direct georeferencing of airborne and spaceborne line scanner images. *Int. Arch. Photogramm. Remote Sens. Spat. Inf. Sci.* **2002**, *34*, 148–153.
33. Stow, D.A.; Riggan, P.J.; Storey, E.J.; Coulter, L.L. Measuring fire spread rates from repeat pass airborne thermal infrared imagery. *Remote Sens. Lett.* **2014**, *5*, 803–812. [CrossRef]
34. Kalacska, M.; Lucanus, O.; Arroyo-Mora, J.P.; Laliberté, É.; Elmer, K.; Leblanc, G.; Groves, A. Accuracy of 3D Landscape Reconstruction without Ground Control Points Using Different UAS Platforms. *Drones* **2020**, *4*, 13. [CrossRef]
35. Wotton, B.M.; Gould, J.S.; McCaw, W.L.; Cheney, N.P.; Taylor, S.W. Flame temperature and residence time of fires in dry eucalypt forest. *Int. J. Wildland Fire* **2012**, *21*, 270–281. [CrossRef]
36. Lee, M.; Ifimov, G.; Naprstek, T.; Arroyo-Mora, J.P.; Craig, G.; Leblanc, G. *Canadian Wildland Fire Momitoring System (CWFMS): Calibration, Pre-Processing and Georeferencing of MWIR Airborne Data*; LTR-FRL-2019-0114; National Research Council: Ottawa, ON, Canada, 2019.
37. Paul, G. *Principles of GNSS, Inertial, and Multisensor Integrated Navigation Systems*, 2nd ed.; Artech: Norwood, MA, USA, 2013.
38. Naprstek, T.; Arroyo-Mora, J.P.; Johnston, J.; Leblanc, G. ReSVA: A MATLAB method to co-register and mosaic airborne video-based remotely sensed data. *Methods X Under Rev.* **2021**.
39. Berk, A.; Conforti, P.; Kennett, R.; Perkins, T.; Hawes, F.; Van Den Bosch, J. MODTRAN® 6: A major upgrade of the MODTRAN® radiative transfer code. In Proceedings of the 2014 6th Workshop on Hyperspectral Image and Signal Processing: Evolution in Remote Sensing (WHISPERS), Lausanne, Switzerland, 24–27 June 2014; pp. 1–4.
40. Ji, Y.; Xu, W.; Li, Y.; Ma, D. Research on the measurement of mid-wavelength infrared average atmospheric characteristics. In Proceedings of the 2013 2nd International Conference on Measurement, Information and Control, Harbin, China, 16–18 August 2013; pp. 115–118.
41. Warren, M.A.; Taylor, B.H.; Grant, M.G.; Shutler, J.D. Data processing of remotely sensed airborne hyperspectral data using the Airborne Processing Library (APL): Geocorrection algorithm descriptions and spatial accuracy assessment. *Comput. Geosci.* **2014**, *64*, 24–34. [CrossRef]
42. Coulter, L.L.; Stow, D.A.; Baer, S. A frame center matching technique for precise registration of multitemporal airborne frame imagery. *IEEE Trans. Geosci. Remote Sens.* **2003**, *41*, 2436–2444. [CrossRef]
43. Coulter, L.L.; Stow, D.A. Assessment of the Spatial Co-registration of Multitemporal Imagery from Large Format Digital Cameras in the Context of Detailed Change Detection. *Sensors* **2008**, *8*, 2161–2173. [CrossRef] [PubMed]
44. Aguilar, M.A.; Aguilar, F.J.; Mar Saldaña, M.d.; Fernández, I. Geopositioning Accuracy Assessment of GeoEye-1 Panchromatic and Multispectral Imagery. *Photogramm. Eng. Remote Sens.* **2012**, *78*, 247–257. [CrossRef]
45. Wooster, M.J.; Roberts, G.; Smith, A.M.S.; Johnston, J.; Freeborn, P.; Amici, S.; Hudak, A.T. Thermal Remote Sensing of Active Vegetation Fires and Biomass Burning Events. In *Thermal Infrared Remote Sensing: Sensors, Methods, Applications*; Kuenzer, C., Dech, S., Eds.; Springer: Dordrecht, The Netherlands, 2013; pp. 347–390. [CrossRef]
46. ESA; EUMETSAT. Sentinel-3 Mission Status: 7 September 2017. Available online: https://sentinel.esa.int/documents/247904/2802412/Sentinel-3-Mission-Status-Report-36-07-sep-2017.pdf (accessed on 19 April 2021).
47. Penney, G.; Richardson, S. Modelling of the Radiant Heat Flux and Rate of Spread of Wildfire within the Urban Environment. *Fire* **2019**, *2*, 4. [CrossRef]

sensors

MDPI

Article

The Wildland Fire Heat Budget—Using Bi-Directional Probes to Measure Sensible Heat Flux and Energy in Surface Fires

Matthew B. Dickinson [1,*], Cyle E. Wold [2], Bret W. Butler [2], Robert L. Kremens [3], Daniel Jimenez [2], Paul Sopko [2] and Joseph J. O'Brien [4]

1 Forestry Sciences Lab, USDA Forest Service, Northern Research Station, 359 Main Road, Delaware, OH 43015, USA

2 Missoula Fire Sciences Laboratory, USDA Forest Service, Rocky Mountain Research Station, 5775 US Highway 10 W, Missoula, MT 59808, USA; cyle.wold@usda.gov (C.E.W.); bret.butler@usda.gov (B.W.B.); dan.jimenez@usda.gov (D.J.); paul.sopko@usda.gov (P.S.)

3 Chester F. Carlson Center for Imaging Science, Rochester Institute of Technology, One Lomb Memorial Drive, Rochester, NY 14623, USA; kremens@cis.rit.edu

4 Center for Disturbance Science, USDA Forest Service, Southern Research Station, 320 Green Street, Athens, GA 30602, USA; joseph.j.obrien@usda.gov

* Correspondence: matthew.b.dickinson@usda.gov; Tel.: +614-556-2271

check for updates

Citation: Dickinson, M.B.; Wold, C.E.; Butler, B.W.; Kremens, R.L.; Jimenez, D.; Sopko, P.; O'Brien, J.J. The Wildland Fire Heat Budget—Using Bi-Directional Probes to Measure Sensible Heat Flux and Energy in Surface Fires. *Sensors* **2021**, *21*, 2135. https://doi.org/10.3390/s21062135

Academic Editor: Robert S. Allison

Received: 31 January 2021
Accepted: 15 March 2021
Published: 18 March 2021

Publisher's Note: MDPI stays neutral with regard to jurisdictional claims in published maps and institutional affiliations.

Copyright: © 2021 by the authors. Licensee MDPI, Basel, Switzerland. This article is an open access article distributed under the terms and conditions of the Creative Commons Attribution (CC BY) license (https://creativecommons.org/licenses/by/4.0/).

Abstract: Sensible energy is the primary mode of heat dissipation from combustion in wildland surface fires. However, despite its importance to fire dynamics, smoke transport, and in determining ecological effects, it is not routinely measured. McCaffrey and Heskestad (A robust bidirectional low-velocity probe for flame and fire application. Combustion and Flame 26:125–127, 1976.) describe measurements of flame velocity from a bi-directional probe which, when combined with gas temperature measurements, can be used to estimate sensible heat fluxes. In this first field application of bi-directional probes, we describe vertical and horizontal sensible heat fluxes during the RxCADRE experimental surface fires in longleaf pine savanna and open ranges at Eglin Air Force Base, Florida. Flame-front sensible energy is the time-integral of heat flux over a residence time, here defined by the rise in gas temperatures above ambient. Horizontal flow velocities and energies were larger than vertical velocities and energies. Sensible heat flux and energy measurements were coordinated with overhead radiometer measurements from which we estimated fire energy (total energy generated by combustion) under the assumption that 17% of fire energy is radiated. In approximation, horizontal, vertical, and resultant sensible energies averaged 75%, 54%, and 64%, respectively, of fire energy. While promising, measurement challenges remain, including obtaining accurate gas and velocity measurements and capturing three-dimensional flow in the field.

Keywords: wildland fire; fire behavior; fire energy; sensible heat flux; convective heat flux; fire radiated energy (FRE); surface fire; residence time; bi-directional probe; flame velocity; gas temperature; RxCADRE Project

1. Introduction

The wildland fire heat budget balances heat generated by combustion (which is reduced by inefficiencies), heat sinks associated with heating fuels to ignition, and heat dissipation ([1]). Based on limited measurements, sensible heat flux (kW/m^2) and energy (kJ/m^2), its time-integral, have been found to dissipate more energy from wildland fires than other modes, including radiation ([2,3]), and we expect that radiated energy, the next largest contributor, will be limited to about 10–20% of fire energy (i.e., total energy generated by combustion [1–5]). Sensible heat flux is the transport of heat in hot gases across a reference plane. The term has roots in the meteorological literature (e.g., [6]) while it is often referred to as convective or advective flux in the engineering literature (e.g., [7,8]). Based on flow velocities and gas temperatures, sensible heat fluxes in wildland fires have been measured using videography ([9,10]), particle velocimetry ([11,12]), one-dimensional anemometry ([13,14]), and two- and three-dimensional anemometry ([15–17]). Sensible heat flux and energy have not been estimated in or just above wildland fire flames because instrumentation has not been sufficiently fire-hardened ([18,19]). Originating in building fire research, McCaffrey and Heskestad ([20]) describe an instrument for measuring flow velocities in flames based on pressure differentials between dynamic (facing on-coming flow) and static (sheltered) orifices. The instrument has the merit that it can be made resistant to the harsh environment of flames.

We can, in a coarse way, define sensible heat flux and energy by considering a control volume that encloses the fuel, flames, heated air and combustion gas products (plume) and soil heated by the fire. All the energy generation from combustion is contained in this volume. A heat budget for the control volume roughly balances heat sources, sinks, and dissipation integrated over the time period from ignition through the cool-down to ambient temperatures (see [1]):

$$Wh_C(1-\varphi) = WQ_P + E_G + E_V + E_L + E_R + E_S \quad (1)$$

where W is the areal fuel consumption on a moisture and ash-free basis (kg/m^2); h_C is the high heat of combustion (i.e., includes heat of condensation of water generated by the combustion process, kJ kg^{-1}); φ is the fractional reduction in h_C because of incomplete combustion, $h_C(1-\varphi)$ is the effective heat of combustion ([21,22]); Q_P is the fuel pre-heating and pyrolysis enthalpy (kJ/kg), defined so as not to include the fuel moisture vaporization enthalpy, see below); E_G is the areal energy density transferred into the soil (kJ/m^2); E_V is the energy transferred to vegetation (e.g., tree stems, branches, and foliage); E_L is the fire latent energy density (kJ/m^2), the condensation energy in water vapor generated from both fuel moisture and the combustion process; E_R is the fire radiative energy density (kJ/m^2), i.e., the radiated flux time-integrated over the period from ignition to cool-down; and E_S is sensible energy density transported by the buoyancy-driven rise of heated combustion products and directly heated air and smoke. Both sides of Equation (1) have units of areal energy density, kJ/m^2. The left side of Equation (1) is the energy available to do work near the fire front which we term 'fire energy'. The heat sink and dissipation modes on the right-hand side are ordered by their expected relative magnitude in flame fronts, although we acknowledge that the ordering is partly conjecture ([1]). All quantities involving fuel mass (e.g., consumption, heats of combustion) are on an ash- and moisture-free basis.

The integrated heat budget is coarse in the sense that heat dissipation modes are not independent and their magnitudes are dependent on the control volume ([1]). For instance, sensible energy is reduced by radiation from the hot flame and plume. Sensible energy is likely to be a larger proportion of fire energy if the control volume is occupied primarily by flame. Heat and momentum from plumes above surface fires is transferred to forest canopies ([23]) but in a control volume that primarily contains the flaming front, that transfer is ignored. Net heat flux at the soil surface will be determined by gas-phase conduction in the burning fuel bed, radiative and convective heat transfer to the soil surface (as mediated by burning fuels and ash), and heat losses by radiation and convection.

In this paper, we describe measurements of sensible heat flux and energy (Equation (1)) based on in situ instruments in wildland surface fires. Sensible heat fluxes and energy have a particular relevance to fire propagation ([24]), plume dynamics ([6,17]), and various ecological fire effects ([25]), including tree crown injury ([26–28]), faunal exposures ([29,30]), and airborne transport of microbes ([31]). This study is a first application of bi-directional probes for measuring gas flow velocities in and just above wildland fire flames. We estimate that sensible energy accounted for more than 50% of fire energy (left-hand side of Equation (1)). We discuss measurement uncertainties and improvements that can be made in future experiments. Our paper is an important step towards closing the wildland fire heat budget ([1]) which provides a standard for assessing measurements and is expected to lead to advances in understanding and predicting fire dynamics, plume dynamics, and fire effects.

2. Materials and Methods

2.1. Study Site and Fire Behavior

Data were collected in early November 2012 within burn blocks in an 8-km x 4-km area of Eglin Air Force Base in northwestern Florida during the Prescribed Fire Combustion and Atmospheric Dynamics Research Experiment (RxCADRE), a coordinated measurements campaign described in Ottmar et al. ([32]) and associated papers. Burn blocks were characterized by either an herbaceous and shrub fuel mix maintained as open range through mowing, fire, and herbicide application (hereafter termed non-forested) or fire-maintained pine savanna with fuel beds including needle cast, turkey oak litter, herbaceous and shrub vegetation, and woody material (hereafter termed forested). Non-forested blocks included large (L1G and L2G) and small burn blocks (S3, S4, S5, S7, S8, and S9) while there was a single large forested block (L2F). Burn blocks, fuels, and fire behavior are described in and Butler et al. ([19]), Dickinson et al. ([33]), Hudak et al. ([34]), and Ottmar et al. ([32]) and are summarized in Table 1. Near-source plumes, including sensible heat fluxes, are described in Clements et al. ([35]).

2.2. Sensible Heat Flux and Energy

Sensible heat fluxes from the flow of hot gases are estimated from gas temperatures measured with fine thermocouples and velocities measured with temperature-compensated bi-directional probes and fast-response pressure transducers. We calculate the horizontal and vertical perturbation heat fluxes and energy and their resultant. In this context, perturbation refers to a departure from the pre-fire background state. Perturbation values are assumed from here forward. Accordingly, horizontal velocity and temperature are as follows:

$$u' = u - \overline{u} \tag{2}$$

$$T' = T - \overline{T} \tag{3}$$

where u and T refer to the instantaneous values, and the overbar refers to the pre-fire average. Streamwise horizontal velocity is positive if its towards the front of the FBP and negative if from the rear. The horizontal sensible heat flux (H'_u) is:

$$H'_u = \rho C_P u' T' \tag{4}$$

where ρ is gas density (kg/m^3), C_P is specific heat capacity (J/kg K), u' is the streamwise velocity (m/s), and T' is the temperature (K). Gas density and heat capacity are temperature dependent (see Appendix A). Vertical (upward) velocity (analogous to Equation (2)) is positive while downward flow is negative. The vertical sensible heat flux is analogous to Equation (4). The FBPs were positioned so that the fire would generally approach from the front (defined by positioning of the incident radiant and total heat flux sensors).

Time-integration of horizontal and vertical sensible heat flux provides sensible energy (kJ/m^2). The integration is limited by the residence time (t_R), shown here for horizontal sensible energy (E_{Su}):

$$E_{S_u} = t \sum_{1}^{t_R} H'_u \tag{5}$$

where the time-step (t) is 0.1 s. Residence time is defined below. Vertical sensible energy is analogous to Equation (5).

We are also interested in the resultant sensible energy which we define as the resultant of the horizontal and vertical sensible energies. Preferable would have been to calculate the resultant of instantaneous horizontal and sensible heat fluxes ([17]), but the separation of horizontal and vertical probes may invalidate this approach. Instead, we estimate resultant sensible energy (E_{Sr}) from horizontal and vertical energies as:

$$E_{S_r} = \left[(E_{S_u})^2 + (E_{S_w})^2 \right]^{\frac{1}{2}} \tag{6}$$

where, again, density and heat capacity are temperature dependent and w refers to vertical.

Table 1. Averaged characteristics of surface fires and flames in forested and non-forested burn blocks from the RxCADRE 2012 fires. Fireline intensity (I) and fuel consumption (W_1) are inferred from nadir radiometer measurements using equations in Kremens et al. (2012, see Supplementary Material) while whole-block estimates of consumption (W_2) are from Hudak et al. (2016). Estimates of flame height (H_F), depth (D_F), residence time (t_R), and fire rate of spread (ROS) are from video analysis ([19,36]). Sample sizes and standard deviations are provided (in parentheses) where applicable.

Burn block	Fuel	Date	I (kW/m)	W_1 ((Mg/ha)	W_2 ((Mg/ha)	H_F ((m)	D_F ((m)	t_R ((s)	ROS ((m/s)
L2F	Forested	11/11/2012	907 (9, 670)	5.0 (9, 2.6)	6.4	0.9 (5, 0.5)	1.3 (5, 0.7)	9 (8, 7)	0.04 (2, 0.05)
L1G	Non-forested	11/04/2012	529 (9, 316)	1.3 (9, 0.5)	1.5	0.7 (6, 0.5)	1.1 (5, 0.8)	11 (7, 7)	0.24 (4, 0.30)
L2G	Non-forested	11/10/2012	739 (12, 358)	1.5 (12, 0.6)	3.1	0.5 (9, 0.2)	0.8 (9, 0.4)	11 (8, 6)	0.89 (3, 0.38)
S3	Non-forested	11/01/2012	479 (5, 79)	1.7 (5, 0.2)	2.6				
S4	Non-forested	11/01/2012	234 (4, 172)	1.6 (4, 0.7)	2.0				
S5	Non-forested	11/01/2012	564 (5, 269)	2.2 (5, 0.6)	2.2	0.4 (4, 0.0)	0.8 (4, 0.3)	11 (4, 4)	0.36 (2, 0.28)
S7	Non-forested	11/07/2012	1179 (4, 641)	3.3 (4, 1.8)	1.8				
S8	Non-forested	11/07/2012	512 (4, 318)	1.9 (4, 0.7)	2.8				
S9	Non-forested	11/07/2012	861 (5, 115)	1.8 (5, 0.9)	1.4				

We calculate perturbation energies, that is, we remove the contribution of pre-fire sensible heat fluxes and focus on the fire residence time, because we are interested in balancing energy generation from fuel combustion (left-hand side of Equation (1)) and energy sinks and dissipation mechanisms. Accordingly, we forego the use of Reynolds (moving) averaging through the residence time required to isolate turbulent kinetic energy from total sensible energy ([17,37]). Instead, the perturbation sensible energies are estimates of fire totals. As described below, the constraints for estimating total sensible energy from a two-dimensional system of probes include the requirement that fires are spreading with the average flow in a direction in line with the horizontal flow measurement.

2.3. Instruments and Measurements

Estimating sensible heat fluxes and energy requires gas velocities and temperatures (Equations (2)–(4)). The core instruments are deployed in a Fire Behavior Package (FBPs) described in Butler et al. ([38]) and shown in Figure 1. The FBP includes one vertically- and one horizontally-oriented bi-directional pressure probe with +/−60 degree directional sensitivity ([20]) and a fine bare Type-K thermocouple (nominally 0.025 mm bead diameter) for temperature measurement. The probe characteristic dimension ([20]) is 12.7 mm. The ends of the fine thermocouple are welded to their corresponding leads and are not visible in Figure 1. The bi-directional probes are separated in space by approximately 36 cm while the thermocouple is positioned between the probes. The vertical probe connects to the FBP container on the left side (when the viewer faces the front of the FBP) and the horizontal probe joins the container on the top (Figure 1). Tubing transmits pressure signals from the dynamic and static sides of each probe separately to differential pressure sensors (Omega Engineering model PX137-0.3DV) which are temperature compensated with a pressure range of approximately ± 2000 Pa. Apart from instruments used to measure gas velocity and temperature, the FBP includes a Medtherm® Dual Sensor Heat Flux sensor (Model 64-20T) that measures incident radiant and total (convective plus radiant) heat deposition onto the face of the sensor and a custom narrow angle radiometer (NAR, [39]) to characterize flame emissive power. The Dual Sensor and NAR are mounted flush with the FBP's container and are oriented towards oncoming fires as best as can be predicted prior to ignition. Data from Dual Sensor and NAR are reported elsewhere ([19]). Each FBP additionally contains a Campbell Scientific® model CR1000 datalogger, two battery packs, and electronics required for each instrument. The sampling interval for all measurements is 10 Hz. The container is covered with two outer layers of fire-shelter material with an inner core of ceramic fiber insulation to prevent excessive heating. The FBP is elevated on a fire-hardened tripod to, nominally, 50 cm aboveground ([19]). As such, approximately, the vertical probe is at 50 cm, the horizontal probe is at 81 cm, and the thermocouple is at 66 cm height aboveground.

Lacking local fuel consumption data, we infer fire energy (left-hand-side of Equation (1)) with data from overhead (nadir) radiometers ([1,33]). We estimate fire energy from fire radiated energy density (FRED, also known as fire radiated energy [FRE]) and an assumption, based on measurements, that 17% of energy generated by combustion was radiated ([33,34]). Two radiometer configurations were used, one elevated to 5.5 m on a tower with a 52 degree field of view and 22.5 m^2 area of regard ([33]) and the second elevated to 7.7 m with a 60 degree field of view and 62 m^2 area of regard ([40]). Radiometer height does not affect the energy estimate ([1]) other than the area over which it is determined. FBPs were positioned just outside the area of regard of the radiometers and oriented across the area of regard and towards the expected approach of the flame front.

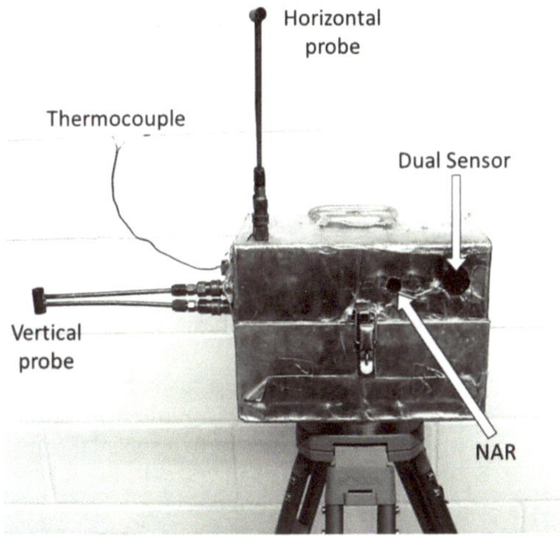

Figure 1. The Fire Behavior Package (FBP) including horizontally- and vertically oriented bi-directional probes and thermocouple. Data from the Dual Sensor and narrow-angle radiometer (NAR) are reported elsewhere ([19]). Nominally, the FBP is positioned 50 cm aboveground and the distance between the probes is approximately 36 cm.

Gas heat capacity and pressure (Equation (4)) and the calibration process required to estimate flow velocity are temperature dependent (Appendix A). We did not adjust for differences between air and flame and plume gases in their physiochemical properties. We used a lookup table to adjust air heat capacity by temperature ([41]). Flow velocity, assuming incompressible flow, is derived by calibration from wind tunnel data (Appendix A). By convention, direction is determined by the sign of the differential pressure measurement with upward being positive on the vertically oriented bi-directional probe and flow towards the front of the Dual Sensor being positive on the horizontally oriented bi-directional probe (see Figure 1).

Time limits to the integrals used to determine sensible energy are determined using 3-s averaged gas temperature measurements because turbulence results in high-frequency fluctuation in temperatures and flows ([37,42]). The start of the integral is determined by the timestep at which a 4-s window moving back in time from peak temperature last contains a temperature rise above threshold. We tested 50, 100, and 200 C rise above background as thresholds. The end of the integral is the timestep at which a window moving forward in time from peak temperature encounters its last rise above threshold within the window. The limits to the integral define what we call residence time in this paper.

2.4. Statistics

Where correlations are reported, they are nonparametric Spearman rank-order correlation coefficients. Comparisons among thresholds used to define residence times are compared by ANOVA on log-transformed data. Within temperature thresholds, comparisons between horizontal and vertical flow velocities and sensible heat fluxes are by paired t-test. Regressions between horizontal, vertical, and resultant sensible energy and fire energy were linear on natural-log transformed data. Statistics were calculated with SAS 9. The standard for judging whether a difference between groups was significant was $p \leq 0.05$.

3. Results

Vertical, horizontal, and resultant sensible heat fluxes and energies (Equations (5) and (6)) are based on flow velocity and direction and gas temperature measurements and inferred temperature-dependent gas density and heat capacity (Equations (2)–(4)). Individual collections were included in the final dataset if we knew from video analysis that the fire approached the FBP from within 60 degrees of perpendicular to the face of the incident heat flux sensor and axis of the horizontal probe (see Dickinson et al., 2019). Where we did not have video information, we further included datasets where sensible heat fluxes in the horizontal and vertical dimensions, and their resultant, were positive on average. We excluded datasets that did not meet the above conditions or which had known equipment problems. Ultimately, we report data from 55 out of 97 datasets.

3.1. Residence Times and Gas Temperatures

After experimenting with a range of thresholds, we ultimately used a 50 C temperature rise above background to determine the residence times over which sensible heat fluxes were integrated (Figure 2). The choice was based on the objective of capturing as much of the perturbation (fire-induced) sensible heat flux and energy as possible. The temperature-rise rule provided a consistent estimate that accounted for variation in ambient (pre-fire) air temperature. We expect that there was minimal heat flux lost by excluding near-ambient temperatures. In practice, the estimates of horizontal, vertical, and resultant sensible energy did not differ among the 50, 100, and 200 C thresholds (ANOVA F-value < 0.9, $p > 0.4$ for all comparisons) although the number of experiments that met the residence time criteria declined as the temperature threshold increased. Residence time used for integration should not be confused with flame residence times which were determined from video analysis and are shorter in duration (Table 1). The frequency distribution of residence times (Figure 2) follows from the wide range of gas temperature regimes in the flames and plumes (Figures 3 and 4).

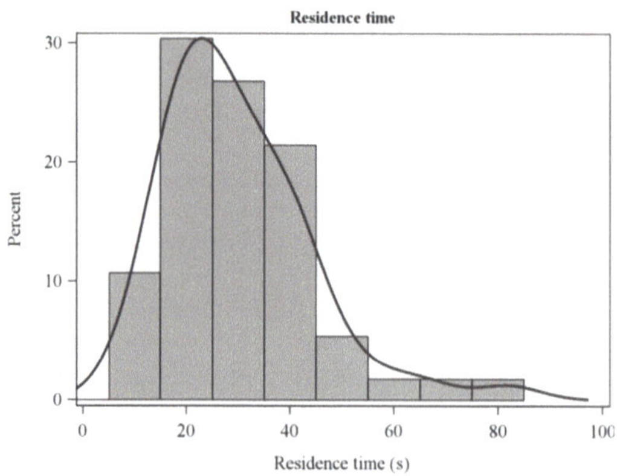

Figure 2. Frequency distribution of residence times defined by a 3-s moving-averaged temperature rise greater than 50 °C above background.

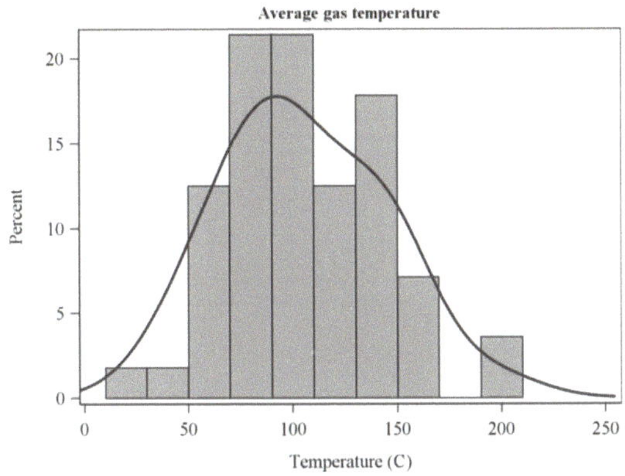

Figure 3. Frequency distribution of average gas temperatures over residence times measured with fine thermocouples.

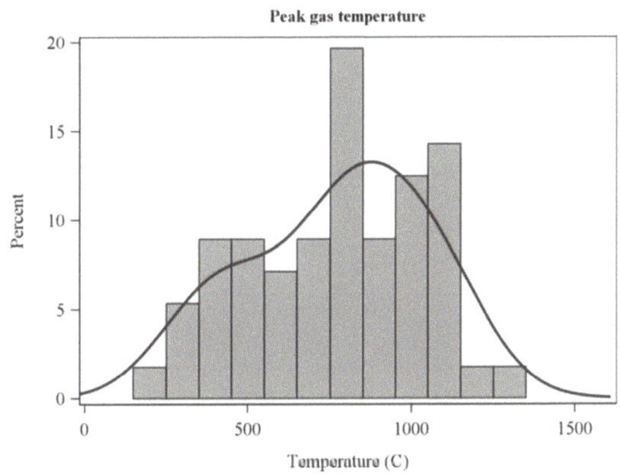

Figure 4. Frequency distribution of peak gas temperatures within residence times measured with fine thermocouples.

3.2. Flow Velocity and Horizontal and Vertical Sensible Heat Flux

The frequency distribution of average horizontal and vertical flow velocities showed a wide range across collections (Figure 5) and were not correlated which each other (R = −0.1, p = 0.6). Average horizontal velocities were larger than vertical velocities (Figure 5, Table 2). Peak flow velocities (Figure 6) show high instantaneous values, particularly for horizontal flow. Average sensible heat fluxes were greater in the horizontal than vertical directions (Table 2, Figure 7). As for flow velocities, horizontal and vertical sensible heat fluxes were not correlated (R = 0.24, p = 0.07). Peak horizontal and vertical sensible heat fluxes were often large (Figure 8) but fluctuated dramatically in our 10 Hz data (Figure 9). Residence times were strongly correlated with vertical sensible energy (R = 0.6, p > 0.0001) but were weakly correlated to horizontal sensible energy (R = 0.26, p = 0.06). The increase in residence time with sensible energy can be seen in Figure 9 in a comparison between timeseries with the lowest and median resultant sensible energies and that from the greatest resultant sensible energy.

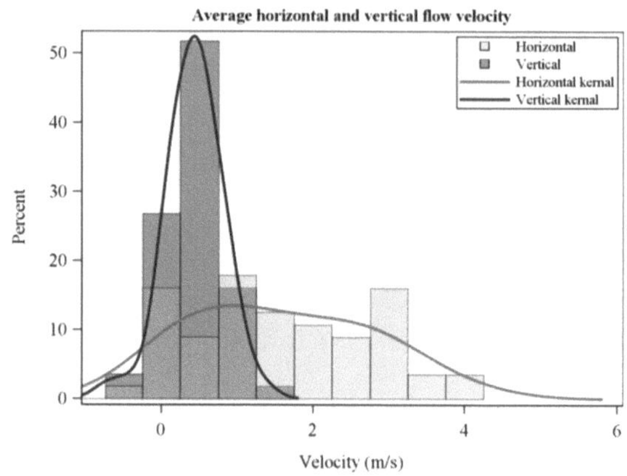

Figure 5. Horizontal and vertical velocities averaged over residence times.

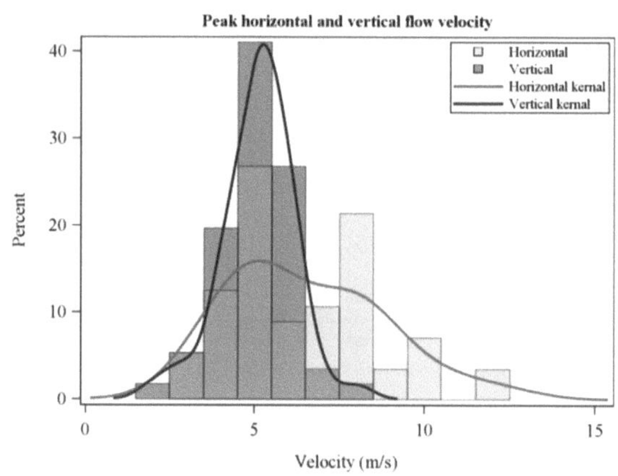

Figure 6. Peak horizontal and vertical flow velocities over residence times.

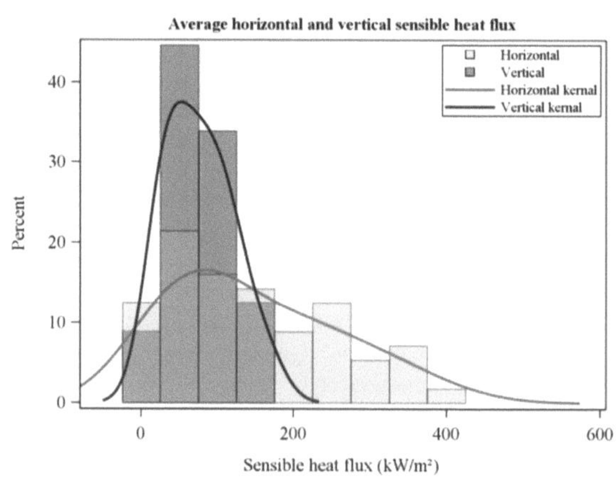

Figure 7. Average horizontal and vertical sensible heat fluxes over residence times.

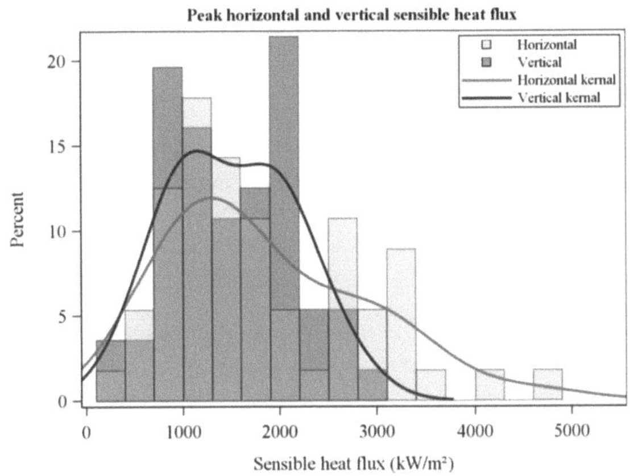

Figure 8. Peak horizontal and vertical sensible heat fluxes over residence times.

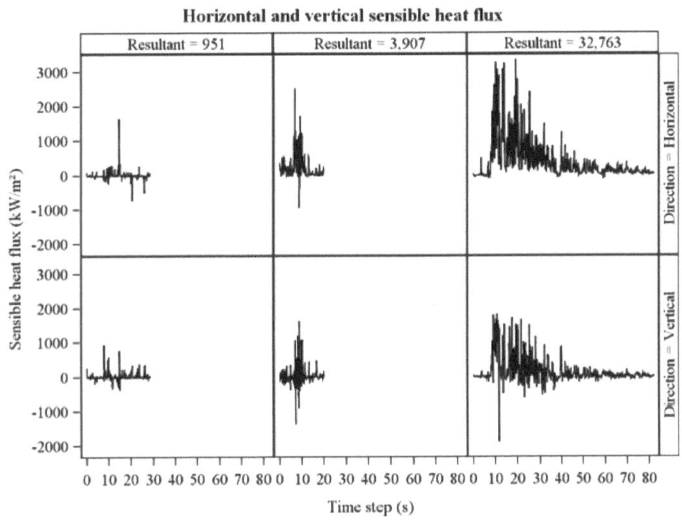

Figure 9. Time series of horizontal and vertical sensible heat flux (rows) across a range in resultant sensible energy (columns). Time series are from the datasets with the lowest (**left**), median (**center**), and highest (**right**) resultant sensible energies (kJ/m^2).

Table 2. Comparisons between horizontal and vertical flow velocities and sensible heat fluxes averaged over residence times and sensible energies. Overall means and, in parentheses, standard deviations and ranges of average values along with t-test statistics are reported.

Dependent Variable	N	Horizontal Mean (SD, Range)	Vertical Mean (SD, Range)	*t*-Value	*p*
Flow velocity	55	1.6 (1.2, −0.4–4.1)	0.4 (0.4, −0.7–1.3)	7.1	<0.0001
Sensible heat flux	55	148 (110, 4–423)	76 (43, 11–171)	4.6	<0.0001
Sensible energy	55	4477 (4978, 171–30753)	2334 (1941, 208–11298)	3.0	0.003

3.3. Horizontal, Vertical, and Resultant Sensible Energy

Horizontal sensible energies, the time-integral of sensible heat fluxes (Figures 5 and 6), were larger than vertical sensible energies (Table 2, Figure 10). Horizontal, vertical, and resultant sensible energies were positively related, with considerable variability, to fire energy inferred from fire radiated energy density (Table 3, Figure 11). The slopes in Table 3 are estimates of the fraction of fire energy dissipated by sensible energy (Equation (1)). Presence of flame, or the hot plume above flames, at the FBP is indicated by peak sensible heat fluxes in Figure 9. Clearly, residence times for integration (Figure 2) include fire-generated sensible heat flux from before and after flame arrival at the FBP.

Table 3. Linear regression statistics for horizontal, vertical, and resultant sensible energies as a function of fire energy (see Figure 11). Dependent and independent variables were log-transformed.

Dependent Variable	N	Intercept (ln[kJ/m^2])	Slope (Dimensionless)	R^2	p
Horizontal	32	2.00	0.75	0.18	0.015
Vertical	32	3.12	0.54	0.16	0.003
Resultant	32	3.25	0.64	0.26	0.003

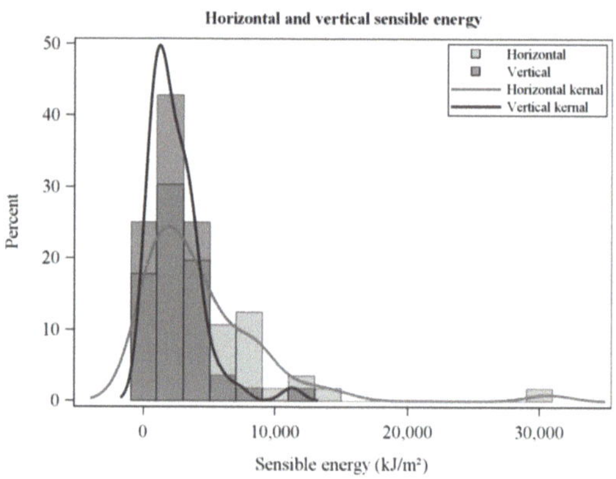

Figure 10. Horizontal and vertical sensible energies resulting from the time-integration of sensible heat fluxes over residence times.

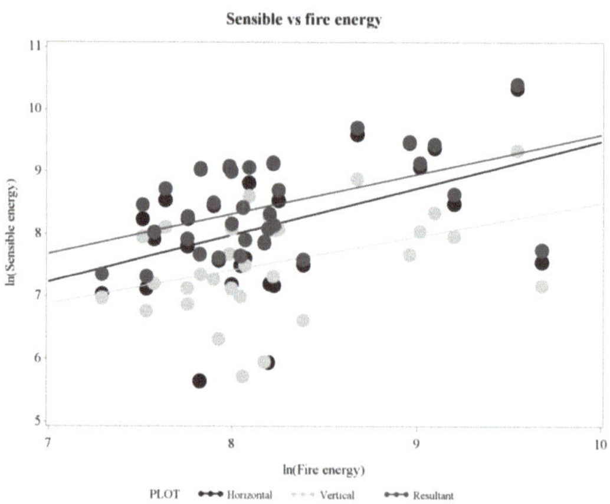

Figure 11. Horizontal, vertical, and resultant sensible energies (kJ/m^2) as a function of fire energy (kJ/m^2) estimated from overhead radiometers on natural log-transformed axes.

4. Discussion

Horizontal, vertical, and resultant energies were estimated to be 75%, 54%, and 64%, respectively, of fire energy generated by the RxCADRE surface fires in longleaf pine savanna and open ranges (Table 3, Figure 11). Clearly, because of their magnitude, sensible heat flux and energy are critical measurements for understanding fire and plume dynamics and fire effects and for closing the fire heat budget (Equation (1)). We believe that these are the first measurements of near-source sensible heat fluxes and energy from wildland fires using in situ (in and near-flame) instruments. Measurements were based on flow estimated from differential pressures using bi-directional probes ([20]) and gas temperatures from fine thermocouples. Measurements of heat dissipation from fires (Equation (1)) have heretofore been limited to radiation measurements ([1–3,5]) and one measurement of energy transferred into the soil ([2]) which we infer was ~5% of fire energy. Fractional radiated energy from field experiments are variable but average somewhere between 14 and 17% of fire energy ([5]). The relative proportion of sensible versus radiant energy should increase with the size of flames (e.g., from surface fires to crown fires) through increases in flame emissivity ([43,44]). As well, radiated fraction has been shown to decrease with increases in fuel moisture ([45]). We recognize that there is substantial variability associated with our estimates of fractional energy (Figure 11) which we attribute, in part, to the fact that we inferred fire energy from radiation measurements and an unrealistic assumption of a constant radiated fraction. In order to balance the integrated wildland fire heat budget (Equation (1)) given the physical complexity of the problem, it is likely that measurements should be coordinated with physics-based fire modeling ([46–48]) within a well-defined control volume. Constraining the control volume is important. For instance, consider that an estimate of sensible energy from in situ instruments positioned in flames will likely be a larger fraction of fire energy than an estimate derived from a measurement in the lower plume because of the progressive loss of heat by radiation.

We recognize limitations that should be addressed in future designs of instruments for measuring sensible heat fluxes in wildland fires. McCaffrey and Heskestad ([20]) describe a polynomial relationship between the calibration coefficient and Reynolds number (Re) that asymptotes at Re > 1000, a value substantially exceeded in our experiments (Re ~8000) where high flow velocity (13 m/s, Figure 6) is combined with temperatures used to define integration limits (50 C + ambient). Given generally positive relationships between gas temperatures and flow velocities, we expect that this situation occurred infrequently and, given low temperatures, would have had a small effect on sensible heat fluxes. In contrast, moderate temperatures combined with high flow velocities would also have yielded Reynolds numbers in the asymptotic range (Re ~3000) which suggests that we may have underestimated flow velocities under these conditions and, thus, underestimated sensible heat fluxes. We discuss instrument design improvements below and note that a probe with a smaller characteristic dimension than ours (12.7 mm) would reduce Reynolds numbers and may be appropriate for wildland fire measurement.

Our gas temperature measurements are from fine, exposed bead thermocouples that we know are biased estimators of gas temperatures ([49,50]). As radiation loss declines and heat gain by convection increases as thermocouple diameter declines, fine thermocouples are more faithful indicators of gas temperatures than thick thermocouples ([49]). Our 0.025 mm diameter thermocouples were near the lower limit for practical use in the field with finer thermocouples being too delicate. Based on Figure 1 in Walker and Stocks [49], we estimate that the error at a peak temperature of 1350 C would be about 11 C for our thermocouples which, at a peak velocity of 13 m/s translates to a reduction in peak sensible heat flux of about 1%. The error is not large, but only applies to peak temperatures. Clements and Seto ([17]) measured maximum horizontal Reynolds averaged heat fluxes of around 120 kW/m^2 at 1.9 m above ground using sonic anemometry during a surface fire, a value substantially lower than the peak sensible heat fluxes we measured (Figures 8 and 9). We attribute the difference to the Reynolds averaging, the greater height of their measurements, and the thicker thermocouples that they used to measure gas temperatures (nominally 0.08 mm). Clements and Seto ([17]) report 30 s Reynolds averaged values which are intended to isolate turbulent kinetic energy and not perturbation energies that we report which only remove pre-fire sensible heat flux and are appropriate for balancing the time-integrated wildland fire heat budget (Equation (1)). The error in peak sensible heat flux resulting from their 0.08 mm diameter thermocouples was probably not the largest source of the difference with our study. Based on data in Walker and Stocks [49], error in peak sensible heat fluxes from thermocouple error in their study was on the order of 3%. Similar considerations are relevant to Clements et al. ([16,35]). Gas temperature measurement is clearly a limitation for achieving accurate estimates of sensible heat fluxes and energy. Shielded-aspirated thermocouples, as opposed to bare thermocouples, maximize heat gain from convection and neutralize heat loss from radiation in steady-state conditions ([51,52]). A limitation is that aspirated thermocouples average over a sampling volume and are, thus, not true estimators of instantaneous gas temperatures. A blending of measurement with fine thermocouples and thermocouple heat budget modeling ([53]) may offer the best means of increasing accuracy of instantaneous gas temperature measurement.

Due to the separation between horizontally- and vertically oriented bi-directional probes (Figure 1), more appropriate for large flames, we are unsure of the error in flow direction based on a resultant of the two instantaneous measurements and we have not reported them here. Proximity of multi-dimensional flow measurements will deliver data that will provide more accurate estimates of flow direction ([54]). Currently, the thermocouple used in the FBP (Figure 1) is located between the two probes, a design that is not ideal, but, we assume, supports estimates of sensible energy that are approximately correct after horizontal and vertical sensible heat fluxes are integrated over residence times.

Bi-directional probes oriented vertically and horizontally clearly miss flow in the third, that is, crossways direction. As such, even though the probes have a wide acceptance angle ([20]), flux and energy are underestimated. We attempted to minimize this problem by only accepting fires for which we had evidence that average flows were upwards and towards the front of the FBP (Figure 1). We do not pretend that we eliminated the problem, however, given that we had no information on crosswise flow and recognize that the turbulent flow that characterizes flames and plumes is three dimensional ([37]). The high variability in sensible heat fluxes shown in Figure 9 reflect this turbulence and the intricate structure of turbulent diffusion flames ([55]). Three-dimensional measurement would clearly be beneficial. We point the reader to a probe assembly involving two precisely oriented bi-directional probes that provides three-dimensional flow except under specific conditions ([56]). Adding a third (crossways) bi-directional probe may be preferable to avoid limiting conditions ([54]). Clearly, a worthy challenge for fire science is to develop accurate methods for measuring three-dimensional flow in situ.

A measurement limitation that is not easily addressed is the fact that sensible heat flux and energy measurements from bi-directional probes have an uncertain footprint because they are sensing at least some heat advected with the flow from approaching flame fronts and, after flame front passage, from heat sources behind the flaming front. Moreover, flame fronts are not steady state and, at best, approach a quasi-steady condition ([57]). As such, it is not possible to know accurately the ground area over which the sensible energy was generated. The indeterminant footprint is indicated by the spikes in sensible heat fluxes before and after peak fluxes (originating from flames and hot gasses in the near-source plume) with a particularly long tail of sensible heat fluxes from the high energy fire on the right-hand side of Figure 9. We can speculate that high-frequency infrared imaging, particularly if done in three dimensions ([58]) could better describe the spatiotemporal field of sensible heat fluxes and energy ([9]). Until spatial measurement approaches are developed, increased replication of relatively local measurements is the only option for better characterizing sensible heat fluxes and energy on an aerial basis.

We assume that sensible energy is largely kinetic and that potential energy is negligible in a flame front wherein flame velocities resulting from buoyancy-driven flow are at a quasi-steady state. We speculate that this assumption would be least tenable during periods where flame velocities were accelerating or decelerating and heat flux rates were increasing or decreasing rapidly. It is not clear how we could address the effects of unsteady fire behavior without measurements of the spatiotemporal field of flow velocities and sensible heat fluxes using imaging methods ([9,58]).

Although smoldering combustion of duff in the longleaf pine savanna (L2F) contributed about 25% to fire energy ([34]), we assume that the bulk of sensible heat flux and energy in RxCADRE surface fires was from flaming combustion. Flame residence times were on the order of 10 s in these fires (Table 1) while our residence times for integration ranged up to nearly 100 s (Figure 2). In fires with substantial duff and downed woody fuel combustion, an assessment of their contribution to sensible heat flux may be possible. Where warranted, contributions from residual combustion might be assessed by first estimating the contribution from flames using a residence time capturing peak fluxes and, then, estimate contributions from residual combustion by extending the trailing limit of the residence time. Ward et al. ([59]) combined flow with chemical emissions measurements to characterize the mass balance of the primary combustion products which they used to partition fuel consumption into contributions from flaming and smoldering. Mass balance measurements may offer an important contribution to balancing the heat budget (Equation (1)).

In situ physical measurements of energy transport in flames and energy deposition in the wildland fire environment are critical for advancing fire science ([18]) yet are relatively sparse in the literature ([42,58,60]). With the advent of modern numerical computation, the physical complexity and computational requirements of wildland fire behavior and effects models has increased, including models designed to simulate fire behavior ([46,47,61–63]), plume transport ([64,65]), and fire effects ([66–68]). More physically realistic models and better basic understanding of fire dynamics require continued measurement development particularly of the basic heat and chemical processes occurring in fires exemplified by our study and others in the field ([17,69–74]) and laboratory ([75–80]). In situ measurements also support remote sensing. Remote sensing measurements directly related to fire heat dissipation are currently limited to infrared radiation which can be quantified at high spatial and temporal resolution over large spatial extents ([81]). Sensible heat fluxes and flow fields derived from coincident nadir radar and infrared measurements may, with more development, provide similar temporal and spatial coverage ([9,82]). There is a need to add value to remotely-sensed measurements through in situ measurements that improve understanding of their physical connections with combustion and energy transport processes in fires ([1,5,18]) and, in turn, their links with plume transport ([65]) and fire effects ([36]).

We describe in situ measurements of sensible heat flux and energy in and near wildland surface fires. Closing the wildland fire heat budget (Equation (1), [1]) remains elusive yet is a scientific quest worth pursuing because doing so will help fire scientists better test their understanding of energy transport and heat deposition in the fire environment ([19,58]), will support smoke plume ([65]) and fire effects model development ([25–27,36]), and will improve remotely sensed measurements of fire behavior and energy ([2,18]). Closing the fire heat budget requires continued measurement development and continued coordinated measurement campaigns over a wide range of fuel and fire behavior characteristics (e.g., [32,65]). Coordinated measurements should target all components of the heat budget, including the heat source ([13–15,70]), heat sinks, latent energy ([16]), soil heating ([2]), radiation ([1–3,5]), and sensible energy ([16,17]). In addition, combining measurement campaigns with fire physics modeling is imperative ([83]) for both advancing the science and furthering the development of wildland fire management decision support systems.

5. Conclusions

Measurements in and above flames from in situ instruments confirm that sensible energies are the dominant mode of heat dissipation from flaming combustion in wildland surface fires (Equation (1)). Measurement challenges remain, including obtaining accurate gas temperatures and three-dimensional flow velocities across the varying spatiotemporal field of spreading fires. Continued development of sensible heat flux and energy measurements is warranted by its magnitude and its importance as a core mechanism driving fire spread, plume rise, and fire effects.

Author Contributions: Conceptualization, M.B.D., R.L.K. and B.W.B.; methodology, B.W.B., C.E.W., D.J., P.S., R.L.K. and M.B.D.; software, C.E.W. and R.L.K.; validation, B.W.B., C.W., R.L.K. and M.B.D.; formal analysis, B.W.B., C.W., R.L.K. and M.B.D.; investigation, B.W.B., C.E.W., D.J., P.S., J.J.O., R.L.K. and M.B.D.; resources, B.W.B., M.B.D., J.J.O.; data curation, B.W.B., C.E.W. and M.B.D.; writing—original draft preparation, M.B.D.; writing—review and editing, M.B.D., B.W.B., C.E.W., R.L.K.; visualization, M.B.D.; supervision, B.W.B.; project administration, B.W.B. and M.B.D.; funding acquisition, B.W.B., J.J.O. and M.B.D. All authors have read and agreed to the published version of the manuscript.

Funding: Data collection was funded by the Northern Research Station, Rocky Mountain Research Station, and the Joint Fire Science Program grant number 11-2-1-11.

Institutional Review Board Statement: Not applicable.

Informed Consent Statement: Not applicable.

Data Availability Statement: Radiometer and FBP data ([84,85]), and instrument locations ([86,87]), and other RxCADRE data are available on the USDA Forest Service Research Data Archive.

Acknowledgments: We thank the Eglin AFB fire management staff, particularly Kevin Hiers, Brett Williams, and the fire crews for logistics and operations support. We also thank the many people whose work and collaboration made the RxCADRE project and the measurements reported here possible, including Roger Ottmar, Casey Teske, Mark Vosburgh, Ben Hornsby, and Louise Loudermilk. Thanks to Scott Goodrick and anonymous reviewers for feedback on earlier drafts.

Conflicts of Interest: The authors declare no conflict of interest.

Appendix A

Flow velocity, assuming incompressible flow, is derived by calibration from wind tunnel data. Each probe was calibrated separately at ambient temperature over a range in velocity from 2.5–14 m/s. Calibration is illustrated here for horizontal flow (u):

$$u_c = \sqrt{\frac{2p}{\rho_c}} \tag{A1}$$

where p is the sensor differential pressure (Pa or kg/m s^2) and ρ_c is the air density at calibration temperature and pressure (kg/m^3). Differential pressure is determined from:

$$p = P_p V \tag{A2}$$

where P_p is the pressure sensitivity of the sensor (volts/Pa) and V is voltage (volts). The air density is determined from:

$$p_c = \frac{P_c M_w}{RT_c} \tag{A3}$$

where P_c is the atmospheric pressure at the time of calibration (Pa), M_w is the molecular weight of dry air (28.9647 × 1000 kg/mol), R is the universal gas constant (8.31 kg m^2/K mols s^2), and T_c is the air temperature at the time of calibration (K). Using Equation (A1) to determine flow velocity from field data:

$$u_f = \sqrt{\frac{2p}{\rho_a}} \tag{A4}$$

where u_f is horizontal velocity in the field and ρ_a is ambient air density. Ambient air density is determined by:

$$\rho_a = \frac{P_a M_w}{RT_a} \tag{A5}$$

where T_a is instantaneous temperature from the thermocouple and P_a is the ambient atmospheric pressure. Velocity in Equation (A4) will only be correct if $T_a = T_c$ and $P_a = P_c$, therefore, a correction must be found to convert u_f to the true air flow velocity, u_c. Ratioing Equations (A3) and (A5) gives:

$$\frac{\rho_c}{\rho_a} = \frac{P_c M_w}{RT_c} / \frac{P_a M_w}{RT_a} \tag{A6}$$

And

$$\rho_c = \frac{P_c T_a}{P_a T_c} \tag{A7}$$

Substituting Equation (A7) into Equation (A1) provides calibrated air flow from field measurements:

$$u_c = \sqrt{\frac{2pP_a T_c}{\rho_a P_c T_a}} \tag{A8}$$

The calibration relationship in Equation (A8) is linear from 2.5–14 m/s.

References

1. Kremens, R.L.; Dickinson, M.B.; Bova, A.S. Radiant flux density, energy density and fuel consumption in mixed-oak forest surface fires. *Int. J. Wildl. Fire* **2012**, *21*, 722–730. [CrossRef]
2. Wooster, M.J.; Roberts, G.; Perry, G.L.W.; Kaufman, Y.J. Retrieval of biomass combustion rates and totals from fire radiative power observations: FRP derivation and calibration relationships between biomass consumption and fire radiative energy release. *J. Geophys. Res. Atmos.* **2005**. [CrossRef]
3. Freeborn, P.H.; Wooster, M.J.; Hao, W.M.; Ryan, C.A.; Nordgren, B.L.; Baker, S.P.; Ichoku, C. Relationships between energy release, fuel mass loss, and trace gas an aerosol emissions during laboratory biomass fires. *J. Geophys. Res. Atmos.* **2008**. [CrossRef]
4. Byram, G.M. Combustian of Forest Fuels. In *Forest Fire Control and Use*; Davis, K.P., Ed.; McGraw-Hill: New York, NY, USA, 1959; pp. 61–89.
5. Johnston, J.M.; Wooster, M.J.; Paugam, R.; Wang, X.; Lynham, T.J.; Johnston, L.M. Direct estimation of Byram's fire intensity from infrared remote sensing imagery. *Int. J. Wildl. Fire* **2017**. [CrossRef]
6. Clements, C.B.; Zhong, S.; Goodrick, S.; Li, J.; Potter, B.E.; Bian, X.; Heilman, W.E.; Charney, J.J.; Perna, R.; Jang, M.; et al. Observing the dynamics of wildland grass fires: FireFlux - A field validation experiment. *Bull. Am. Meteorol. Soc.* **2007**. [CrossRef]
7. Thomas, P.H. The size of flames from natural fires. *Symp. Combust.* **1963**, *9*, 844–859. [CrossRef]
8. Holman, J.P. *Heat Transfer*, 9th ed., McGraw-Hill: New York, NY, USA, 2002.
9. Clark, T.L.; Radke, L.; Coen, J.; Middleton, D. Analysis of small-scale convective dynamics in a crown fire using infrared video camera imagery. *J. Appl. Meteorol.* **1999**. [CrossRef]
10. Coen, J.; Mahalingam, S.; Daily, J. Infrared imagery of crown-fire dynamics during FROSTFIRE. *J. Appl. Meteorol.* **2004**. [CrossRef]
11. Lozano, J.; Tachajapong, W.; Weise, D.R.; Mahalingam, S.; Princevac, M. Fluid dynamic structures in a fire environment observed in laboratory-scale experiments. *Combust. Sci. Technol.* **2010**, *182*, 858–878. [CrossRef]
12. Morandini, F.; Silvani, X.; Susset, A. Feasibility of particle image velocimetry in vegetative fire spread experiments. *Exp. Fluids* **2012**, *53*, 237–244. [CrossRef]
13. Ward, D.E.; Hardy, C.C.; Ottmar, R.D.; Sandberg, D.V. A sampling system for measuring emissions from west coast prescribed fires. In Proceedings of the Pacific Northwest International Section of the Air pollution Control Association, Vancouver, BC, Canada, 15–17 November 1982; pp. 1–10.
14. Ward, D.E. Smoke and fire characteristics for cerrado and deforestation burns in Brazil: BASE-B experiment. *J. Geophys. Res.* **1992**. [CrossRef]
15. Susott, R.A.; Ward, D.E.; Babbitt, R.E.; Latham, D.J. The measurement of trace emissions and combustion characteristics for a mass fire. In *Global Biomass Buring: Atmospheric, Climatic, and Biosphere Implications*; Levine, J.S., Ed.; MIT Press: Cambridge, MA, USA, 1991; pp. 245–257.
16. Clements, C.B.; Potter, B.E.; Zhong, S. In situ measurements of water vapor, heat, and CO2 fluxes within a prescribed grass fire. *Int. J. Wildl. Fire* **2006**. [CrossRef]
17. Clements, C.B.; Seto, D. Observations of Fire–Atmosphere Interactions and Near-Surface Heat Transport on a Slope. *Boundary-Layer Meteorol.* **2014**. [CrossRef]
18. Kremens, R.L.; Smith, A.M.S.; Dickinson, M.B. Fire metrology: Current and future directions in physics-based measurements. *Fire Ecol.* **2010**. [CrossRef]
19. Butler, B.; Teske, C.; Jimenez, D.; O'Brien, J.; Sopko, P.; Wold, C.; Vosburgh, M.; Hornsby, B.; Loudermilk, E. Observations of energy transport and rate of spread from low-intensity fires in longleaf pine habitat - RxCADRE 2012. *Int. J. Wildl. Fire* **2016**. [CrossRef]

20. McCaffrey, B.J.; Heskestad, G. A robust bidirectional low-velocity probe for flame and fire application. *Combust. Flame* **1976**. [CrossRef]
21. Dietenberger, M. Update for combustion properties of wood components. *Fire Mater.* **2002**. [CrossRef]
22. Babrauskas, V. Effective heat of combustion for flaming combustion of conifers. *Can. J. For. Res.* **2006**. [CrossRef]
23. Kiefer, M.T.; Zhong, S.; Heilman, W.E.; Charney, J.J.; Bian, X. A Numerical Study of Atmospheric Perturbations Induced by Heat From a Wildland Fire: Sensitivity to Vertical Canopy Structure and Heat Source Strength. *J. Geophys. Res. Atmos.* **2018**, *123*, 2555–2572. [CrossRef]
24. Dupuy, J.L.; Maréchal, J. Slope effect on laboratory fire spread: Contribution of radiation and convection to fuel bed preheating. *Int. J. Wildl. Fire* **2011**. [CrossRef]
25. O'Brien, J.J.; Hiers, J.K.; Varner, J.M.; Hoffman, C.M.; Dickinson, M.B.; Michaletz, S.T.; Loudermilk, E.L.; Butler, B.W. Advances in Mechanistic Approaches to Quantifying Biophysical Fire Effects. *Curr. For. Reports* **2018**, *4*, 161–177. [CrossRef]
26. Wagner, C.E. Van Height of Crown Scorch in Forest Fires. *Can. J. For. Res.* **1973**. [CrossRef]
27. Michaletz, S.T.; Johnson, E.A. A heat transfer model of crown scorch in forest fires. *Can. J. For. Res.* **2006**. [CrossRef]
28. Kavanagh, K.L.; Dickinson, M.B.; Bova, A.S. A way forward for fire-caused tree mortality prediction: Modeling a physiological consequence of fire. *Fire Ecol.* **2010**. [CrossRef]
29. Dickinson, M.B.; Norris, J.C.; Bova, A.S.; Kremens, R.L.; Young, V.; Lacki, M.J. Effects of wildland fire smoke on a tree-roosting bat: Integrating a plume model, field measurements, and mammalian dose-response relationships. *Can. J. For. Res.* **2010**. [CrossRef]
30. Dickinson, M.B.; Lacki, M.J.; Cox, D.R. Fire and the endangered Indiana bat. In Proceedings of the 3rd Fire in Eastern Oak Forests Conference, Carbondale, IL, USA, 20–22 May 2008; Hutchinson, T.F., Ed.; U.S. Department of Agriculture, Forest Service, Northern Research Station: Newtown Square, PA, USA, 2009; pp. 51–75.
31. Kobziar, L.N.; Thompson, G.R. Wildfire smoke, a potential infectious agent. *Science* **2020**. [CrossRef]
32. Ottmar, R.D.; Hiers, J.K.; Butler, B.W.; Clements, C.B.; Dickinson, M.B.; Hudak, A.T.; O'Brien, J.J.; Potter, B.E.; Rowell, E.M.; Strand, T.M.; et al. Measurements, datasets and preliminary results from the RxCADRE project - 2008, 2011 and 2012. *Int. J. Wildl. Fire* **2016**. [CrossRef]
33. Dickinson, M.B.; Hudak, A.T.; Zajkowski, T.; Loudermilk, E.L.; Schroeder, W.; Ellison, L.; Kremens, R.L.; Holley, W.; Martinez, O.; Paxton, A.; et al. Measuring radiant emissions from entire prescribed fires with ground, airborne and satellite sensors—RxCADRE 2012. *Int. J. Wildl. Fire* **2016**, *25*, 48. [CrossRef]
34. Hudak, A.T.; Dickinson, M.B.; Bright, B.C.; Kremens, R.L.; Loudermilk, E.L.; O'Brien, J.J.; Hornsby, B.S.; Ottmar, R.D. Measurements relating fire radiative energy density and surface fuel consumption - RxCADRE 2011 and 2012. *Int. J. Wildl. Fire* **2016**. [CrossRef]
35. Clements, C.B.; Lareau, N.P.; Seto, D.; Contezac, J.; Davis, B.; Teske, C.; Zajkowski, T.J.; Hudak, A.T.; Bright, B.C.; Dickinson, M.B.; et al. Fire weather conditions and fire-atmosphere interactions observed during low-intensity prescribed fires - RxCADRE 2012. *Int. J. Wildl. Fire* **2016**. [CrossRef]
36. Dickinson, M.B.M.B.; Butler, B.W.B.W.; Hudak, A.T.A.T.; Bright, B.C.B.C.; Kremens, R.L.R.L.; Klauberg, C. Inferring energy incident on sensors in low-intensity surface fires from remotely sensed radiation and using it to predict tree stem injury. *Int. J. Wildl. Fire* **2019**, *28*. [CrossRef]
37. Clements, C.B.; Zhong, S.; Bian, X.; Heidman, W.E.; Byun, D.W. First observations of turbulence generated by grass fires. *J. Geophys. Res. Atmos.* **2008**. [CrossRef]
38. Butler, B.W.; Jimenez, D.; Forthofer, J.; Shannon, K.; Sopoko, P. A portable system for characterizing wildland fire behavior. In Proceedings of the Proceedings of the 6th International Conference on Forest Fire Research, Coimbra, Portugal, 15–18 November 2010; Viegas, D.X., Ed.; Imprensa da Universidade de Coimbra: Coimbra, Portugal, 2010; pp. 1–13.
39. Butler, B.W.; Cohen, J.; Latham, D.J.; Schuette, R.D.; Sopko, P.; Shannon, K.S.; Jimenez, D.; Bradshaw, L.S. Measurements of radiant emissive power and temperatures in crown fires. *Can. J. For. Res.* **2004**. [CrossRef]
40. O'Brien, J.J.; Loudermilk, E.L.; Hornsby, B.; Hudak, A.T.; Bright, B.C.; Dickinson, M.B.; Hiers, J.K.; Teske, C.; Ottmar, R.D. High-resolution infrared thermography for capturing wildland fire behaviour: RxCADRE 2012. *Int. J. Wildl. Fire* **2016**. [CrossRef]
41. Lide, D.R. *CRC Handbook of Chemistry and Physics*, 90th ed.; CRC Press: Boca Raton, FL, USA, 2009; ISBN 9781420090840.
42. Frankman, D.; Webb, B.W.; Butler, B.W.; Jimenez, D.; Harrington, M. The effect of sampling rate on interpretation of the temporal characteristics of radiative and convective heating in wildland flames. *Int. J. Wildl. Fire* **2013**. [CrossRef]
43. Pastor Ferrer, E.; Rigueiro, A.; Zárate López, L.; Gimenez, A.; Arnaldos Viger, J.; Planas Cuchi, E. Experimental methodology for characterizing flame emissivity of small scale forest fires using infrared thermography techniques. In *Forest Fire Research & Wildland Fire Safety*; Viegas, D.X., Ed.; Millpress: Rotterdam, Holland, 2002; ISBN 90-77017-72-0.
44. Johnston, J.M.; Wooster, M.J.; Lynham, T.J. Experimental confirmation of the MWIR and LWIR grey body assumption for vegetation fire flame emissivity. *Int. J. Wildl. Fire* **2014**. [CrossRef]
45. Smith, A.M.S.; Tinkham, W.T.; Roy, D.P.; Boschetti, L.; Kremens, R.L.; Kumar, S.S.; Sparks, A.M.; Falkowski, M.J. Quantification of fuel moisture effects on biomass consumed derived from fire radiative energy retrievals. *Geophys. Res. Lett.* **2013**. [CrossRef]
46. Mell, W.; Jenkins, M.A.; Gould, J.; Cheney, P. A physics-based approach to modelling grassland fires. *Int. J. Wildl. Fire* **2007**. [CrossRef]
47. Linn, R.; Reisner, J.; Colman, J.J.; Winterkamp, J. Studying wildfire behavior using FIRETEC. *Int. J. Wildl. Fire* **2002**. [CrossRef]

48. Morvan, D.; Dupuy, J.L. Modeling the propagation of a wildfire through a Mediterranean shrub using a multiphase formulation. *Combust. Flame* **2004**. [CrossRef]
49. Walker, J.D.; Stocks, B.J. Thermocouple errors in forest fire research. *Fire Technol.* **1968**. [CrossRef]
50. Van Wagner, C.E.; Methven, I.R. Two recent articles on fire ecology. *Can. J. For. Res.* **1978**. [CrossRef]
51. Blevins, L.G.; Pitts, W.M. Modeling of bare and aspirated thermocouples in compartment fires. *Fire Saf. J.* **1999**. [CrossRef]
52. Philpot, C.W. *Temperatures in a Large Natural-Fuel Fire*; Research Note PSW-90; US Forest Service, Pacific Southwest Forest and Range Experiment Station: Berkeley, CA, USA, 1965.
53. Bova, A.S.; Dickinson, M.B. Beyond "fire temperatures": Calibrating thermocouple probes and modeling their response to surface fires in hardwood fuels. *Can. J. For. Res.* **2008**. [CrossRef]
54. Grumstrup, T.P.; Forthofer, J.M.; Finney, M.A. Measurement of three-dimensional flow speed and direction in wildfires. In *Advances in Forest Fire Research*; Viegas, D.X., Ed.; Imprensa da Universidade de Coimbra: Coimbra, Portugal, 2018; pp. 542–548. ISBN 9789892616506.
55. Drysdale, D. *An Introduction to Fire Dynamics*, 3rd ed.; John Wiley & Sons, Ltd: Hoboken, NJ, USA, 2011; ISBN 9781119975465.
56. Newman, J.S. Multi-Directional Flow Probe Assembly for Fire Application. *J. Fire Sci.* **1987**. [CrossRef]
57. Rothermel, R.C. *A Mathematical Model for Predicting Fire Spread in Wildland Fuels*; Research Paper INT-115; USDA Forest Service, Intermountain Forest & Range Experiment Station: Ogden, UT, USA, 1972.
58. Morandini, F.; Toulouse, T.; Silvani, X.; Pieri, A.; Rossi, L. Image-Based Diagnostic System for the Measurement of Flame Properties and Radiation. *Fire Technol.* **2019**. [CrossRef]
59. Ward, D.E.; Hao, W.M.; Susott, R.A.; Babbitt, R.E.; Shea, R.W.; Kauffman, J.B.; Justice, C.O. Effect of fuel composition on combustion efficiency and emission factors for African savanna ecosystems. *J. Geophys. Res. Atmos.* **1996**. [CrossRef]
60. Butler, B.W.; Jimenez, D.; Forthofer, J.M.; Shannon, K.S.; Sopko, P. A portable system for characterizing wildland fire behavior. In Proceedings of the VI Internationl Conference on Forest Fire Research; Viegas, D.X., Ed.; University of Coimbra: Coimbra, Portugal; p. 13.
61. Call, P.T.; Albini, F.A. Aerial and surface fuel consumption in crown fires. *Int. J. Wildl. Fire* **1997**, *7*, 259–264. [CrossRef]
62. Morvan, D.; Dupuy, J.L. Modeling of fire spread through a forest fuel bed using a multiphase formulation. *Combust. Flame* **2001**. [CrossRef]
63. Sullivan, A.L. Wildland surface fire spread modelling, 1990–2007. 1: Physical and quasi-physical models. *Int. J. Wildl. Fire* **2009**. [CrossRef]
64. Kochanski, A.K.; Jenkins, M.A.; Mandel, J.; Beezley, J.D.; Clements, C.B.; Krueger, S. Evaluation of WRF-SFIRE performance with field observations from the FireFlux experiment. *Geosci. Model Dev.* **2013**. [CrossRef]
65. Liu, Y.; Kochanski, A.; Baker, K.R.; Mell, W.; Linn, R.; Paugam, R.; Mandel, J.; Fournier, A.; Jenkins, M.A.; Goodrick, S.; et al. Fire behaviour and smoke modelling: Model improvement and measurement needs for next-generation smoke research and forecasting systems. *Int. J. Wildl. Fire* **2019**, *28*. [CrossRef]
66. Campbell, G.S.; Jungbauer, J.D.; Bristow, K.L.; Hungerford, R.D. Soil temperature and water content beneath a surface fire. *Soil Sci.* **1995**, *159*. [CrossRef]
67. Massman, W.J.; Frank, J.M.; Mooney, S.J. Advancing investigation and physical modeling of first-order fire effects on soils. *Fire Ecol.* **2010**. [CrossRef]
68. Chatziefstratiou, E.K.; Bohrer, G.; Bova, A.S.; Subramanian, R.; Frasson, R.P.M.; Scherzer, A.; Butler, B.W.; Dickinson, M.B. FireStem2D—A Two-Dimensional Heat Transfer Model for Simulating Tree Stem Injury in Fires. *PLoS ONE* **2013**. [CrossRef]
69. Stocks, B.J.; Lynham, T.J.; Lawson, B.D.; Alexander, M.E.; Van Wagner, C.E.; McAlpine, R.S.; Dubé, D.E. The Canadian Forest Fire Danger Rating System: An Overview. *For. Chron.* **1989**. [CrossRef]
70. Ward, D.E.; Radke, L.F. Emissions Measurements from Vegetation Fires: A Comparative Evaluation of Methods and Results. In *Fire in the Environment: The Ecological, Atmospheric, and Climatic Importance of Vegetation Fires*; Crutzen, P.J., Goldammer, J.G., Eds.; John Wiley & Sons: Chischester, England, 1993; pp. 53–76, ISBN 0-471-93604-9.
71. Hiers, J.K.; O'Brien, J.J.; Mitchell, R.J.; Grego, J.M.; Loudermilk, E.L. The wildland fuel cell concept: An approach to characterize fine-scale variation in fuels and fire in frequently burned longleaf pine forests. *Int. J. Wildl. Fire* **2009**. [CrossRef]
72. Stocks, B.J.; Alexander, M.E.; Wotton, B.M.; Stefner, C.N.; Flannigan, M.D.; Taylor, S.W.; Lavoie, N.; Mason, J.A.; Hartley, G.R.; Maffey, M.E.; et al. Crown fire behaviour in a northern jack pine - Black spruce forest. *Can. J. For. Res.* **2004**. [CrossRef]
73. Lachlan McCaw, W.; Gould, J.S.; Phillip Cheney, N.; Ellis, P.F.M.; Anderson, W.R. Changes in behaviour of fire in dry eucalypt forest as fuel increases with age. *For. Ecol. Manage.* **2012**. [CrossRef]
74. Silvani, X.; Morandini, F. Fire spread experiments in the field: Temperature and heat fluxes measurements. *Fire Saf. J.* **2009**. [CrossRef]
75. Catchpole, W.R.; Catchpole, E.A.; Butler, B.W.; Rothermel, R.C.; Morris, G.A.; Latham, D.J. Rate of spread of free-burning fires in woody fuels in a wind tunnel. *Combust. Sci. Technol.* **1998**. [CrossRef]
76. Sun, L.; Zhou, X.; Mahalingam, S.; Weise, D.R. Comparison of burning characteristics of live and dead chaparral fuels. *Combust. Flame* **2006**. [CrossRef]
77. White, R.H.; Zipperer, W.C. Testing and classification of individual plants for fire behaviour: Plant selection for the wildlandurban interface. *Int. J. Wildl. Fire* **2010**. [CrossRef]

78. Dickinson, M.B.; Hutchinson, T.F.; Dietenberger, M.; Matt, F.; Peters, M.P. Litter species composition and topographic effects on fuels and modeled fire behavior in an oak-hickory forest in the Eastern USA. *PLoS ONE* **2016**. [CrossRef]
79. Safdari, M.S.; Rahmati, M.; Amini, E.; Howarth, J.E.; Berryhill, J.P.; Dietenberger, M.; Weise, D.R.; Fletcher, T.H. Characterization of pyrolysis products from fast pyrolysis of live and dead vegetation native to the Southern United States. *Fuel* **2018**. [CrossRef]
80. Safdari, M.S.; Amini, E.; Weise, D.R.; Fletcher, T.H. Comparison of pyrolysis of live wildland fuels heated by radiation vs. convection. *Fuel* **2020**. [CrossRef]
81. Riggan, P.J.; Tissell, R.G.; Lockwood, R.N.; Brass, J.A.; Pereira, J.A.R.; Miranda, H.S.; Miranda, A.C.; Campos, T.; Higgins, R. Remote measurement of energy and carbon flux from wildfires in Brazil. *Ecol. Appl.* **2004**. [CrossRef]
82. Rodriguez, B.; Lareau, N.P.; Kingsmill, D.E.; Clements, C.B. Extreme Pyroconvective Updrafts During a Megafire. *Geophys. Res. Lett.* **2020**. [CrossRef]
83. Linn, R.R.; Winterkamp, J.L.; Furman, J.H.; Williams, B.; Hiers, J.K.; Jonko, A.; O'Brien, J.J.; Yedinak, K.M.; Goodrick, S. Modeling Low Intensity Fires: Lessons Learned from 2012 RxCADRE. *Atmosphere (Basel)* **2021**, *12*, 139. [CrossRef]
84. Jimenez, D.M.; Butler, B.W. RxCADRE 2012: In-Situ Fire Behavior Measurements. Fort Collins, CO: Forest Service Research Data Archive. 2016. Available online: https://doi.org/10.2737/RDS-2016-0038 (accessed on 17 March 2021).
85. Dickinson, M.B.; Kremens, R.L. RxCADRE 2008, 2011, and 2012: Radiometer Data. Fort Collins, CO: Forest Service Research Data Archive. 2015. Available online: https://doi.org/10.2737/RDS-2015-0036 (accessed on 17 March 2021).
86. Hudak, A.T.; Dickinson, M.B.; Rodriguez, A.J. RxCADRE 2008, 2011, and 2012: Radiometer Locations. Forest Service Research Data Archive. 2015. Available online: https://doi.org/10.2737/RDS-2015-0035 (accessed on 17 March 2021).
87. Hudak, A.T.; Dickinson, M.B.; Rodriguez, A.J.; Bright, B.C. RxCADRE 2012: Instrument and Infrared Target Survey Locations and Attributes. Fort Collins, CO: Forest Service Research Data Archive. 2016. Available online: https://doi.org/10.2737/RDS-2016-0014 (accessed on 17 March 2021).

sensors

MDPI

Article

Top-Down Estimation of Particulate Matter Emissions from Extreme Tropical Peatland Fires Using Geostationary Satellite Fire Radiative Power Observations

Daniel Fisher [1,2,*], Martin J. Wooster [1,2,*], Weidong Xu [1,2], Gareth Thomas [2,3] and Puji Lestari [4]

1 Leverhulme Centre for Wildfires, Environment and Society, Department of Geography, King's College London, Aldwych, London WC2B 4BG, UK; weidong.xu@kcl.ac.uk
2 NERC National Centre for Earth Observation (NCEO), Leicester LE1 7RH, UK; gareth.thomas@stfc.ac.uk
3 RAL Space, STFC Rutherford Appleton Laboratory, Harwell Campus, Didcot OX11 0QX, UK
4 Faculty of Civil and Environmental Engineering, ITB, JL. Ganesha No.10, Bandung 40132, Indonesia; pujilest@indo.net.id
* Correspondence: daniel.fisher@kcl.ac.uk (D.F.); martin.wooster@kcl.ac.uk (M.J.W.)

Received: 2 October 2020; Accepted: 8 December 2020; Published: 10 December 2020

Abstract: Extreme fires in the peatlands of South East (SE) Asia are arguably the world's greatest biomass burning events, resulting in some of the worst ambient air pollution ever recorded ($PM_{10} > 3000\ \mu g \cdot m^{-3}$). The worst of these fires coincide with El Niño related droughts, and include huge areas of smouldering combustion that can persist for months. However, areas of flaming surface vegetation combustion atop peat are also seen, and we show that the largest of these latter fires appear to be the most radiant and intensely smoke-emitting areas of combustion present in such extreme fire episodes. Fire emissions inventories and early warning of the air quality impacts of landscape fire are increasingly based on the fire radiative power (FRP) approach to fire emissions estimation, including for these SE Asia peatland fires. "Top-down" methods estimate total particulate matter emissions directly from FRP observations using so-called "smoke emission coefficients" [C_e; $g \cdot MJ^{-1}$], but currently no discrimination is made between fire types during such calculations. We show that for a subset of some of the most thermally radiant peatland fires seen during the 2015 El Niño, the most appropriate C_e is around a factor of three lower than currently assumed (~16.8 ± 1.6 $g \cdot MJ^{-1}$ vs. 52.4 $g \cdot MJ^{-1}$). Analysis indicates that this difference stems from these highly radiant fires containing areas of substantial flaming combustion, which changes the amount of particulate matter emitted per unit of observable fire radiative heat release in comparison to more smouldering dominated events. We also show that even a single one of these most radiant fires is responsible for almost 10% of the overall particulate matter released during the 2015 fire event, highlighting the importance of this fire type to overall emission totals. Discriminating these different fires types in ways demonstrated herein should thus ultimately improve the accuracy of SE Asian fire emissions estimates derived using the FRP approach, and the air quality modelling which they support.

Keywords: tropical peatlands; landscape fire; emissions; FRP; total particulate matter

1. Introduction

In the second half of the 20th Century, changing land management practices in the lowland tropical swamp forests of equatorial South East (SE) Asia led many to be degraded from their natural forest-covered, persistently waterlogged state [1–5]. The carbon-rich peat surface now becomes dry enough to burn at times of low rainfall in many of these deforested and drained areas, leading to substantial increases in the region's susceptibility to extreme landscape fire—particularly during the periodic droughts brought on by El Niño–Southern Oscillation events [6–9]. During the extreme 2015 El Niño, an overwhelming number of landscape fires occurred in the tropical peatlands of Kalimantan and Sumatra during September and October [10]. Truly extreme ambient atmospheric concentrations of particulate matter (PM) were generated across parts of Indonesia in particular, representing one of the worst sustained air pollution events ever recorded (see Figure 1, main image). PM_{10} concentrations sometimes exceeded 3000 $\mu g \cdot m^{-3}$, as reported in [10], and very substantial impacts on human mortality have been suggested as a result of this air pollution [11,12]. However, such estimates are built on as yet uncertain fire-emitted PM totals along with dose-response functions derived for Western populations that are generally subject to far lower, yet more long-term, PM exposures (e.g., those from urban air pollution).

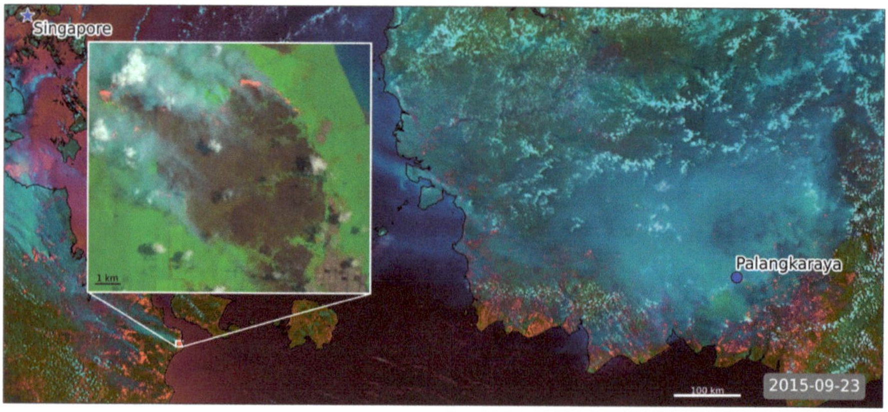

Figure 1. Satellite imagery of Kalimantan and South Sumatra taken on 23 September 2015 during the extreme landscape fires exacerbated by the strong El Niño-related drought. Main image: Visible Infrared Imaging Radiometer Suite (VIIRS) false colour composite (R: 3.7 μm; G: 0.865 μm; B: 0.488 μm) where clouds appear white, smoke blue—grey, and actively burning fires—bright red. Inset image: Landsat false colour composite (R: B6 (SWIR); G: B5 (NIR); B: B4 (VIS)) for the same day as the main image, focused on the longest of the Sumatran plumes present in the main image whose source region is boxed in white on the VIIRS imagery. Again, clouds appear white, smoke grey, and actively burning fires orange/red; also note the large burned area, which appears dark brown. The fire shown in the Landsat scene is plume/fire matchup Fire 7 in Figure 2.

Emissions from fires on peatlands and other landscapes are mostly derived using so-called "bottom-up" approaches, which usually require the combination of satellite-derived burned area data with information on fuel consumption per unit area. The latter can be difficult to obtain, especially perhaps where peatland fires burn down into the carbon-rich soil to varying degrees—generating fires that can persist in the smouldering phase for weeks or months and can consume very large amounts of biomass per unit area [13]. Such "bottom-up" approaches are also only able to be applied after the fire event has occurred, making them unsuitable to support near-real-time emissions estimates and air quality early warning applications [14,15]. An alternative fire emissions estimation approach, and one that can be applied in real-time, is the so-called "top-down" method linking satellite-derived fire radiative energy (FRE; MJ) observations directly to total particulate matter (TPM; g) emissions through a so-called "smoke emissions coefficient" [C_e; $g \cdot MJ^{-1}$]. This is possible as FRE is correlated with fuel mass combusted [16] and also removes the need to estimate fuel consumption per unit area, relying only on knowledge of an appropriate C_e factor.

Tropical peat soils consist of ~54–60% carbon [17], and organic carbon is the primary particulate matter emission component resulting from its combustion [10]. Research foci related to peatland fire emissions have up to now primarily been focused on smouldering fires in this organic soil [8,18–20], which, whilst involving relatively slow combustion rates per unit area, produces substantially more PM emissions per unit of dry matter burned than do flaming or mixed phase (i.e., smouldering and flaming) fires in vegetation (e.g., [21]). These characteristics of smouldering peat are reflected in the TPM C_e used to produce current "top-down" fire emissions estimates in peatland fires, these being far higher than those of tropical forest vegetation fires (see Section 2). However, it is currently unclear whether such high smoke emission coefficients are singularly appropriate for all fires occurring on tropical peatlands, because some areas may involve flaming as well as smouldering fuel consumption. Indeed, data collected by the Advanced Himawari Imager (AHI) carried onboard the Himawari-8 satellite, along with data from the Visible Infrared Imaging Radiometer Suite (VIIRS) and Landsat Advanced Land Imager (ALI) satellite instruments (Section 4.2 and Figure 1), show evidence of highly radiant peatland fires, which appear to involve substantial flaming combustion of surface vegetation in addition to smouldering peat [22]. The likelihood of smoke being injected above the atmospheric boundary layer during flaming combustion is greater than for smouldering combustion [23], and this in turn likely supports more significant transboundary transport of the polluted air (e.g., to Singapore and mainland Malaysia [24,25] in the region studied herein). Accurately identifying and estimating the PM emissions for such flaming peatland fires is therefore likely to be important for effective modelling of air pollution transport, and for better gauging their ultimate air quality impacts. The objectives of the current work are therefore to (i) examine the PM emission characteristics of more intensely radiant tropical peatland fires in particular, through the development and application of a more appropriate top-down TPM C_e for such events, derived here using AHI fire radiative power (FRP; MW) products [26] and collocated observations of aerosol optical depth (AOD) obtained from VIIRS; and (ii) assess the impact of this updated C_e compared to use of existing values from studies that have not considered discriminating between smouldering and mixed phase fires. Our ultimate aim is to use the findings from this work to improve real-time air pollution forecasting during peatland fire episodes, which we demonstrate include flaming combustion of surface vegetation as well as smouldering combustion of subsurface organic soils.

2. Landscape Fire Emission Estimation Overview

"Bottom-up" methodologies convert observations of terrestrial phenomena into estimates of atmospheric emissions using assumptions and/or additional terrestrial parameters, for example, fuel consumption per unit area measures [27,28]. The two most widely used "bottom-up" emissions inventories are the Global Fire Emissions Database (GFED) [28], and the Global Fire Assimilation System (GFAS) [27]. In both systems, information derived from Earth observation (EO) on either burned area (BA (m^2); GFED) or FRP (GFAS) are converted into estimates of smoke emission via application of species-specific emission factors (EF; $g{\cdot}kg^{-1}$) applied to the bottom-up generated estimates of fuel consumption (kg). Fuel consumption estimates in GFED are produced by multiplying satellite-derived burned area measures by an assumed (modelled) fuel consumption per unit area, whereas in GFAS they are generated via application of biome-specific coefficients linking FRE and dry matter fuel consumption totals, which themselves have been generated using past comparisons of satellite-derived FRE data from GFAS and GFED fuel consumption values [27].

To support real-time air quality forecasting, we focus on an approach based on FRP measures, as they can be derived at the temporal resolution of the satellite rather than over multiple days as is typical of BA-based approaches [29]. Whilst significant advancements have been made during the last two decades in the active fire detection and fire characterisation algorithms that underlie FRP approaches (e.g., [29,30]), and landcover specific EF databases are becoming more detailed in their contents [31–33], there remain uncertainties in the FRP approach to fire emissions estimation. These can stem from (i) the relatively limited sampling frequency provided by the polar-orbiting satellites that currently dominate provision of FRP [34]; (ii) the fact that some of any surface-fire emitted FRP maybe intercepted by overlying tree canopies [35]; (iii) that some fires are too small or weakly burning to be detected in the most common spaceborne FRP data products (see comparisons by [36]); and (iv), in the case of tropical peatlands, from the rather limited (and until recently largely laboratory-derived) emission factors for peat burning [37]. Perhaps the most significant uncertainties are associated with the conversion between FRP and fuel consumption rate or totals. The conversion factor for this process has generally been derived in one of two ways. First, using measurements of FRP and fuel consumption in small laboratory-scale vegetation fires [16], although these may not be fully appropriate for application to satellite-derived FRP observations—especially in forested environments due to the issues outlined in [38], such as interception of radiated heat by forest canopies. Additionally, such conversion factors may not be applicable to types of fire involving at least some subsurface combustion [13], since fires burning underground will likely show a different amount of surface-emitted FRP per kg of fuel burned than do "normal" surface fires. Second, via comparisons between FRP data and GFED-derived fuel consumption totals (e.g., see [27]), which thus still leaves issues such as non-detection of smaller burned areas in GFED [39]—the under detection of peat burn area, in particular [40], and the use of difficult-to-estimate fuel loads [41] and combustion completeness [42] parameters within GFED. The last of these issues seems especially problematic for subsurface peat fires, where there remain rather limited measurements of peat depth-of-burn (DOB) [13], and where an accurate method to estimate the DOB of any particular satellite-detected fire remains lacking.

For these reasons, we focus on adapting the fully top-down "Fire Radiative Energy eMissions" (FREM) approach of [38,43] to tropical peatland fires. The FREM approach uses species- and biome-specific C_e to directly link FRP data to the emissions of any particular species within the smoke, thus removing the need for the interim fuel consumption estimation step. The method is "top-down" as it relies on atmospheric and terrestrial remote sensing observations only, as detailed in [37,38,44,45]. In contrast to the Moderate Resolution Image Radiometer (MODIS) based approach of [44], the FREM approach also attempts to minimise the number of assumptions required when deriving each biome-specific C_e by exploiting the almost continuous, very high temporal resolution FRP data available from geostationary sensors. Using these data, it is possible to derive an estimate of the total FRE released by a fire during the period when it produced a particular smoke plume. Each plume contains a certain amount of TPM, which itself can be estimated using satellite-derived AOD products [38,43–45] and further detailed in Section 3.3. Developing a set of matchup fires for which FRE and TPM values are determined enables the derivation of specific C_e for the biome and fire-type of interest, and it is our aim here to do this exclusively for the type of more intensely radiant fires seen on tropical peatlands and shown in Figure 1.

3. Top-Down Estimation of Particulate Matter Emissions

3.1. Algorithm Requirements and Plume Digitisation

As introduced in [38] and presented in Section 2, in order to establish an appropriate TPM C_e for a particular biome, collocated satellite-derived observations of smoke plume AOD and FRE are required for a set of matchup fires statistically representative of the biome's fire events. To derive C_e for the type of highly radiant peatland fires we study here, we found it necessary to further modify parts of the original methodology to address three specific issues:

(i) In [38], entire plumes were manually digitised from the satellite imagery to create the southern African fire matchups. The radiant heat output (FRP) of the largest fires investigated in SE Asia is more than an order of magnitude higher than those in that original study, however, and their extensive smoke plumes often merge and/or have indistinct boundaries—making accurate delineation of a fires entire plume often impossible. There is also far more significant potential for cloud contamination of the plume observations in the SE Asian environment (see Figures 1 and 2).

(ii) In [38], it was assumed that each plume analysed had been produced between the start of the most recent diurnal cycle of the associated fire and the time of the polar orbiting satellite overpass used to generate the AOD product. However, certain of the SE Asian fires did not show obvious FRP minima during the night, meaning that the start time of the temporal integration period over which FRE was calculated could not be determined in the same way.

(iii) The extreme optical thickness of the peatland fire plumes means parts of them are often incorrectly masked as meteorological cloud by satellite AOD products, or given an unrealistically constant maximum AOD (this includes the standard MODIS AOD products employed by [38]), potentially resulting in low biased C_e estimates.

To deal with issue (i) and select an appropriate matchup between the TPM in a given plume and the FRE for the fire that produced it, only a spatial subset of each plume was digitised (as shown in Figure 2), rather than the entire plume. This avoided the need to identify the complete plume extent, greatly reduced issues of cloud contamination, and enabled us to exclude areas of the plume that had merged with smoke from other fires. In parallel with these advantages, only using part of the plume also resolved issue (ii), since we could then integrate the FRP values only over the time that the part of the plume used for the TPM calculation had been generated. Identifying this time period then became one of the most complex steps in the calculation, and the approach used to do so is discussed in Section 3.2. To resolve issue (iii), a modified AOD retrieval approach, able to provide AOD estimates for very optically thick smoke, was developed, as discussed in Section 3.3.

Using this modified approach, C_e was then generated for the highly radiant SE Asian peatland fires using a suitable set of potential target plumes identified using the NASA Worldview webtool (https://worldview.earthdata.nasa.gov). MODIS and VIIRS true colour composite browse imagery were comprehensively searched over the islands of Kalimantan and Sumatra from July through October 2015, and ultimately thirteen plumes (from tens of potential plumes) were identified that contained digitizable subsets (for detailed plume information see Table 1 contained in Section 4.1). These subsets were considered of sufficient length for reliable calculation of the time taken for the plume to be produced (see Section 3.2, typically tens of km), had sufficient high quality AOD retrievals for reliable estimation of TPM in optically thick conditions (see Section 3.3), and had collocated FRP observations coming from Himawari-8 AHI active fire pixels with each fire confirmed as being located on peatland (see Figure 2) using the SE Asian peatland shape files available from Global Forest Watch (http://data.globalforestwatch.org/datasets/). All data for the 13 digitized plumes were reprojected to a UTM grid with a 750 m pixel resolution, chosen to reflect the nominal resolution of VIIRS M-band pixels (see Section 3.3) at or close to nadir.

Table 1. Key statistics for the thirteen matchup fires shown in Figure 2 and used to derive the total particulate matter (TPM) smoke emissions coefficient (C_e) in Figure 3. Data for the plume subsets and the fires contained within the plume polygon depicted in Figure 2 are shown.

Sample ID	Date (2015)	FRE (10^7 MJ)	TPM (10^7 g)	Mean Plume AOD	Plume Area (10^8 m^2)	Plume Length (km)	Plume Velocity (ms^{-1})	Time (s)	Landcover Concession Type
1	07/06	0.98	16.6	0.65	10.9	35.9	13.1	2753	oil palm
2	08/07	0.167	2.1	0.38	2.39	26.4	10.3	2553	none
3	08/07	0.07	5.7	0.57	4.34	37.6	7.6	4951	none
4	08/07	0.80	9.9	0.66	6.38	25.1	4.7	5297	none
5	09/11	0.32	10.3	1.94	2.43	21.5	10.6	2030	fibre
6	09/22	0.57	19.9	2.02	3.67	30.6	11.5	2665	none
7	09/23	1.11	26.0	2.11	5.24	32.4	21.4	1512	none
8	09/23	0.32	7.4	1.78	1.99	23.6	11.1	2137	fibre
9	09/24	0.35	3.5	0.63	2.43	22.5	10.3	2186	fibre
10	10/03	1.28	17.8	1.82	4.53	23.9	12.3	1937	fibre
11	10/04	1.62	25.0	2.41	5.04	26.3	10.5	2497	fibre
12	10/20	1.12	14.4	2.08	4.23	13.3	21.5	618	fibre
13	10/20	0.07	0.6	0.15	1.89	22.9	14.1	1634	none

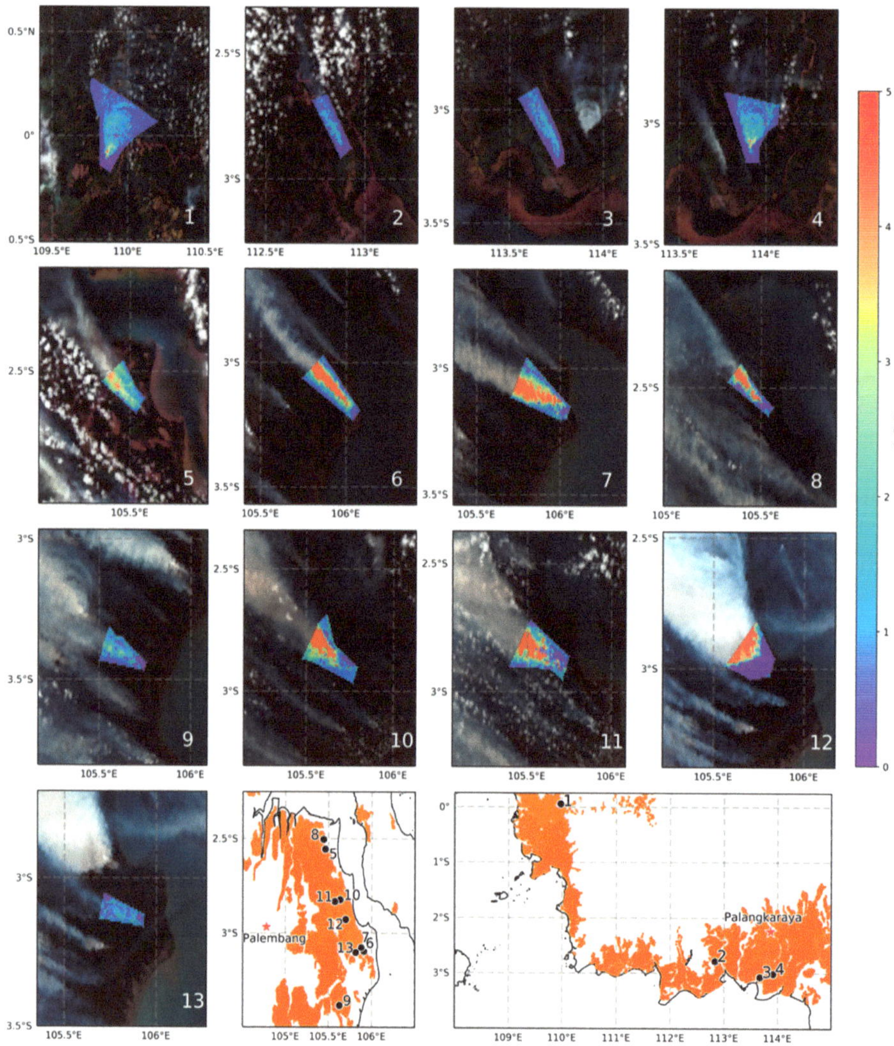

Figure 2. Plumes and overlain aerosol optical depth (AOD) from the thirteen highly radiant peatland fires on Sumatra and Kalimantan burning between July and October 2015, which provided our matchup dataset. The two inset maps show the location of each plume on Sumatra (first map) and Kalimantan (second map), with peatland areas shaded according to Global Forest Watch (http://data.globalforestwatch.org/datasets/).

3.2. Temporal Integration of FRP to FRE Using Plume Velocity Estimates

To estimate each fire's FRE, we used the FRP measures for each of the active fire pixels detected within the digitised fire and plume polygon, as discussed in Section 3.1. These FRP measures, expressed in watts $\times 10^6$ (MW), were calculated from level-1b Himawari imagery [26] at the locations of active fire pixel detections made using an adaption of the geostationary fire thermal anomaly (FTA) algorithm prototyped in [46] and described in detail in [30]. The FRP measures themselves were calculated using the MIR-radiance method of [16,47]. The FRE (in joules $\times 10^6$; MJ) for each fire was then calculated from the temporal integration of its FRP measures. This was easily achieved due to the

10-min imaging frequency of the Himawari-8 data that avoided the need for interpolation during long gaps between FRP observations. The time interval for each FRP temporal integration was set to be that over which the matching plume subset (Figure 2) was produced. This was calculated from the plume subset length (m) and horizontal velocity (m·s^{-1}). The former was determined by creating a vector $\boldsymbol{p}$ from the plume's approximate distal end (identified as the approximate midpoint of the distal end during digitisation) to the mean UTM-projected location of the Himawari-8 active fire pixels contained within the plume polygon. The total length of the plume subset is then defined as $\|\boldsymbol{p}\|$. Plume velocity was derived from consecutive Himawari-8 0.55-μm images using the OpenCV Farneback optical flow algorithm [48] (see Appendix B) applied to the seven AHI images preceding each of the VIIRS overpasses used to provide the plume AOD measure (see Section 3.3). Seven images, when paired, provided a 1-h subset of Himawari-8 observations from which to estimate the flow. This was deemed suitable as most plume subsets, when evaluating their flows, were produced by the associated fire in less than an hour. From the image flow estimates, plume velocity, w (m·s^{-1}) is estimated as:

$$w = \frac{\max_i \left\| \left(\overline{x}_{flow,\ i}, \overline{y}_{flow,\ i} \right) \right\|}{s} \tag{1}$$

where i defines the set of geostationary image pairs between which the flow is calculated, $\overline{x}_{flow,\ i}$ is the mean subpixel shift (i.e., flow) in the UTM projected geostationary image x-axis with the mean calculated only from pixels contained within the plume subset polygon, and $\overline{y}_{flow,\ i}$ is the same but for the y-axis. The denominator s is the total number of seconds between temporally adjacent Himawari image acquisitions, which occur every 10 min (i.e., 600 s).

The flow measures, $\overline{x}_{flow,\ i}$ and $\overline{y}_{flow,\ i}$, are subject to two screening steps prior to calculating their mean: (i) to exclude any flow pixels not physically representative of the plume-indicated wind direction, the angle between all flow vectors and $\boldsymbol{p}$ is calculated, and only flow vectors within $\pm\frac{\pi}{8}$ radians of the plume vector are retained; (ii) some nonplume flow pixels are sometime contained within the plume polygon, typically providing flow estimates for the land surface (i.e., 0) and are excluded. Finally, the image pair with the maximum of the screened flows is retained as it was found to be the most reliable indicator of plume speed based on comparisons to ECMWF Re-Analysis 5 (ERA5) wind speeds (see Appendix B, Figure A2).

Total time for the plume subset to be produced was calculated as $t = \|\boldsymbol{p}\| \cdot w^{-1}$. It is this duration t that was used as the time period over which the matching FRP time-series for the fire was integrated to estimate FRE. The end of the integration period was the time of the VIIRS overpass providing the AOD estimate, whilst the start of the integration period was t seconds earlier.

3.3. TPM Estimation

AOD retrievals for each plume subset (Figure 2) were derived from the VIIRS multispectral 750-m spatial resolution "M-band" observations via two approaches. The first for pixels showing optically thick conditions (i.e., extreme AODs, here >2.0) and the second for all other pixels. For the latter, AOD values from the standard VIIRS aerosol product were used, based on land pixels whose 555 nm AOD was derived using a modified dark target (DT) algorithm (originally developed for MODIS) [49]. The modified algorithm uses five M-bands (M1: 0.412 μm; M2: 0.445 μm; M3: 0.488 μm; M5: 0.672 μm; M11: 2.25 μm) to estimate AOD, and a key enhancement over its use with MODIS is that five aerosol types are evaluated in the retrieval's optimisation step (compared to the two assessed in the MODIS DT product), and these include both low- and high-absorption smoke aerosols taken from AERONET inversion climatologies [50]. The VIIRS AOD product is available from the NOAA CLASS data centre as an aggregated environmental data record (EDR) at a nominal 6-km spatial resolution, or an unaggregated pixel-level intermediate product (IP). The IP product was used herein because its higher spatial resolution better captures the AOD variability within each plume. Retrieval confidence flags are produced at the IP level and used in EDR generation, ranging from 1 (good) through 4 (non-retrieved), and AODs higher than 2.0 are considered invalid, even though they are common in biomass burning plumes [51].

For pixels showing extreme AODs, no publicly available satellite-derived standard AOD product currently exists that retrieves AODs greater than 5, even though such values exist in the SE Asian biomass burning plumes of 2015 [51]. Therefore, to provide data for these pixels we deployed the Oxford-RAL Aerosol and Cloud retrieval algorithm (ORAC) for all pixels showing optically thick conditions (AOD > 2.0). ORAC is able to retrieve AOD even in such extreme AOD situations [52], but was not used for the optically thin (AOD < 2.0) condition since when applied to data from single-view sensors such as VIIRS (as opposed to dual-view sensors such as ATSR), the retrievals retain significant and unwanted sensitivity to changes in the underlying land surface reflectance [52]. We applied ORAC to data from VIIRS channels M3 (0.488 μm), M4 (0.555 μm), M5 (0.673 μm), M7 (0.865 μm), M8 (1.24 μm), and M11 (2.25 μm), parameterising the retrievals with the smoke optical properties reported in [50] for tropical forest fires in the Amazon. These are unlikely to be a perfect match for the smoke coming from SE Asia fires, which, for example, will have very likely been contributed to by organic soil burning, so wherever possible we compared the ORAC-retrieved AODs from SE Asian plumes to those derived from the surface AERONET network (which itself provides data suitable for retrieving AOD in even optically thick conditions). A bias was noted between the ORAC and AERONET AOD retrievals made in the extreme AOD conditions considered here, believed most likely to arise from differences between the assumed and true optical properties of the smoke particulates. This bias was easily adjusted for by using a single linear scaling factor (see Appendix A).

Even with the two AOD retrieval methods noted above, complete plume subset coverage was not always achieved, and quality indicators were used to screen out low quality AOD retrievals. Specifically, only VIIRS IP data with quality of 2 or better and ORAC retrievals with costs of ≤10 (a limit applied previously [53]) were used, with the ORAC cost providing an indication of the consistency between the retrieved and observed atmospheric state [52]. The AODs of any non-retrieved pixels were estimated using a radial basis function interpolation approach. This was found to provide the most representative AOD estimates when evaluated against alternative interpolation approaches on artificially removed AOD pixels (see Appendix C).

Excess AOD was calculated for every pixel in each plume subset by differencing the pixel's AOD value from that derived for the upwind plume background (defined from an additional polygon drawn immediately upwind of the fire during the digitisation process). The total particulate matter (TPM, g) in the plume subset was then calculated from the excess AOD totalled over all pixels using:

$$TPM = A \cdot AOD_e \cdot \beta^{-1} \tag{2}$$

where A is the area (m^2) of the plume subset polygon, AOD_e (unitless) is the summed excess AOD for the plume subset, and β is the mass attenuation coefficient ($m^2 \cdot g^{-1}$) for the constituent smoke.

The selection of an appropriate mass attenuation coefficient depends on numerous and largely unknown factors associated with a fires environment, and the usual solution is to assume a globally or regionally appropriate value, such as the 4.6 $m^2 \cdot g^{-1}$ applied in [45] or the regional mean of 3.5 ± 1.0 $m^2 \cdot g^{-1}$ used for Southern African fires in [38]. However, peatland fires generally involve organic soil combustion as well as combustion of vegetation, and thus a different mass extinction coefficient may be warranted compared to smoke coming from a purely surficial vegetation fire [54]. However, since the fires we focus upon are highly radiant and clearly involve substantial flaming combustion of vegetation (Section 4.2 and Figures 1, 5, and 6), we retain use of the most widely applied biomass burning smoke mass extinction coefficient of 4.6 $m^2 \cdot g^{-1}$ [44]. Further investigations of this parameter based on field or laboratory measures is suggested as a future research priority. Using this smoke mass extinction coefficient allowed us to generate TPM values for the plume subsets of our matchup fires, and combining these with the per-fire FRE values we generated the smoke emissions coefficient C_e using ordinary least squares (OLS) regression forced through the origin.

4. Results and Discussion

4.1. Derivation of TPM Emission Coefficient (C_e)

Figure 3 shows the data from the thirteen matchup fires of Figure 2, used to estimate C_e as 16.8 ± 1.6 g·MJ^{-1} through OLS regression, with Table 1 reporting various metrics for each matchup. For comparison, in Table 2 are the existing TPM C_e published for SE Asian peatland fires, either derived from the multiplication of an FRE to fuel consumption conversion factor with a TPM emission factor [27], or calculated directly via a "top-down" method relating FRE to TPM emissions [37]. Values range from 69.3 g·MJ^{-1} in the former case to 52.4 g·MJ^{-1} in the latter, and are far higher than the 16.8 ± 1.6 g·MJ^{-1} determined herein. The highest C_e (69.3 g·MJ^{-1}) was derived using a largely "bottom-up" approach where a dry matter fuel consumption rate conversion factor of 5.87 kg·MJ^{-1} was established though comparison of GFAS-derived FRP values against burned-area-derived peatland fuel consumption estimates coming from GFED [27], and the 11.8 g·kg^{-1} TPM emissions factor from deforestation fires [33]. A recent update to the EF from [32] provides an explicit, though indirectly estimated, TPM EF value for peat of 27.5 g·kg^{-1}, resulting in an even larger C_e estimate of 161 g·MJ^{-1}. The "top-down" C_e of 52.4 g·MJ^{-1} was calculated from a set of 19 collocated smoke plumes and fires burning on SE Asian peatlands, with the TPM measurements derived from MODIS deep blue AOD products [55] and the FRE calculated from MODIS FRP data provided by successive overpasses of the Terra and Aqua satellites and an assumed FRE diurnal distribution [37].

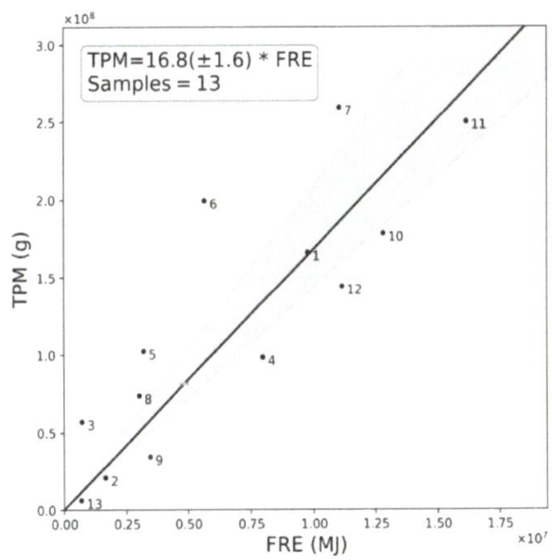

Figure 3. Comparison of total particulate matter (TPM) in the plume subset and matching fire radiative energy (FRE) released from the causal fire, shown for the thirteen matchup fires of Figure 2, with each numbered as shown therein. The ordinary least squares (OLS) linear best fit through the origin is used to generate a TPM smoke emissions coefficient (C_e; 16.8 g·MJ^{-1}) appropriate for this type of highly radiant SE Asian peatland fire, with the grey shaded area indicating the 95% confidence interval [14.3 g·MJ^{-1}, 20.9 g·MJ^{-1}].

Table 2. Existing peatland and tropical forest fire TPM smoke emissions coefficients (C_e) linking TPM emissions to FRE release, for comparison to that derived herein (Figure 3).

Source	Peatlands (g·MJ^{-1})	Tropical Forests (g·MJ^{-1})
[27]	69.3	11.3
[37]	52.4	30.0
[44]	-	15–32

4.2. Discussion of TPM Emission Coefficient (C_e) Differences

The TPM C_e reported in [27,37] for SE Asian peatland fires are very substantially larger than the 16.8 g·MJ^{-1} value generated herein (Figure 3). A key cause of this difference is suggested to be the fact that unlike our highly radiant peatland fires, which contain substantial areas of flaming combustion (Figure 1), many fires on peatlands tend to be solely smouldering, which generates far more particulate matter emissions per unit of dry matter burned than do flaming fires [21].

A second cause is related to the typical surface temperatures found at locations of burning peat, which are typically far cooler than those of surface vegetation fires. In [16], it is shown how the MIR-radiance method of FRP retrieval is adapted to a temperature range relevant to smouldering and flaming-phase vegetation combustion. Whilst the approach greatly benefits from not requiring knowledge of the actual fire effective temperature, it is tuned for temperatures exceeding 650 K (considered a good lower limit for smouldering phase activity [56]). Below this temperature, the radiative power output becomes increasingly underestimated unless the FRP coefficient is rederived for a cooler temperature range [16]. Such FRP underestimation at lower temperatures is in some cases advantageous, because in standard vegetation fires such temperatures are mostly associated with recently burned but still cooling ground that is not consuming fuel. However, by burning underground, subsurface peat fires can still consume fuel whilst generating surface temperatures far lower than 650 K [57], meaning the equivalent FRP to fuel consumption rate conversion factor would need to be increased compared to that for standard surface smouldering or flaming fires. This may be one reason that during comparisons between MODIS-derived FRP measures and GFED-derived fuel consumptions, [27] found an "FRE-to-fuel consumption" conversion factor for savanna fires rather similar to that of the small-scale laboratory vegetation burns of [16], but for peatland fires the estimate was more than an order of magnitude higher. To demonstrate, Figure 4 shows infrared surface temperature data of SE Asian peatland fires acquired by us during the October 2015 Kalimantan fieldwork described in [10]. The data come from two areas of burning peat. The first area is a largely cleared location next to a drainage canal imaged from a low altitude UAV (Figure 4a,b). Ambient surface temperatures are below 313 K (40 °C) and elevated temperatures at the areas of combustion are mostly lower than the aforementioned 650 K (377 °C) threshold used by default in the MIR-radiance method. These elevated surface temperatures appear mostly to be generated via heat conduction from below at locations where smouldering peat consumption is proceeding subsurface, often below a layer of white ash. The second area is a more forested location (Figure 4c,d), where the thermal imagery was acquired from the ground. Here, some surface woody material is being consumed by smouldering combustion, but the vast majority of elevated surface temperatures, and certainly the hottest ones, are from areas of peat and appear to be actively generated by convective heat transfer related to the passage of hot smoke through vents and cracks in the peat surface. Substantial areas exceed the 650 K (377 °C) threshold, though most are still cooler than this—as confirmed by the normalised frequency distribution shown in Figure 4e.

A third cause for the difference between the C_e estimate determined herein and the far higher values determined by [27,37] is that when examining wide-scale peat fires with moderate to low spatial resolution spaceborne data, the pixels successfully triggering the active fire (AF) detection algorithm typically comprise many individual subsurface fires, which together can present a sufficiently significant surface thermal signature to be detected. However, there will likely be many fire regions with insufficient high temperature areas to trigger the AF detection algorithm, meaning their FRP will not contribute to the FRP total, but may well contribute significantly to smoke production or estimated BA-derived fuel consumptions. Such FRP underestimation caused by undetected AF pixels will likely be greatest in the case of subsurface smouldering peat fires, where far more of the pixel area must be covered by the elevated temperature material for the AF detection algorithm to identify the pixel as containing a fire, compared to fires at flaming temperatures.

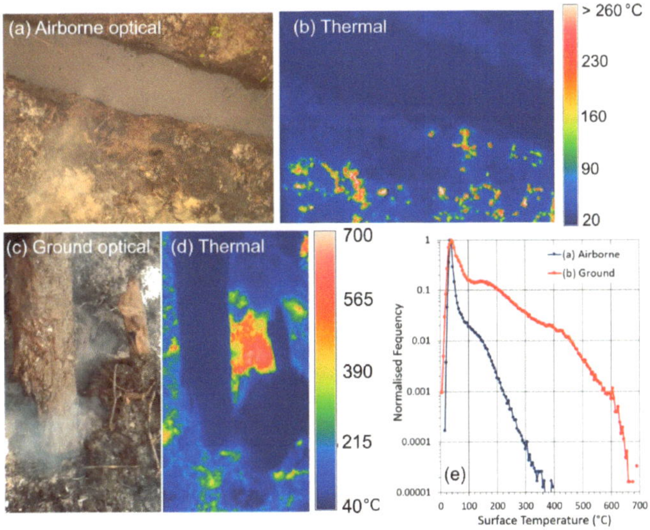

Figure 4. SE Asian peatland fire infrared surface temperatures and matched optical imagery. Data come from infrared camera measurements conducted on 14 October 2015 during the Central Kalimantan field campaign described in [10], and were acquired close to location 3 detailed therein. (**a**) Example airborne optical image of subsurface peat burning in a largely treeless area next to a drainage canal, captured from a UAV flying ~20 m height above ground, along with (**b**) the matching thermal (8–14 µm) brightness temperature image. (**c**) Example ground-based optical image of peat burns in a more forested area, where hot smoke is escaping from vents and holes and there is more woody material to burn. (**d**) Matching thermal brightness temperature image. (**e**) Normalised frequency of surface temperatures collected in the two different areas from multiple examples of the type of airborne and ground-based thermal imagery shown in (**b**,**d**), and where the more forested locations (**c**,**d**) show significantly higher temperatures, albeit still primarily coming from subsurface combustion.

The above factors provide an indication as to why the C_e values established previously for peat fires, assumed to comprise a significant subsurface smouldering combustion component, are so much higher than those established for the potentially more flaming-dominated fires assessed herein. Confirmation of these fires' flaming nature comes from several sources. Firstly, exploiting the large temperature differential between flaming and smouldering combustion in SE Asian peatland fires (~400 K, [57]) allows the spectral ratio method demonstrated in [58,59] to discriminate between them. This was conducted using the ratio of the excess (i.e., above background) mid- and long-wave infrared (MWIR and LWIR, respectively) spectral radiances available in the AHI FRP product at the location of detected active fire pixels [26]. Comparing this against an empirically defined threshold value of 5.4 (W m^2 sr^{-1} µm^{-1} (W m^2 sr^{-1} µm^{-1})$^{-1}$), equivalent to a combustion temperature of 700 K, allowed us to identify whether each AF pixel was dominated by areas of combustion cooler or hotter than this (validation of the defined threshold against known flaming locations is provided in Appendix E). Assessing AHI-detected fires burning across the SE Asian study region on a 1-arcminute grid aggregated over the months of September and October 2015 (see Appendix D, Figure A4), showed that approximately ~51% of the peatland grid cells have a median spectral ratio of >5.4 (i.e., hotter than 700 K) compared to ~65% of grid cells not located on peatlands, a statistically significant difference ($p < 0.01$) in terms of temperature. Aggregating the data from peatlands into hourly bins (see Appendix D, Figure A5) returns median spectral ratio values of >5.4 from 12:00 to 20:00 local time, an indication that flaming combustion is more prevalent during the day, as might be expected since daytime conditions tend to promote more intense fire activity [9,24]. These findings together indicate that this spectral ratio thresholding is suitable for discriminating between more flaming and more smouldering dominated areas of combustion, and that smouldering combustion is more prevalent in peatland areas and at night. Figure 1 (inset) also proves the presence of substantial areas of flaming combustion, as do the spectral ratio values for the 13 peatland fires of Figure 2, which are shown in Figure 5a, with the median spectral ratio of all but two of the fires exceeding the defined threshold of 5.4. A further comparison of the spectral ratio values for the AHI active fire pixels of our matchup fires was made against all active fire pixels detected in peatland and non-peatland areas in the study region across September and October 2015 (Figure 6a). The median spectral ratio for the matchup fires (7.2) is significantly higher than that for the wider peatland (4.8) and non-peatland (5.7) areas, indicating that the matchup fires have substantially more flaming combustion. Note that the median spectral ratio across all peatland fires is less than the threshold of 5.4, indicating the predominance of smouldering combustion in peatland areas (but not in our highly radiant matchup fires).

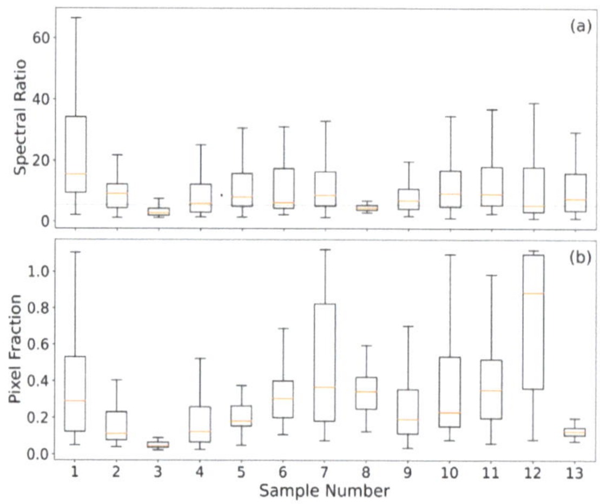

Figure 5. Results of the investigation as to the nature of the thirteen matchup peatland fires focused on herein and shown in Figure 2. (**a**) Boxplots of the "excess above background" 3.9 to 11 μm spectral radiance ratio for the Himawari active fire pixels making up each fire, along with that associated with a 700 K fire (5.4 W m^2 sr^{-1} μm^{-1} (W m^2 sr^{-1} μm^{-1})$^{-1}$); shown by the dashed vertical line. (**b**) Boxplots representing the fraction of each Himawari pixel that would be required to be covered by a subsurface smouldering peat fire having an effective surface temperature of 412 K (mean of the surface temperatures determined in [57]). Himawari-8 active fire pixels are those detected within the associated plume polygons shown in Figure 2 in the hour prior to the VIIRS afternoon overpass. In both (**a**) and (**b**), the boxplots follow standard conventions showing the median (red horizontal lines), interquartile range (box), and minimum and maximum values (whiskers).

A second assessment is made in Figure 5b, where for active fire pixels associated with each matchup fire we estimate the required subpixel fraction f that would need to be combusting to generate the observed FRP, assuming a 412 K (139 °C) surface temperature (the mean smouldering peat fire temperature determined in [57]):

$$f = \frac{FRP \cdot \left(\frac{\sigma}{a} \cdot \beta(\lambda, T)\right)^{-1}}{A_s} \tag{3}$$

where f is the fractional area (unitless); FRP is the fire radiative power (W), $\frac{\sigma}{a}$ is the AHI specific FRP coefficient (sr μm); $\beta(\lambda,\ T)$ is the Planck function used to define spectral radiance (W m^2 sr^{-1} μm^{-1}) at wavelength λ (μm), here 3.9 μm and temperature T (K), here 412 K; and A_s is the area (m^2) of the AHI pixel (4 km^2 at the subsatellite point). Given the relatively large AHI pixel areas, the resulting fractional areas coming from application of (3) to our matchup fires appear unrealistically high, with median pixel fractions of >20% for the majority of these fires. This contrasts against the same estimate of subpixel fraction calculated for all active fire pixels detected on peatland and non-peatland areas (Figure 6b), which indicates a far more reasonable median pixel fraction of ~0.05% (peatland) and ~0.04% (non-peatland). This difference suggests (i) that our matchup fires must contain substantial areas of combustion far hotter than 412 K (which would then make their subpixel fractional areas far lower and more realistic), and (ii) that they are among the most radiant fires occurring during the 2015 El Niño period in the study region.

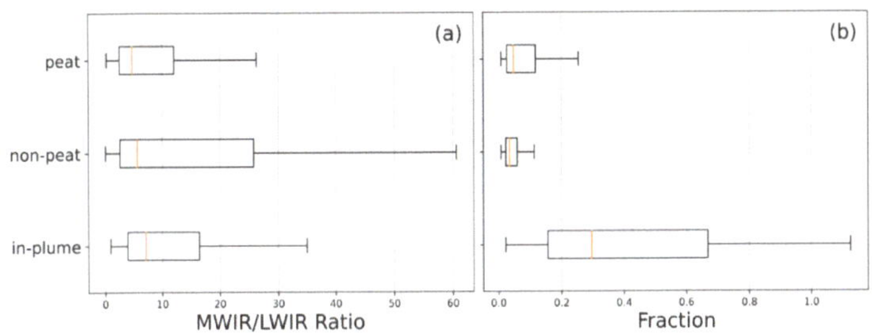

Figure 6. Spectral ratio (**a**) and subpixel active fire fraction (**b**) for AHI-detected active fire pixels contained in our matchup fires against all AHI-detected active fire pixels for September and October 2015 identified in the study region (latitude range: 10° N:10° S; longitude range: 100° E:120° E). Active fire pixels are separated into those burning on peatlands and on non-peatlands using the Global Forest Watch peat map shown in Figure 2. As with Figure 5, the boxplots in (**a**,**b**) follow standard conventions.

Together, the data from these three assessments support the interpretation that our matchup fires are far more radiant and contain far more flaming activity than do the majority of the peatland fires identified by AHI during September–October 2015, and since more than 50% of them are located on Indonesian governmental peatland concessions for wood fibre (6 samples) or palm oil (1 sample) plantations (see Table 1), substantial surface vegetation and vegetation litter is indeed likely to be available at these sites to support flaming combustion of the type shown in Figure 1. Thus, it follows that the C_e established for our targeted highly radiant fires is associated with peatland burns containing a significant flaming phase component.

4.3. Consideration of Contributions to Uncertainties

As is apparent in Figure 1, wildfire plumes are rather transparent in the SWIR and MWIR spectral regions, because these wavelengths are significantly longer than typical smoke particle diameters. However, the extreme optical thickness of the plumes of our matchup fires likely does provide some reduction on the MWIR radiance measures used to establish FRP [60], though the magnitude and variability of this is currently uncertain and the FRP values are thus only atmospherically corrected for the effect of atmospheric water vapour and trace gases [26]. This, along with active fire detection errors of omission, results in an increased C_e since the measured FRP is low biased. As discussed in Section 4.2, C_e values derived using AHI might be expected to be lower than those produced using MODIS (assuming AOD values were correctly specified in both cases), given the reduced capability of AHI to detect less radiant fires due to its coarser nadir point pixel size [26,46] (up to 66% active fire detection omission rate [26]). Such active fire errors of omission are, however, to some extent accounted for in our approach—since whilst some active fire pixels in our matchup fires that are burning below the AHI minimum FRP detection threshold (~40 MW [26]) may remain undetected, their TPM contribution is captured in the plume and incorporated into the derived C_e. Thus, as long as the C_e derived herein is applied to AHI data and to fires with characteristics similar to those of the matchups focused on here, errors will be minimised. This does, however, highlight the need for an approach to effectively determine whether a fire is combusting predominantly in the flaming or smouldering phase, since our derived C_e has been focused on the more radiant peatland fires that contain substantial flaming activity. The spectral ratio method applied herein demonstrates one potentially suitable approach for doing so.

Further uncertainties are introduced during the establishment of the temporal integration period under the situation where smoke plumes are constantly evolving, with significant changes in their appearance, sometimes occurring between AHI image acquisitions, which make tracking their motion challenging, especially for some of the limited plume extents used herein. Such problems are minimised by the very high 10-min temporal resolution of AHI, which allows a smaller window size to be used to capture image motion, and by the strict postprocessing applied in the extraction of reliable motion estimates. Furthermore, comparisons made against ERA5 profiles of wind speed, shown in Appendix B, demonstrate that the horizontal wind speed estimates produced using our motion tracking approach to the AHI-detected plumes are reasonable, with differences of $<2\ m{\cdot}s^{-1}$ in all instances.

The AOD estimates used to generate the TPM measure of our matchup fires have four main sources of identified uncertainty: the interpolation of missing AOD values; the assumed ambient background AOD; the issues of estimating AOD in conditions of optically thick smoke; and inconsistent wind speed over the plume cross section, such that AODs observed within the plume are not fully resultant from the fire during the FRP integration time period. Estimating appropriate AODs for missing pixels is perhaps the most challenging problem of the four. In [37], this issue was tackled using the maximum near-fire-observed AOD value, but since the cause of the missing data is often related to the fact that a pixel's AOD exceeds the maximum possible with the retrieval approach, or that the smoke is optically thick and misidentified as cloud [51], assigning the maximum "unsaturated" AOD value found elsewhere in the plume will certainly lead to a low biased AOD estimate. The consequence would be a low biased in-plume TPM estimate and resulting C_e coefficient. By contrast, the radial basis function interpolation method deployed herein is able to provide interpolated AOD estimates even exceeding the maximum observed at retrieved pixels, by providing an estimate influenced by the gradients of the proximal AOD values. Various approaches can be employed to generate background AOD estimates, [38] for example, use of a regional value, but this seems inappropriate for conditions studied herein where very significant spatial variations in ambient aerosol load are seen over small spatial scales. We assumed that the AOD measured immediately upwind of the plume (as determined by the plume velocity direction) is that which is representative of the background AOD, since it should not be contributed to by the fire's smoke emission. In [51], the potential for accurately retrieving AOD > 5 optical depths is shown to be problematic, but even if similar limitations apply to the optically thick AOD retrievals conducted with the ORAC retrieval approach used herein, only an average of 6% of pixels across the 13 plumes shown in Figure 2 have optical depths above 5 (a maximum of 15.4% of pixels in plume 11), and so the overall effect will be limited. The effect of differing wind speed across the plume cross section is difficult to ascertain, and has not been considered in previous work where temporal plume subsetting methods similar to those employed here are used [44,45]. For plumes that are not injected above the boundary layer, the inclusion of aerosol material from outside the FRP integration period in the TPM calculation is more likely due to increased turbulence [61]. For plumes that are injected above the boundary layer, there is less turbulence [61] and so it may be more reasonable to assume that the AOD contained within a plume cross section is more closely associated with an instant of the producing fire. Given that most of the fires evaluated here are very large and have well-defined plumes, above boundary layer injection seems probable, and we therefore assume that the TPM contained in the plume is well correlated with the FRE, as indicated by Figure 3.

4.4. Significance of Flaming Phase Dominated Fires

Since the SE Asian peatland fire regime is often assumed to be dominated by persistent smouldering combustion [18,62], it is worthwhile to consider the significance of the highly radiant flaming-phase peatland fires that comprise our matchup dataset and how they contribute to the overall smoke emission totals of the region. To do this, we studied the signatures of the most radiant matchup fire in our dataset, that associated with Plume 12 in Figure 2. This Sumatran fire burned from 18 to 27 October 2015 according to our AHI-derived FRP time series (Figure 7), and released 1.76×10^{10} MJ of fire radiative energy. Applying our "top-down" C_e to this FRE total results in a total emitted particulate matter mass of ~0.3 Tg for this fire alone, which represents around 9% of particulate emissions from all Sumatran fires burning in September and October 2015 (assuming ~3 Tg of $PM_{2.5}$ released during this period and that $PM_{2.5}$ typically comprises 90% or more of emitted PM mass totals [10,21], giving ~3.4 Tg of total particulates). Thus, this single fire alone markedly contributes to regional TPM emission totals, and our matchup dataset shows thirteen such intensely burning highly radiant peatland fires, whose TPM emission rates are each far higher than those of the far more common low intensity smouldering peat fires. Combined, the TPM emissions of the twelve other fires studied totals to ~0.4 Tg of particulate matter mass. To demonstrate the inappropriateness of applying existing C_e values (Table 2) to these types of high intensity peatland fire, along with the updated "bottom-up" value of 161 g·MJ^{-1} derived from [32], we calculated the equivalent TPM total for the fire associated with Plume 12. TPM totals of 0.93, 1.22, and 2.84 Tg were calculated for this single fire from the alternative coefficients, which are clearly too high given the total from all Sumatran fires is estimated to be ~3.4 Tg [10].

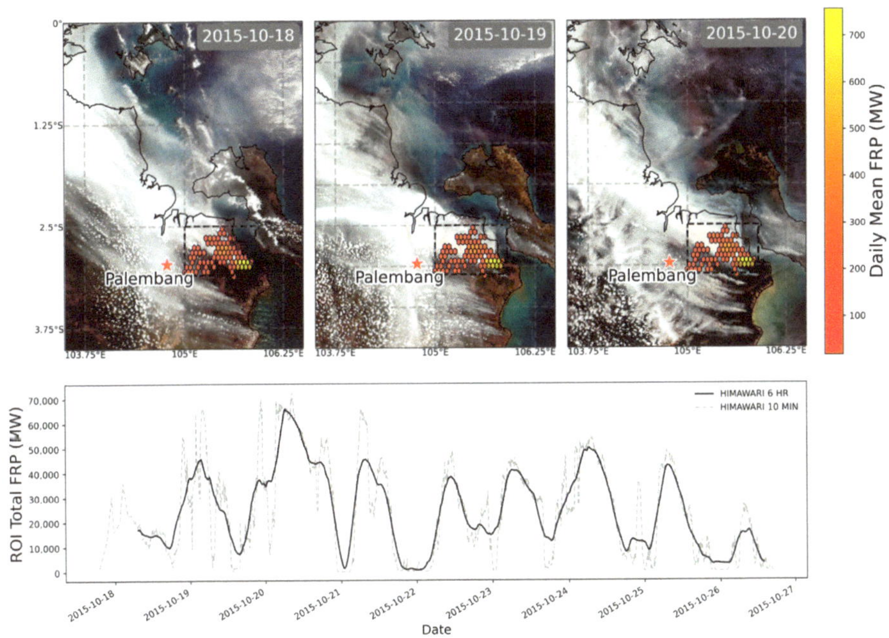

Figure 7. Data of the highly radiant Sumatran peatland fire associated with Plume 12 of Figure 2, and confirmed as showing flaming as well as smouldering activity in Figure 5. The upper plot shows the smoke plume and the spatial distribution of mean Himawari-8 derived FRP for three days of the fire. The lower plot shows the nine-day time series (each tick represents a 24-h period) of total Himawari-derived FRP for the fire (contained in the black bounding box marked in the images), with the dashed line showing the original 10-min resolution data and the solid line a 6-h sliding window. The data from this fire used in the matchup process to derive C_e (Figures 2 and 3) came from the 20 October 2015 record (Table 1).

The high Himawari-derived FRP values of the matchup fires shown in Figure 2 mean they are consuming significant amounts of fuel per unit time. This is probably occurring in spreading flaming fire fronts of the type shown in Figure 1, and in fuels consisting of both surface vegetation as well as organic peat soil. In [10], it was also found that the particulate matter emissions factor for this type of flaming vegetation fire atop burning peat is significantly higher than for smouldering peat fires alone, further increasing the amount of particulate matter they release per unit time. Overall, our results demonstrate that indiscriminate application of a single C_e, for example, from [27] or [37], based solely on the fact that the landcover type is peatland is inadvisable.

5. Conclusions

Our analysis focused on a series of the most radiant fires burning in Indonesian peatlands during the 2015 El Niño (September and October 2015). These fires were responsible for the highest fuel consumption rates and thus the highest rates of PM emission during this event. We generate a matchup dataset of these fires and use it to generate a TPM C_e estimate of 16.8 ± 1.6 g·MJ^{-1}, which can then be applied to all FRP data of this type of fire to estimate TPM emissions per unit time (g·s^{-1}), even in real-time. Our derived C_e is far lower than those currently published for SE Asian peatland fires, which typically exceed 50 g·MJ^{-1} [27,63]. The difference is primarily attributed to the fact that the highly radiant fires focused upon herein include significant flaming activity in surface fuels, such as forest and plantation vegetation, as well as smouldering combustion within the peat. Deriving a single C_e value for all SE Asian peatland fires, which area-wise tend to be dominated by subsurface smouldering combustion, is the likely reason for the far higher smoke emissions coefficients coming from past studies. Our results have important consequences for the estimation of particulate matter emissions for SE Asian peatland fires using the FRP approach, since we show that application of an inappropriate C_e derived for smouldering fires can result in highly biased PM emission estimates when applied to the more intensely radiant fires that can also be found on SE Asian peatlands. Indeed, these latter types of fire can be highly important to overall PM emissions totals and are thus important to consider accurately. We estimate that the single most radiant Sumatran fire assessed herein produced approximately 9% of all $PM_{2.5}$ emissions originating from Sumatran fires in September and October 2015. Such large fires can greatly affect regional air quality, including in highly populated parts of Indonesia and neighbouring countries such as Singapore [24,25].

Our work demonstrates for the first time the importance of considering combustion phase both when establishing C_e for a particular biome in which very different types of fire (smouldering dominated vs. mixed phase or flaming dominated) can occur, and during application of the resulting coefficient to subsequent FRP observations. The FREM approach and our optimised C_e coefficient appear well suited for use in fire emissions inventorying and air quality early-warning, and indeed the highly radiant, surficial flaming fires we focus on herein are more energetic and thus more likely to inject pollutants into the free atmosphere that may assist their transboundary transport [64]. Further work is recommended to determine the most appropriate smoke mass extinction coefficient for peatland fires and for best discriminating areas of flaming- and smouldering-dominated combustion to ensure the most appropriate C_e is applied to each type of fire.

Author Contributions: Conceptualization, D.F., M.J.W., and G.T.; methodology, D.F., M.J.W., W.X., G.T., and P.L.; analysis, D.F., W.X., and G.T.; writing—original draft preparation, D.F., M.J.W., and P.L.; writing—review and editing, D.F. and M.J.W. All authors have read and agreed to the published version of the manuscript.

Funding: Funding for this research came from NERC National Capability funding to the National Centre for Earth Observation (NE/Ro16518/1) and from the Leverhulme Centre for Wildfires, Environment and Society through the Leverhulme Trust (RC-2018-023).

Acknowledgments: We wish to thank to Takeshi Murata of the National Research and Development Institute of Information and Communications Technology Integrated Data System Research Development Office for use of the National Institute of Information and Communications Technology (NICT) Science Cloud, who enabled access to Himawari-8 data. VIIRS data were obtained via the NOAA Comprehensive Large Array-data Stewardship System.

Conflicts of Interest: The authors declare no conflict of interest.

Appendix A

To assess the satellite-derived AODs used herein, we made use of the AERONET database, which was reprocessed in 2018 with the AERONET Version 3 algorithm. This provided a number of improvements to the cloud screening and instrument anomaly detection processes [65], including a new very high AOD restoration step that retained previously excluded cases of high aerosol loading typically associated with biomass burning plumes [66]. The uncertainty in the V3 AOD is ~0.01 (±0.015) optical depths at visible wavelengths [65], very similar to that in the V2 algorithm. As shown in Figure A1, a number of AERONET stations are located in the SE Asian region of interest, and for the period August through October 2015 those located at Jambi, Kuching, Makassar, Palangkaraya, Pontianak, and Singapore were examined for the occurrence of overlying smoke and clouds (manually assessed using VIIRS true colour composites). When an overlying smoke plume was found to coincide with an AERONET station location, the AERONET database was queried for a temporally collocated Level 2.0 (cloud screened and quality assured) AOD sample, with a time difference between the VIIRS overpass and the AERONET sampling of up to ±30 min permitted. To account for the differences in spatial sampling characteristics between AERONET and the VIIRS imagery, all AERONET samples within ±30 min of the VIIRS overpass were averaged and the VIIRS AODs retrieved with the ORAC optimal estimation aerosol and cloud retrieval algorithm [67] and from the official VIIR IP AOD product [49] were averaged over a 10 km^2 window centred on the station location. For a spatiotemporal collocation to be included in the assessment, at least three AERONET and five "good" quality VIIRS AOD retrievals (at 500 and 650 nm) had to be available for averaging; for ORAC this is indicated by a retrieval cost of <10 and for the IP data a retrieval flag of 0.

This procedure identified 50 suitable samples across the AERONET station subset under consideration, comprising 22 samples with smoke features with optical depths ≥2.0 and 28 samples with optical depths <2.0. As AERONET stations do not retrieve AOD at 550 nm, it is interpolated from the measurement at 650 nm using the wavelength specific angstrom exponent. From the VIIRS observations, a total of 43 (out of 50) collocated ORAC retrievals were available for intercomparison to the AERONET station AODs and these are shown in Figure A1. The collocated retrievals have AODs ranging from ~0.1 to ~5.0 for AERONET and ~0.5 to ~9.0 for ORAC, and for the samples with AERONET optical depths <2.0 a mean AOD difference of 1.41 ± 1.06 is observed with the ORAC retrievals from VIIRS, whereas for the samples with AERONET optical depths ≥2.0 differences were 1.96 ± 1.29. There appears significantly more spread for the data obtained in less optically thick smoke conditions (Figure A1), agreeing with the fact that ORAC AOD retrievals for single view sensors such as VIIRS retain significant sensitivity to the underlying land surface reflectance in such conditions, introducing more uncertainty [52]. Both optically thick and thin observations show a positive bias of the ORAC-derived AODs compared to AERONET, and the cause seems likely to be differences between the optical properties of the aerosols assumed in ORAC and the true optical properties of the smoke. However, the observed bias is systematic and adjustable using a simple linear correction (slope: 1.25; intercept 0.92) (Figure A1). This adjustment is applied to all ORAC-derived AODs used herein.

For the IP aerosol product, a total of 20 (out of 28) collocated high quality retrievals were available for intercomparison, also shown in Figure A1. A small bias of 0.12 ± 0.23 is observed between these retrievals and those of AERONET, indicating that for the smoke conditions focused on here with the VIIRS IP product the AOD estimates are generally sound, supporting the findings of more extensive assessments reported elsewhere [68,69]. However, the IP product cannot be used to assess very high AOD values, which is the reason that the ORAC-retrievals of AOD are required for such situations. All but one of the IP AODs are associated with AERONET observations of less than 2.0 optical depths; however, eight (~29%) of the AERONET samples with AODs of <2.0 are not retrieved with high quality by the IP algorithm, potentially indicating possible coverage issues.

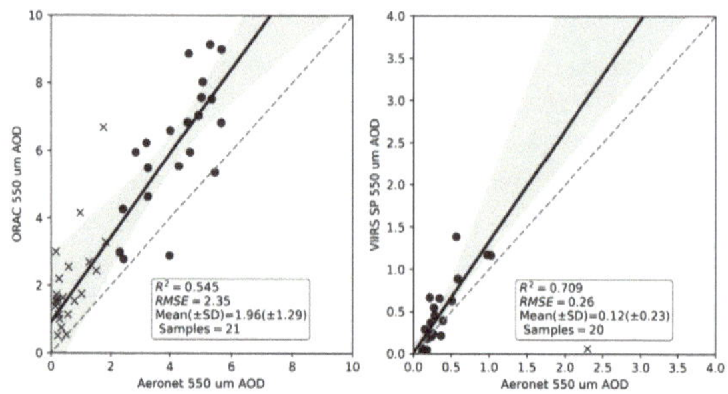

Figure A1. Intercomparison between collocated AOD retrievals made from VIIRS using ORAC [52] and the official VIIRS IP AOD product [49], both conducted at 550 nm and for the set of SE Asian fires focused on herein. ORAC was able to retrieve AODs significantly higher than those present in the VIIRS IP AOD product, so the scales of the two graphs are different. The OLS linear best fit through each set of data is shown in both cases, and at left the crosses relate to points with AOD < 2.0.

Appendix B

The openCV2 optical flow parameters used to estimate motion between Himawari-8 image acquisitions are pyr_scale = 0.5, levels = 1, winsize = 5, iterations = 7, poly_n = 7, poly_sigma = 1.5, and flags = cv2.OPTFLOW_FARNEBACK_GAUSSIAN.

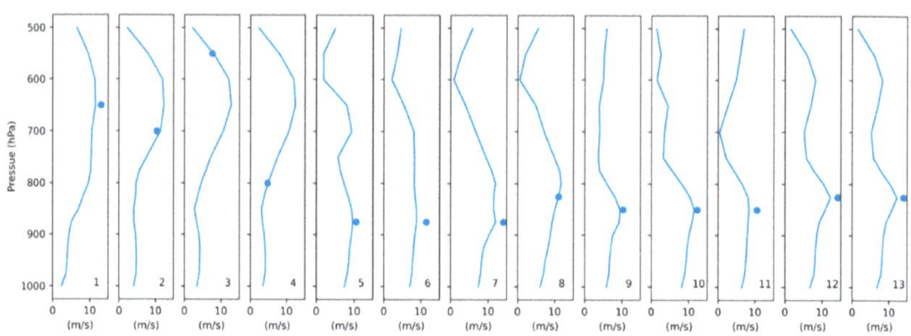

Figure A2. Comparisons of spatially and temporally collocated ERA5 wind-speed magnitudes (calculated from u and v components) against in-plume horizontal velocity retrievals derived from the Farneback optical flow algorithm. The in-plume horizontal velocity is matched to the closest ERA5 velocity across pressure levels, with most plumes injecting near 850 hPa (~1.5 km above mean sea level), the height value assumed in [44].

Appendix C

Three different AOD interpolation approaches were assessed for their performance in providing estimates for missing pixels in the AOD products. The first approach assessed is the most basic and involves replacing missing data points with a mean value derived from the entire plume. The second replaces missing points using estimates generated from a multiple linear regression that associates spectral radiances in the VIIRS M3, M4, and M5 bands with the observed aerosol optical depth. The final approach assessed is radial basis function (RBF) interpolation. To evaluate the three approaches, all thirteen plumes in this study had 25% of their valid AOD retrievals (from the merged ORAC and IP AOD product generated using the approach described in Section 3.3) removed, and values for those locations were refilled using the three approaches considered herein. In Figure A3, the results for each plume and interpolation method are shown, with the evaluation metric being the cumulative distribution of $1 - |pred/obs|$ for all refilled pixels in the plume, where *pred* is the predicted AOD and *obs* is the observed AOD. The RBF approach clearly outperforms the alternative methods, with nearly 80% of the refilled data points across all plumes having an interpolation error of near 0, and ~90% of the interpolated pixels have an error of <20%. The alternative methods do not perform nearly as well, hence the selection of the RBF interpolation approach to replace missing AOD values in this study.

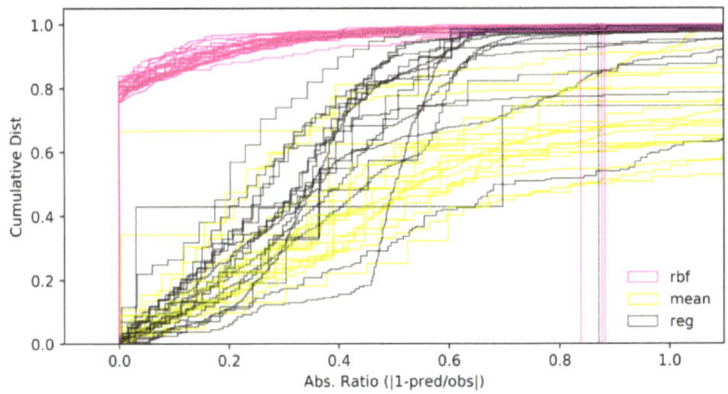

Figure A3. Assessment of three different approaches (radial basis function (RBF), mean value, and regression) for generating AOD estimates for missing pixels. The x-axis shows the error function evaluated for each approach, where *pred* is the predicted AOD for a missing pixel and *obs* is the observed AOD in that pixel. The radial basis function (RBF) performs best overall for the thirteen plumes evaluated, with typically ~80% of the refilled pixels evaluated in each plume having an error near zero.

Appendix D

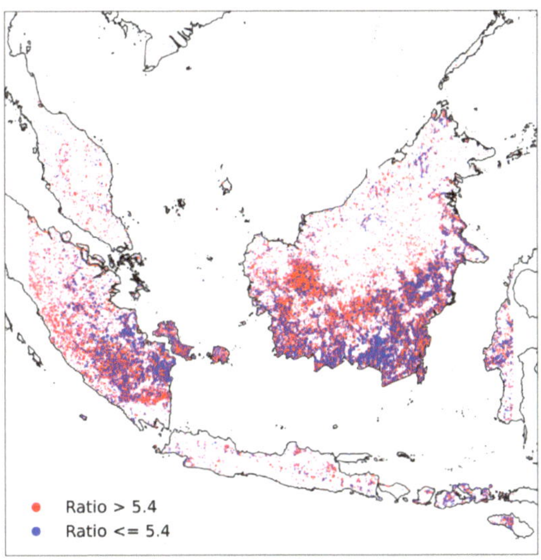

Figure A4. Map of the study region showing gridded (1-arcminute; ~2 km) AHI active fire detection locations for September and October 2015, coloured by the median spectral ratio for each grid cell (defined in Section 4.2).

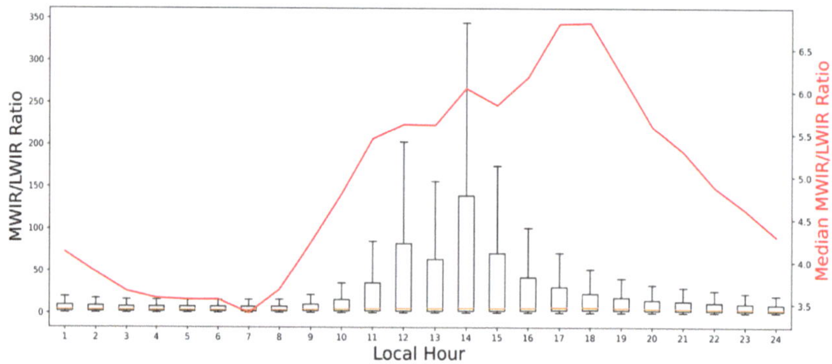

Figure A5. Spectral ratio (defined in Section 4.2) boxplots for September and October 2015 for the region shown in Figure A4, aggregated into hourly bins (local solar time). Red line (RHS, *y*-axis) provides a zoomed view of the median value for each boxplot. The boxplots follow the same conventions described in the Figure 5.

Appendix E

To assess the suitability of the MWIR/LWIR spectral ratio defined in Section 4.2, night-time collocated SWIR observations at 1.6 µm can be used. Shown in Figure A6 are two Planck curves, one assuming 5% of a unit pixel combusting at 500 K (i.e., smouldering) and the other 0.01 of a unit pixel combusting at 1300 K (i.e., flaming). It is apparent that at 500 K the spectral radiance produced at 1.6 µm is close to zero, whereas there is significant spectral radiance at the flaming combustion temperature. At night, in the absence of reflected solar radiation, 1.6 µm hotspot pixels raised above the background signal therefore must contain flaming activity. The distribution of night-time background signal (i.e., instrument noise) for AHI 1.6 µm (B5) imagery (12:00:00 UTC to 20:00:00 UTC) over the study region shown on Figure A8 for 25 September 2015 is shown in Figure A7. As shown in the distribution, it can be assumed that any night-time SWIR pixel with spectral radiances above 0.1 W m^2 sr^{-1} μm^{-1} contains flaming activity. In Figure A8, all thermal anomalies detected by the AHI FRP product on 25 September are shown and overlain are all AHI night-time (12:00:00 UTC to 20:00:00 UTC) 1.6 µm SWIR detections that exceed the 0.1 W m^2 sr^{-1} μm^{-1} threshold. It is apparent that many thermal anomalies detected by the AHI FRP product are not detected in the SWIR imagery, indicating the lack of any intense flaming activity in these fires. To assess the suitability of the MWIR/LWIR spectral ratio, the observed night-time AHI 1.6 µm spectral radiances can therefore be used, with the expectation that for MWIR/LWIR spectral ratios of <5.4 the observed 1.6 µm radiances should be near the sensor noise floor (i.e., <0.1 W m^2 sr^{-1} μm^{-1}) and that for MWIR/LWIR spectral ratios of ≥5.4 the observed 1.6 µm spectral radiances should be above the noise floor (i.e., ≥0.1 W m^2 sr^{-1} μm^{-1}). This analysis is shown in Figure A9, and it can be seen that for MWIR/LWIR spectral ratios of <5.4 the observed 1.6 µm radiances are found to be less than 0.1 W m^2 sr^{-1} μm^{-1}, with a median value of 0.06 W m^2 sr^{-1} μm^{-1} and for spectral ratios of ≥5.4 the 1.6 µm radiances are typically greater than 0.1 W m^2 sr^{-1} μm^{-1} with a median value of 0.15 W m^2 sr^{-1} μm^{-1}. This demonstrates that MWIR/LWIR spectral ratio values of >5.4 are typically associated with thermal anomalies that produce detectable signals at 1.6 µm at night, which is indicative of flaming combustion.

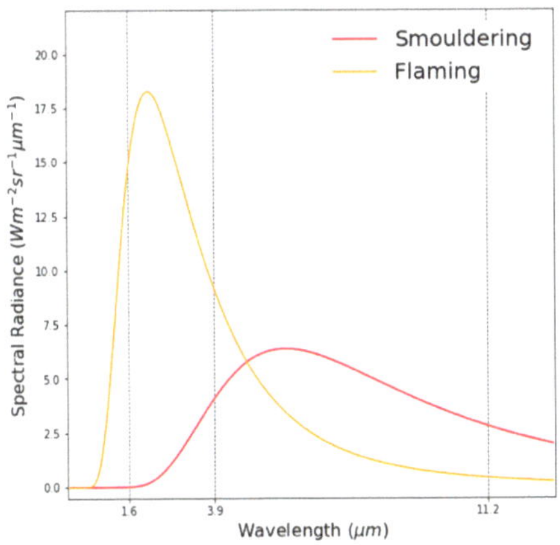

Figure A6. Planck curves for blackbodies at temperatures typical for flaming and smouldering combustion.

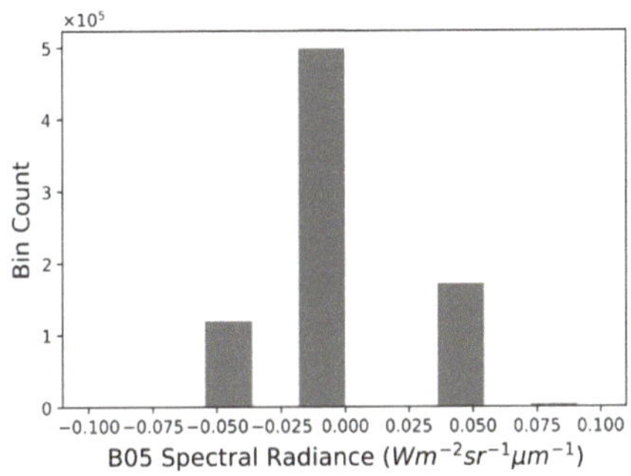

Figure A7. Spectral radiance distribution of nonthermal anomaly containing AHI B05 (1.6 µm) pixels for 25 September 2015 for the region shown in Figure A8.

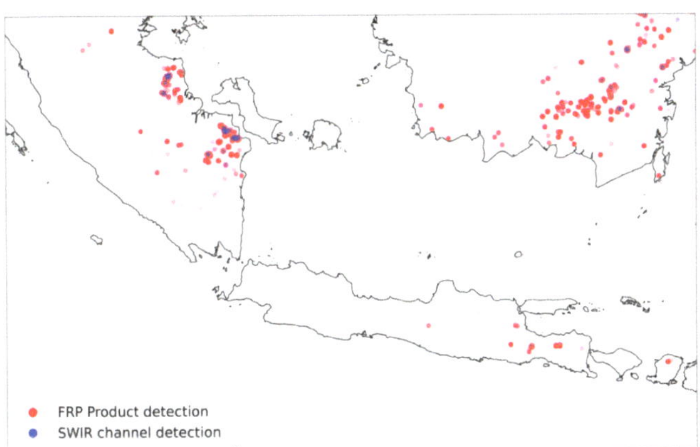

Figure A8. In red are all hotspots detected by the AHI FRP product on 25 September 2015 and in blue are all AHI SWIR pixels with a spectral radiance value >0.1 W m^2 sr^{-1} μm^{-1}.

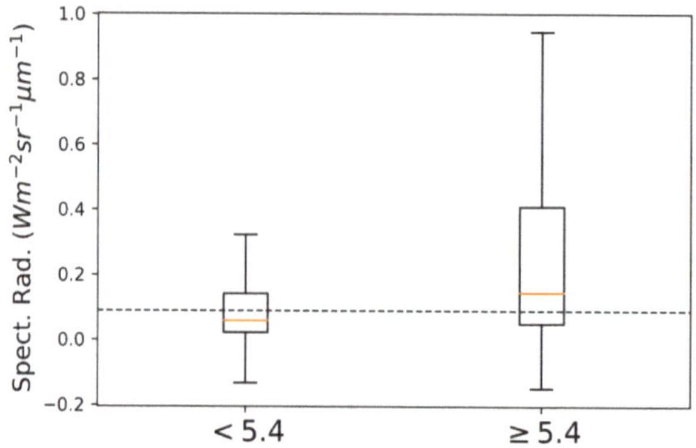

Figure A9. Boxplots of collocated SWIR spectral radiances for all AHI FRP product thermal anomaly containing pixels shown in Figure A8 binned based on whether the pixel meets the MWIR/LWIR spectral ratio threshold of 5.4 W m^2 sr^{-1} μm^{-1} (W m^2 sr^{-1} $\mu m^{-1})^{-1}$ or not. The boxplots follow the same conventions defined in Figure 5.

References

1. Dohong, A.; Aziz, A.A.; Dargusch, P. A review of the drivers of tropical peatland degradation in South-East Asia. *Land Use Policy* **2017**, *69*, 349–360. [CrossRef]
2. Hooijer, A.; Page, S.; Canadell, J.G.; Silvius, M.; Kwadijk, J.; Wösten, H.; Jauhiainen, J. Current and future CO_2 emissions from drained peatlands in Southeast Asia. *Biogeosciences* **2010**, *7*, 1505–1514. [CrossRef]
3. Miettinen, J.; Shi, C.; Liew, S.C. Deforestation rates in insular Southeast Asia between 2000 and 2010. *Glob. Chang. Biol.* **2011**, *17*, 2261–2270. [CrossRef]
4. Page, S.E.; Hooijer, A. In the line of fire: The peatlands of Southeast Asia. *Philos. Trans. R. Soc. B Biol. Sci.* **2016**, *371*, 20150176. [CrossRef] [PubMed]
5. Sloan, S.; Locatelli, B.; Wooster, M.J.; Gaveau, D.L.A. Fire activity in Borneo driven by industrial land conversion and drought during El Niño periods, 1982–2010. *Glob. Environ. Chang.* **2017**, *47*, 95–109. [CrossRef]
6. Huijnen, V.; Wooster, M.J.; Kaiser, J.W.; Gaveau, D.L.A.; Flemming, J.; Parrington, M.; Inness, A.; Murdiyarso, D.; Main, B.; van Weele, M. Fire carbon emissions over maritime southeast Asia in 2015 largest since 1997. *Sci. Rep.* **2016**, *6*, 26886. [CrossRef]
7. Tacconi, L. Preventing fires and haze in Southeast Asia. *Nat. Clim. Chang.* **2016**, *6*, 640–643. [CrossRef]
8. Turetsky, M.R.; Benscoter, B.; Page, S.; Rein, G.; van der Werf, G.R.; Watts, A. Global vulnerability of peatlands to fire and carbon loss. *Nat. Geosci.* **2015**, *8*, 11–14. [CrossRef]
9. Wooster, M.J.; Perry, G.L.W.; Zoumas, A. Fire, drought and El Niño relationships on Borneo (Southeast Asia) in the pre-MODIS era (1980–2000). *Biogeosciences* **2012**, *9*, 317–340. [CrossRef]
10. Wooster, M.J.; Gaveau, D.L.A.; Salim, M.A.; Zhang, T.; Xu, W.; Green, D.C.; Huijnen, V.; Murdiyarso, D.; Gunawan, D.; Borchard, N.; et al. New Tropical Peatland Gas and Particulate Emissions Factors Indicate 2015 Indonesian Fires Released Far More Particulate Matter (but Less Methane) than Current Inventories Imply. *Remote Sens.* **2018**, *10*, 495. [CrossRef]
11. Crippa, P.; Castruccio, S.; Archer-Nicholls, S.; Lebron, G.B.; Kuwata, M.; Thota, A.; Sumin, S.; Butt, E.; Wiedinmyer, C.; Spracklen, D.V. Population exposure to hazardous air quality due to the 2015 fires in Equatorial Asia. *Sci. Rep.* **2016**, *6*, 37074. [CrossRef] [PubMed]
12. Koplitz, S.N.; Mickley, L.J.; Marlier, M.E.; Buonocore, J.J.; Kim, P.S.; Liu, T.; Sulprizio, M.P.; DeFries, R.S.; Jacob, D.J.; Schwartz, J.; et al. Public health impacts of the severe haze in Equatorial Asia in September–October 2015: Demonstration of a new framework for informing fire management strategies to reduce downwind smoke exposure. *Environ. Res. Lett.* **2016**, *11*, 094023. [CrossRef]
13. Simpson, J.; Wooster, M.J.; Smith, T.; Trivedi, M.; Vernimmen, R.; Dedi, R.; Shakti, M.; Dinata, Y. Tropical peatland burn depth and combustion heterogeneity assessed using UAV photogrammetry and airborne LiDAR. *Remote Sens.* **2016**, *8*, 1000. [CrossRef]

14. Kelly, F.J.; Fuller, G.W.; Walton, H.A.; Fussell, J.C. Monitoring air pollution: Use of early warning systems for public health. *Respirology* **2012**, *17*, 7–19. [CrossRef] [PubMed]
15. Monitoring and Early Warning of Smoke Haze by Southeast Asia Regional Centre. Available online: https://public.wmo.int/en/media/news-from-members/monitoring-and-early-warning-of-smoke-haze-southeast-asia-regional-centre (accessed on 18 November 2020).
16. Wooster, M.J.; Roberts, G.; Perry, G.L.W.; Kaufman, Y.J. Retrieval of biomass combustion rates and totals from fire radiative power observations: FRP derivation and calibration relationships between biomass consumption and fire radiative energy release. *J. Geophys. Res. Atmos.* **2005**, *110*. [CrossRef]
17. Setyawati, W.; Damanhuri, E.; Lestari, P.; Dewi, K. Emission factor from small scale tropical peat combustion. *IOP Conf. Ser. Mater. Sci. Eng.* **2017**, *180*, 012113. [CrossRef]
18. Hu, Y.; Fernandez-Anez, N.; Smith, T.E.L.; Rein, G. Review of emissions from smouldering peat fires and their contribution to regional haze episodes. *Int. J. Wildland Fire* **2018**, *27*, 293–312. [CrossRef]
19. Huang, X.; Rein, G. Upward-and-downward spread of smoldering peat fire. *Proc. Combust. Inst.* **2019**, *37*, 4025–4033. [CrossRef]
20. Rein, G.; Cleaver, N.; Ashton, C.; Pironi, P.; Torero, J.L. The severity of smouldering peat fires and damage to the forest soil. *Catena* **2008**, *74*, 304–309. [CrossRef]
21. Reid, J.S.; Koppmann, R.; Eck, T.F.; Eleuterio, D.P. A review of biomass burning emissions part II: Intensive physical properties of biomass burning particles. *Atmos. Chem. Phys.* **2005**, 27. [CrossRef]
22. Atwood, E.C.; Englhart, S.; Lorenz, E.; Halle, W.; Wiedemann, W.; Siegert, F. Detection and Characterization of Low Temperature Peat Fires during the 2015 Fire Catastrophe in Indonesia Using a New High-Sensitivity Fire Monitoring Satellite Sensor (FireBird). *PLoS ONE* **2016**, *11*, e0159410. [CrossRef] [PubMed]
23. Gonzalez-Alonso, L.; Val Martin, M.; Kahn, R.A. Biomass-burning smoke heights over the Amazon observed from space. *Atmos. Chem. Phys.* **2019**, *19*, 1685–1702. [CrossRef]
24. Aouizerats, B.; van der Werf, G.R.; Balasubramanian, R.; Betha, R. Importance of transboundary transport of biomass burning emissions to regional air quality in Southeast Asia during a high fire event. *Atmos. Chem. Phys.* **2015**, *15*, 363–373. [CrossRef]
25. Atwood, S.A.; Reid, J.S.; Kreidenweis, S.M.; Yu, L.E.; Salinas, S.V.; Chew, B.N.; Balasubramanian, R. Analysis of source regions for smoke events in Singapore for the 2009 El Nino burning season. *Atmos. Environ.* **2013**, *78*, 219–230. [CrossRef]
26. Xu, W.; Wooster, M.J.; Kaneko, T.; He, J.; Zhang, T.; Fisher, D. Major advances in geostationary fire radiative power (FRP) retrieval over Asia and Australia stemming from use of Himarawi-8 AHI. *Remote Sens. Environ.* **2017**, *193*, 138–149. [CrossRef]
27. Kaiser, J.W.; Heil, A.; Andreae, M.O.; Benedetti, A.; Chubarova, N.; Jones, L.; Morcrette, J.-J.; Razinger, M.; Schultz, M.G.; Suttie, M. Biomass burning emissions estimated with a global fire assimilation system based on observed fire radiative power. *Biogeosciences* **2012**, *9*, 527. [CrossRef]
28. Van der Werf, G.R.; Randerson, J.T.; Giglio, L.; van Leeuwen, T.T.; Chen, Y.; Rogers, B.M.; Mu, M.; van Marle, M.J.E.; Morton, D.C.; Collatz, G.J.; et al. Global fire emissions estimates during 1997–2016. *Earth Syst. Sci. Data* **2017**, *9*, 697–720. [CrossRef]
29. Giglio, L.; Schroeder, W.; Justice, C.O. The collection 6 MODIS active fire detection algorithm and fire products. *Remote Sens. Environ.* **2016**, *178*, 31–41. [CrossRef]
30. Wooster, M.J.; Roberts, G.; Freeborn, P.H.; Govaerts, Y.; Beeby, R.; He, J.; Lattanzia, A.; Mullen, R. Meteosat SEVIRI Fire Radiative Power (FRP) products from the Land Surface Analysis Satellite Applications Facility (LSA SAF): Part 1—Algorithms, product contents & analysis. *Atmos. Chem. Phys.* **2015**, *15*, 13217–13239. [CrossRef]
31. Akagi, S.K.; Yokelson, R.J.; Wiedinmyer, C.; Alvarado, M.J.; Reid, J.S.; Karl, T.; Crounse, J.D.; Wennberg, P.O. Emission factors for open and domestic biomass burning for use in atmospheric models. *Atmos. Chem. Phys.* **2011**, *11*, 4039–4072. [CrossRef]
32. Andreae, M.O. Emission of trace gases and aerosols from biomass burning—An updated assessment. *Atmos. Chem. Phys.* **2019**, *19*, 8523–8546. [CrossRef]
33. Andreae, M.O.; Merlet, P. Emission of trace gases and aerosols from biomass burning. *Glob. Biogeochem. Cycles* **2001**, *15*, 955–966. [CrossRef]

34. Freeborn, P.H.; Wooster, M.J.; Roberts, G. Addressing the spatiotemporal sampling design of MODIS to provide estimates of the fire radiative energy emitted from Africa. *Remote Sens. Environ.* **2011**, *115*, 475–489. [CrossRef]
35. Roberts, G.; Wooster, M.J.; Lauret, N.; Gastellu-Etchegorry, J.-P.; Lynham, T.; McRae, D. Investigating the impact of overlying vegetation canopy structures on fire radiative power (FRP) retrieval through simulation and measurement. *Remote Sens. Environ.* **2018**, *217*, 158–171. [CrossRef]
36. Zhang, T.; Wooster, M.J.; De Jong, M.C.; Xu, W. How Well Does the 'Small Fire Boost'Methodology Used within the GFED4. 1s Fire Emissions Database Represent the Timing, Location and Magnitude of Agricultural Burning? *Remote Sens.* **2018**, *10*, 823. [CrossRef]
37. Lu, X.; Zhang, X.; Li, F.; Cochrane, M.A. Investigating Smoke Aerosol Emission Coefficients Using MODIS Active Fire and Aerosol Products: A Case Study in the CONUS and Indonesia. *J. Geophys. Res. Biogeosci.* **2019**, *124*, 1413–1429. [CrossRef]
38. Mota, B.; Wooster, M.J. A new top-down approach for directly estimating biomass burning emissions and fuel consumption rates and totals from geostationary satellite fire radiative power (FRP). *Remote Sens. Environ.* **2018**, *206*, 45–62. [CrossRef]
39. Randerson, J.T.; Chen, Y.; van der Werf, G.R.; Rogers, B.M.; Morton, D.C. Global burned area and biomass burning emissions from small fires. *J. Geophys. Res. Biogeosci.* **2012**, *117*. [CrossRef]
40. Gaveau, D.L.A.; Salim, M.A.; Hergoualc'h, K.; Locatelli, B.; Sloan, S.; Wooster, M.; Marlier, M.E.; Molidena, E.; Yaen, H.; DeFries, R.; et al. Major atmospheric emissions from peat fires in Southeast Asia during non-drought years: Evidence from the 2013 Sumatran fires. *Sci. Rep.* **2014**, *4*, 6112. [CrossRef]
41. Boschetti, L.; Eva, H.D.; Brivio, P.A.; Grégoire, J.M. Lessons to be learned from the comparison of three satellite-derived biomass burning products. *Geophys. Res. Lett.* **2004**, *31*. [CrossRef]
42. Veraverbeke, S.; Hook, S.J. Evaluating spectral indices and spectral mixture analysis for assessing fire severity, combustion completeness and carbon emissions. *Int. J. Wildland Fire* **2013**, *22*, 707–720. [CrossRef]
43. Nguyen, H.M.; Wooster, M.J. Advances in the estimation of high Spatio-temporal resolution pan-African top-down biomass burning emissions made using geostationary fire radiative power (FRP) and MAIAC aerosol optical depth (AOD) data. *Remote Sens. Environ.* **2020**, *248*, 111971. [CrossRef]
44. Ichoku, C.; Ellison, L. Global top-down smoke-aerosol emissions estimation using satellite fire radiative power measurements. *Atmos. Chem. Phys.* **2014**, *14*, 6643–6667. [CrossRef]
45. Ichoku, C.; Kaufman, Y.J. A method to derive smoke emission rates from MODIS fire radiative energy measurements. *IEEE Trans. Geosci. Remote Sens.* **2005**, *43*, 2636–2649. [CrossRef]
46. Roberts, G.J.; Wooster, M.J. Fire detection and fire characterization over Africa using Meteosat SEVIRI. *IEEE Trans. Geosci. Remote Sens.* **2008**, *46*, 1200–1218. [CrossRef]
47. Wooster, M.J.; Zhukov, B.; Oertel, D. Fire radiative energy for quantitative study of biomass burning: Derivation from the BIRD experimental satellite and comparison to MODIS fire products. *Remote Sens. Environ.* **2003**, *86*, 83–107. [CrossRef]
48. Farnebäck, G. Two-Frame Motion Estimation Based on Polynomial Expansion. In *Image Analysis*; Bigun, J., Gustavsson, T., Eds.; Springer: Berlin/Heidelberg, Germany, 2003; Volume 2749, pp. 363–370.
49. Jackson, J.M.; Liu, H.; Laszlo, I.; Kondragunta, S.; Remer, L.A.; Huang, J.; Huang, H.-C. Suomi-NPP VIIRS aerosol algorithms and data products. *J. Geophys. Res. Atmos.* **2013**, *118*, 12673–12689. [CrossRef]
50. Dubovik, O.; Holben, B.; Eck, T.F.; Smirnov, A.; Kaufman, Y.J.; King, M.D.; Tanré, D.; Slutsker, I. Variability of Absorption and Optical Properties of Key Aerosol Types Observed in Worldwide Locations. *J. Atmos. Sci.* **2002**, *59*, 590–608. [CrossRef]
51. Shi, Y.R.; Levy, R.C.; Eck, T.F.; Fisher, B.; Mattoo, S.; Remer, L.A.; Slutsker, I.; Zhang, J. Characterizing the 2015 Indonesia Fire Event Using Modified MODIS Aerosol Retrievals. *Atmos. Chem. Phys. Discuss.* **2018**, 1–26. [CrossRef]
52. Thomas, G.E.; Carboni, E.; Sayer, A.M.; Poulsen, C.A.; Siddans, R.; Grainger, R.G. Oxford-RAL Aerosol and Cloud (ORAC): Aerosol retrievals from satellite radiometers. In *Satellite Aerosol Remote Sensing over Land*; Kokhanovsky, A.A., de Leeuw, G., Eds.; Springer Praxis Books; Springer: Berlin/Heidelberg, Germany, 2009; pp. 193–225; ISBN 978-3-540-69397-0.
53. Bulgin, C.E.; Palmer, P.I.; Merchant, C.J.; Siddans, R.; Gonzi, S.; Poulsen, C.A.; Thomas, G.E.; Sayer, A.M.; Carboni, E.; Grainger, R.G.; et al. Quantifying the response of the ORAC aerosol optical depth retrieval for MSG SEVIRI to aerosol model assumptions. *J. Geophys. Res. Atmos.* **2011**, *116*. [CrossRef]

54. Chand, D.; Schmid, O.; Gwaze, P.; Parmar, R.S.; Helas, G.; Zeromskiene, K.; Wiedensohler, A.; Massling, A.; Andreae, M.O. Laboratory measurements of smoke optical properties from the burning of Indonesian peat and other types of biomass. *Geophys. Res. Lett.* **2005**, *32*. [CrossRef]
55. Sayer, A.M.; Munchak, L.A.; Hsu, N.C.; Levy, R.C.; Bettenhausen, C.; Jeong, M.-J. MODIS Collection 6 aerosol products: Comparison between Aqua's e-Deep Blue, Dark Target, and "merged" data sets, and usage recommendations. *J. Geophys. Res. Atmos.* **2014**, *119*, 13965–13989. [CrossRef]
56. Hurley, M.J.; Gottuk, D.T.; Hall, J.R., Jr.; Harada, K.; Kuligowski, E.D.; Puchovsky, M.; Watts, J.M., Jr.; Wieczorek, C.J. *SFPE Handbook of Fire Protection Engineering*; Springer: Greenbelt, MD, USA, 2015.
57. Elvidge, C.D.; Zhizhin, M.; Hsu, F.-C.; Baugh, K.; Khomarudin, M.R.; Vetrita, Y.; Sofan, P.; Hilman, D. Long-wave infrared identification of smoldering peat fires in Indonesia with nighttime Landsat data. *Environ. Res. Lett.* **2015**, *10*, 065002. [CrossRef]
58. Fisher, D.; Wooster, M.J. Multi-decade global gas flaring change inventoried using the ATSR-1, ATSR-2, AATSR and SLSTR data records. *Remote Sens. Environ.* **2019**, *232*, 111298. [CrossRef]
59. Fisher, D.; Wooster, M.J. Shortwave IR Adaption of the Mid-Infrared Radiance Method of Fire Radiative Power (FRP) Retrieval for Assessing Industrial Gas Flaring Output. *Remote Sens.* **2018**, *10*, 305. [CrossRef]
60. Kumar, S.S.; Hult, J.; Picotte, J.; Peterson, B. Potential Underestimation of Satellite Fire Radiative Power Retrievals over Gas Flares and Wildland Fires. *Remote Sens.* **2020**, *12*, 238. [CrossRef]
61. Val Martin, M.; Kahn, R.A.; Logan, J.A.; Paugam, R.; Wooster, M.; Ichoku, C. Space-based observational constraints for 1-D fire smoke plume-rise models: Smoke plume-rise constraints. *J. Geophys. Res.* **2012**, *117*. [CrossRef]
62. Wiggins, E.B.; Czimczik, C.I.; Santos, G.M.; Chen, Y.; Xu, X.; Holden, S.R.; Randerson, J.T.; Harvey, C.F.; Kai, F.M.; Yu, L.E. Smoke radiocarbon measurements from Indonesian fires provide evidence for burning of millennia-aged peat. *Proc. Natl. Acad. Sci. USA* **2018**, *115*, 12419–12424. [CrossRef]
63. Lu, X.; Zhang, X.; Li, F.; Cochrane, M.A. Investigating Smoke Emission Coefficients using MODIS Fire Radiative Energy and Smoke Aerosols. *AGU Fall Meet. Abstr.* **2018**, *51*. [CrossRef]
64. Val Martin, M.; Logan, J.A.; Kahn, R.A.; Leung, F.-Y.; Nelson, D.L.; Diner, D.J. Smoke injection heights from fires in North America: Analysis of 5 years of satellite observations. *Atmos. Chem. Phys.* **2010**, *10*, 1491–1510. [CrossRef]
65. Giles, D.M.; Sinyuk, A.; Sorokin, M.S.; Schafer, J.S.; Smirnov, A.; Slutsker, I.; Eck, T.F.; Holben, B.N.; Lewis, J.; Campbell, J.; et al. Advancements in the Aerosol Robotic Network (AERONET) Version 3 Database—Automated Near Real-Time Quality Control Algorithm with Improved Cloud Screening for Sun Photometer Aerosol Optical Depth (AOD) Measurements. *Atmos. Meas. Tech. Discuss.* **2018**, 1–78. [CrossRef]
66. Smirnov, A.; Holben, B.N.; Eck, T.F.; Dubovik, O.; Slutsker, I. Cloud-Screening and Quality Control Algorithms for the AERONET Database. *Remote Sens. Environ.* **2000**, *73*, 337–349. [CrossRef]
67. Thomas, G.E.; Poulsen, C.A.; Sayer, A.M.; Marsh, S.H.; Dean, S.M.; Carboni, E.; Siddans, R.; Grainger, R.G.; Lawrence, B.N. The GRAPE aerosol retrieval algorithm. *Atmos. Meas. Tech.* **2009**, *2*, 23. [CrossRef]
68. Huang, J.; Kondragunta, S.; Laszlo, I.; Liu, H.; Remer, L.A.; Zhang, H.; Superczynski, S.; Ciren, P.; Holben, B.N.; Petrenko, M. Validation and expected error estimation of Suomi-NPP VIIRS aerosol optical thickness and Ångström exponent with AERONET. *J. Geophys. Res. Atmos.* **2016**, *121*, 7139–7160. [CrossRef]
69. Xiao, Q.; Zhang, H.; Choi, M.; Li, S.; Kondragunta, S.; Kim, J.; Holben, B.; Levy, R.C.; Liu, Y. Evaluation of VIIRS, GOCI, and MODIS Collection 6 AOD retrievals against ground sunphotometer observations over East Asia. *Atmos. Chem. Phys.* **2016**, *16*, 1255–1269. [CrossRef]

Publisher's Note: MDPI stays neutral with regard to jurisdictional claims in published maps and institutional affiliations.

© 2020 by the authors. Licensee MDPI, Basel, Switzerland. This article is an open access article distributed under the terms and conditions of the Creative Commons Attribution (CC BY) license (http://creativecommons.org/licenses/by/4.0/).

sensors

MDPI

Article

Thermal Infrared Sensing for Near Real-Time Data-Driven Fire Detection and Monitoring Systems

Maria João Sousa [1,*], **Alexandra Moutinho** [1] and **Miguel Almeida** [2]

[1] IDMEC, Instituto Superior Técnico, Universidade de Lisboa, Av. Rovisco Pais, 1, 1049-001 Lisboa, Portugal; alexandra.moutinho@tecnico.ulisboa.pt

[2] ADAI, University of Coimbra, Rua Pedro Hispano, 12, 3030-289 Coimbra, Portugal; miguelalmeida@adai.pt

* Correspondence: maria.joao.sousa@tecnico.ulisboa.pt

Received: 15 October 2020; Accepted: 23 November 2020; Published: 28 November 2020

check for updates

Abstract: With the increasing interest in leveraging mobile robotics for fire detection and monitoring arises the need to design recognition technology systems for these extreme environments. This work focuses on evaluating the sensing capabilities and image processing pipeline of thermal imaging sensors for fire detection applications, paving the way for the development of autonomous systems for early warning and monitoring of fire events. The contributions of this work are threefold. First, we overview image processing algorithms used in thermal imaging regarding data compression and image enhancement. Second, we present a method for data-driven thermal imaging analysis designed for fire situation awareness in robotic perception. A study is undertaken to test the behavior of the thermal cameras in controlled fire scenarios, followed by an in-depth analysis of the experimental data, which reveals the inner workings of these sensors. Third, we discuss key takeaways for the integration of thermal cameras in robotic perception pipelines for autonomous unmanned aerial vehicle (UAV)-based fire surveillance.

Keywords: thermal infrared cameras; thermal imaging data; wildfire detection; active fire monitoring; early warning systems; unmanned aerial systems

---

1. Introduction

With the emergent effects of climate change, several regions worldwide have been undergoing an increasing number of more intense and devastating wildfire events, as well as extended fire seasons [1,2]. In this context, given the high spatial and temporal uncertainty intrinsic to these phenomena, environment monitoring is determinant for firefighting activities to mitigate the consequences of these events.

Currently, limited areas can be monitored through automatic systems in watchtowers. However, by being installed near the ground—hence, at low altitudes—these systems have several limitations. Since these are widely based on visible range sensors, clouds can be easily confused with smoke and the sunset or reflections can be mistaken by flames, leading to false alarms. Moreover, as the systems depend on the identification of a smoke column, flames can only be detected when the fire has increased in magnitude, thus often preventing early fire detection. Furthermore, solutions based on satellite data have considerable latency, thus also hindering their application for early detection.

Despite the widespread use of aerial means in prevention of and emergency response to forest fires, piloted aircraft require highly trained personnel and are expensive to operate [3,4], limiting the number of vehicles that are used for firefighting and surveillance tasks. However, given that these scenarios require the coverage of extensive areas where the environment is highly dynamic, the availability of aerial means is paramount on tactical and operational levels for situation awareness.

To address this issue, in recent years, there has been active research towards developing systems based on unmanned aerial vehicles (UAVs) for fire detection [5,6] and operational support [7–9], but it has intensified lately as a result of the difficulties faced in responding to large-scale wildfires, which has reinforced the need to detect fires in an early stage, as well as to provide near real-time monitoring.

In parallel, the recent advances in robotic perception have opened a path towards autonomous robotics by taking advantage of novel sensors and intelligent systems to enable autonomous navigation and exploration [10]. The breakthroughs in sensor technology and embedded computing have resulted in the progressive decrease in the weight/size ratio and equipment cost while incorporating powerful computing capabilities. These emerging technologies are enabling real-time processing of high-dimensional data, which allows UAVs to be equipped with advanced thermal and optical image sensors as well as computing platforms for on-board data processing.

In this context, the general aim of this investigation is to contribute to the development of UAV-based systems for fire detection and monitoring, namely in what concerns the integration of thermal cameras in robotic perception for these tasks.

The main objective of this study is to provide a comprehensive understanding of the behavior of thermal imaging sensors in fire detection scenarios by relating the sensor response and the image processing employed in this type of device. In that sense, this work explores sensor data from two different thermal cameras by linking the raw sensor data to the mapping functions employed to obtain images encoded in pseudocolor. For this purpose, several fire experiments in controlled conditions were performed in laboratory and field trials covering distinct operating regimes of this type of sensor.

The contributions of this article are threefold: (1) an overview of state-of-the-art image processing algorithms widely used in thermal imaging cameras; (2) an in-depth analysis of the behavior of thermal cameras targeted at fire identification scenarios based on controlled fire experiments; (3) discussion of the implications of the insights exposed and the potential developments in robotic perception towards autonomous UAV-based fire detection and monitoring systems.

Considering the growing availability of low-cost, compact thermal cameras for UAV-borne applications, the uptake of this technology will progressively increase in the near future [11]. As this topic is an active area of research, the relevant insights outlined in this article provide important considerations to guide future research. Hence, also contributing to the practical implementation of thermal-enabled fire detection and monitoring systems, which bring great potential for minimizing the impacts of fire events.

1.1. Related Work

The advent of the evolution of remote sensing technologies has led to the continuous improvement of fire hazard identification and risk assessment systems. Over the years, several types of platforms have been explored for these solutions, such as satellites, high-altitude aircraft, and remotely piloted aircraft systems, enabling the assessment of the progression of fire events [12].

Although satellite-based imagery is used by emergency response agencies to monitor large-scale wildfires that burn over extensive periods, the wait interval for a satellite overpass induces a considerable time delay, which prevents its application in time-sensitive fire detection scenarios, such as emergency evacuations or search-and-rescue operations [13]. Despite its value from a strategic standpoint, for tactical and operational decision support, the availability of updated information is crucial. To address this while avoiding the expensive operation costs of piloted aircraft, UAVs are considered as a viable alternative for remote sensing by providing local coverage with high spatial and temporal resolution.

In previous contributions, wildfire detection applications based on airborne systems have explored visible range, thermal, or multispectral technologies [14]. However, since the radiation emitted from a fire is high in the thermal range, there has been significant interest in the use of thermal infrared bands [12]. In that sense, the following review focuses on contributions employing the thermal range.

Thermal infrared cameras provide sensing capabilities suitable for ongoing environmental monitoring by operating in both daylight and in night conditions. Additionally, in contrast to visible range sensors, for which smoke severely affects the ability to detect and track the perimeters of fire fronts, for thermal infrared cameras, the impact of smoke presents lower interference.

Regarding the development of automatic algorithms, several contributions have presented advances towards this objective. On the one hand, works on image processing algorithms have focused on the extraction of image descriptors to obtain signals representing fire instances to detect the presence of a fire by assessing the time-series data [15]. On the other hand, since thermal images adapt according to the context in the field of view of the camera, considering only the brightness information can lead to false alarms. To address this issue, some approaches combine both spatial and temporal features, i.e., brightness, motion, and flicker [16]. In turn, the integration of infrared cameras has been suggested as a way to complement visual range fire surveillance systems so as to harness the advantages of both visual and thermal features and to yield more accurate early fire detection rates [17–19]. More recently, an off-line processing fire-tracking algorithm based on edge detection has been proposed to process geo-referenced thermal images previously acquired using an airborne thermal camera [20].

Note, however, that not all the previously mentioned research efforts focus specifically or solely on wildfire detection. In a wider scope, considering fire detection outdoors, authors have recognized the difficulty in applying image processing algorithms in this setting due to a significant rate of false positives caused by external factors, such as weather conditions, sunlight reflections, or saturation of the infrared sensors caused by other heated objects [21]. Given the challenges faced in real contexts, it becomes rather complex to overcome these limitations with ad hoc classical computer vision algorithms.

In alternative to computer vision methods, to improve generalization in dynamic scenarios, intelligent systems approaches have been proposed, namely data-driven models based on feature engineering and fuzzy inference systems [22]. The nonlinear approach has been successfully tested in fire experiments under controlled conditions and validated in highly dynamic environments, such as camping sites.

Most previous approaches were designed for data acquisition purposes and algorithm design, whereas only recently autonomous systems are being developed, as a result of the increased data processing capabilities aboard aerial vehicles.

1.2. Proposed Approach

Thermal cameras have a clear potential for wildfire detection and monitoring tasks, but the path towards their integration in robotic perception pipelines, which are essential for autonomous systems, is still rather understudied. To contribute to narrowing this gap, this work conducts a comprehensive study of thermal imaging sensing to extend the understanding of the inner workings of this technology and how it can be leveraged for wildfire surveillance tasks. For that purpose, it is important to identify the main challenges this work addresses in the following.

First, although thermal imaging cameras are increasingly available for a multitude of industrial domains [23], with expanding model ranges and accessible equipment costs, their application is still limited, which can be attributed to two factors. On the one hand, most applications rely on a human-in-the-loop approach, which typically requires specialized training and technical expertise to interpret the image data, but this is generally based on high-level knowledge oriented to the domain of application. On the other hand, as discussed in the literature review, machine vision approaches depend on feature-based approaches derived from image data, which do not take into account the image processing algorithms underlying the output data. However, these abstraction layers hinder the development of automatic algorithms due to the adaptive nonlinear nature that is at the core of the behavior of these systems. Therefore, the knowledge of the underlying processing methods involved in generating the output image is central to understanding how to leverage this technology in a robotic perception framework. In that sense, the first step in this work concerns the overview of state-of-the-art image processing algorithms employed in most commercial off-the-shelf thermal cameras.

Second, the adoption of thermal imaging cameras for wildfire detection scenarios differs considerably from general applications, e.g., industrial inspection or precision agriculture, in the sense that it deals with extreme temperatures. In this context, the importance of quantitative information is less prevalent than that of qualitative data because a fire can be identified by high temperature gradients with respect to ambient conditions, and can thus be detected using the relative temperature differences in the images. To that effect, having radiometric information is not determinant because the intensity levels can translate the relative difference between objects in the scene. Nonetheless, a correct interpretation of the adaptive algorithms is required because the color-encoding schema adapts to the range of measurements in each instance. For these reasons, after covering the processing algorithms in the first part of this work, we demonstrate the implications of their usage in wildfire detection scenarios. To that end, several fire experiments under controlled conditions were conducted to study the behavior of thermal cameras in those situations to characterize how the raw sensor data are mapped to visually interpretable pseudocolor images. In this regard, attention is paid to the identification of the saturation levels of this type of sensor.

Third, we discuss the implications of the insights exposed and the potential developments in robotic perception towards autonomous UAV-based fire monitoring systems, namely through the identification of current roadblocks and possible enabling solutions that manufacturers should integrate for the widespread use of this technology.

After this introduction, the remainder of this article is structured as follows: Section 2 covers the overview of image processing algorithms employed in thermal cameras. Section 3 presents a data-driven method based on thermal imaging for fire situation awareness. Section 4 presents the experimental setups and conditions of the laboratory and field trials. Section 5 analyzes the results and discusses the implications for the integration of thermal imaging in robotic perception of wildfire surveillance. Finally, Section 6 presents the conclusions and offers suggestions for future research.

2. Thermal Imaging

The basic principle of thermal imaging is based on the concept of sensing the radiance emitted from objects of interest. Note that above absolute zero temperature, i.e., 0 K, all bodies emit thermal radiation. In a general definition, what is referred to as thermal range is comprised of radiation with wavelengths in the 10^{-7} to 10^{-4} m interval of the electromagnetic spectrum. Hence, this range includes spectral bands in the ultraviolet, visible, and infrared regions of the spectrum.

Although this range is rather ample, thermal cameras usually only cover a part of it, which varies depending on the model and the application for which it is intended. More specifically, the devices employed in this work are based on uncooled microbolometer detectors that cover part of the infrared band, namely the 7.5–13.5 $\times 10^{-6}$ m spectrum, which is widely common for camera models in the market.

Thermal infrared cameras employ complex signal processing architectures in order to output images that convey relative temperature differences between objects in a scene [24]. This process comprises several stages, which are summarized in Figure 1 and are briefly described in the following.

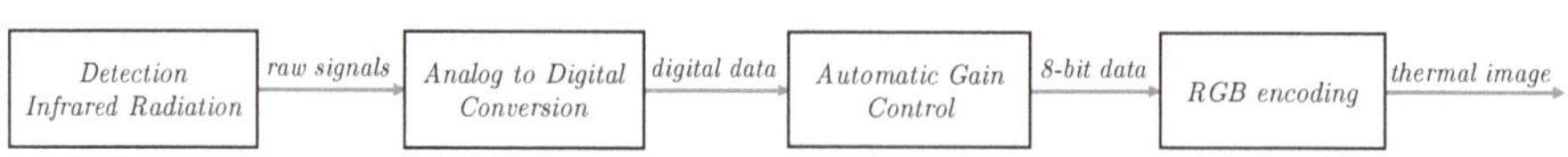

Figure 1. High-level diagram of the process flow in uncooled microbolometer arrays.

In the first stage, incident infrared radiation is absorbed, inducing changes of resistance in each microbolometer detector of the focal plane array, which are translated into a time-multiplexed electrical signal by a readout integrated circuit (ROIC) [25]. Then, the array is calibrated automatically each time it is powered to match all microbolometers to the same input/output characteristic function that relates the measured radiance intensity and the output signal. This is performed through a linearization process and temperature compensation of the signals from individual detectors of the array [26]. Second, with these compensated signals, the measurements are transformed into raw pixel values that translate the intensity values that compose a monochromatic image. Raw image data can be subsequently transformed into pseudocolor images through an automatic gain control procedure and RGB encoding according to a user-specified color palette to facilitate interpretation.

In addition to sensing the radiation being emitted from objects in the field of view, another important aspect concerning thermal cameras is the ability to measure temperature. While this topic is extensively covered in the literature, with respect to how the incident radiation is transformed into approximate temperature readings [26], in this work, we do not delve into this matter for two main reasons. First, with this study, we aim to address the general image processing pipeline that is transversal to most thermal cameras, i.e., irrespective of these having radiometric capabilities or being non-radiometric. Second, although in different contexts, the correction of temperature values of thermographic images can be performed *a posteriori* using off-line post-processing methods, this requires a known reference in the image content [27]. In the case of wildfire surveillance, if we consider the environment to be open and unknown with regard to the temperature, i.e., without access to external absolute temperature readings, on-line thermal correction of the calibration for real-time applications is not possible.

Taking into account these considerations, the main focus of this work is on how raw digital readings are encoded into pseudocolor images, which can be applicable to image data from both radiometric and non-radiometric thermal cameras.

2.1. Sensor Characteristics: Preliminaries

Thermal imaging systems are optical instruments that are able to generate two-dimensional representations of the surrounding environment as a function of the strength of incoming thermal radiation. These exteroceptive sensors transform a digitally encoded array of pixels into an image according to the camera perspective projection, which depends on the focal length of the lens.

Digital images are generated according to the characteristics of the camera sensor, namely spatial resolution, temporal resolution and dynamic range [28]. The spatial resolution is intrinsically related to the size of the focal plane array and defines the number of pixels in an image, as well as the corresponding aspect ratio. The temporal resolution is associated with the operating frequency of the device, i.e., the frame rate at which the camera yields image data. In turn, the dynamic range corresponds to the interval of intensity values represented. Additionally, the resulting images also depend on the automatic gain, which is a fundamental aspect of the manner in which thermal images are encoded, since most camera models employ automatic gain control, as will be discussed further along.

To explore the effects of these characteristics, this work resorts to two distinct thermal cameras, namely the FLIR SC660 (FLIR Systems, Inc., Wilsonville, OR, USA) and FLIR Vue Pro (FLIR Systems, Inc., Wilsonville, OR, USA), which operate in the regions denominated as far infrared (FIR) or long-wave infrared (LWIR). The main specifications of these camera models are presented in Table 1.

Regarding the field of view (FOV) specifications and focal length, note that these cameras have distinct characteristics, which will influence the data recorded. Moreover, spatial resolution, temporal resolution and dynamic range also vary between both models, which is an aspect to take into consideration. In the case of the FLIR SC660, the automatic gain can be adjusted to different configurations to change the dynamic range.

In addition, keeping in mind the payload budget of small UAVs and comparing the weight and size of both cameras, only the FLIR Vue Pro has suitable characteristics to be taken onboard a small UAV. However, given the exploratory basis of this work, the radiometric capabilities of the FLIR SC660 are valuable for the study of these sensors in extreme conditions inherent to fire scenarios.

Table 1. Summary of the specifications of the thermal cameras.

	FLIR SC660	FLIR Vue Pro
Spatial resolution (px)	640 × 480	336 × 256
Temporal resolution (Hz)	30	8.3
Bit resolution (bit)	14	16
Focal length (mm)	19	9.0
Horizontal field of view (FOV) (°)	45	35
Vertical FOV (°)	34	27
Spectral band (10^{-6} m)	7.5∼13.5	7.5∼13.5
Measurement range (°C)	−40∼+1500	−60∼+150
Size (L × W × H) (mm)	299 × 144 × 147	63 × 44.4 × 44.4
Weight (g)	1800	92.1–113.4

Although this work explores these cameras, other alternative models have been released more recently, with higher spatial resolution and gimbal integration, albeit at superior costs. Notwithstanding, from the standpoint of the hardware and firmware available, the update does not have a significant impact on thermal imaging results, since the major benefits of the newer models are the facilitated integration for deployment on UAVs or graphic user interfaces for control from mobile devices. Thus, the analysis presented herein also applies to different thermal cameras.

In the following, the image processing methods for generating thermal images are described through examples explaining the mapping algorithms that perform the transformation of raw digital data into pseudocolor-encoded images.

2.2. Mapping Raw Digital Data to Thermal Images

Thermal cameras are generally single-band in the sense that they only produce monochromatic images, which subsequently undergo a sequence of image processing steps to transform the raw digital data into pseudocolor images to highlight the details of the scene context. This processing pipeline can be implemented either on the device for storage or digital output, or on external software.

Currently, commercial off-the-shelf devices already provide a variety of color palettes to enhance the visual interpretation of the amounts of radiance captured by the sensors. However, to design intelligent algorithms for autonomous systems, the color-encoding schema have to be well suited for the robotic perception approach, which is essential for fulfilling the application requirements. For this reason, this also requires a deeper understanding of the image processing pipeline to leverage the potential of this type of sensor for novel applications.

The raw intensity levels are given by a digital number assigned by the sensor analog-to-digital converter, which can be of 14-bit or 16-bit order depending on the sensor bit resolution. In this work, we will employ both these alternatives, as specified in Table 1, but since these algorithms apply in the same way to both versions, the following examples showcase only the processing of raw data in 14-bit space. Further along in the analyses of the sensor response to fire scenarios both cases will be covered in detail. Note that the bit resolution relates to the temperature range of the camera, and for camera models with different modes, such as the FLIR SC660, the intensity level values will also be influenced by the camera configuration, i.e., the high-gain or low-gain modes.

To obtain a thermal image in pseudocolor, the image processing pipeline is divided into two main steps: (1) application of a data compression technique denoted as the automatic gain control; (2) application of the color palette specified, yielding images with three channels corresponding to the RGB color-space representation.

Automatic gain control (AGC) is a histogram-based technique that performs the transformation between raw data formats to 8-bit image data. This processing method is responsible for data compression, which implies a considerable loss of information. For the 16-bit case, from a range of possible values from 0 to 65,535, the resulting image will be represented with values in the 0 to

255 interval. To counteract the decrease in detail, the AGC algorithms are designed to enhance the image contrast and brightness in order to highlight the scene context.

The following sections cover with illustrative examples the main variants of AGC algorithms implemented in thermal cameras that practitioners should be aware of to leverage this technology. Then, the color-mapping schema and several available color palettes are presented in Section 2.2.3.

2.2.1. Histogram-Based Automatic Gain Control

AGC methods are typically variants of the histogram-based operations widely used in computer vision for contrast enhancement, e.g., histogram equalization [28]. However, in thermal imaging, AGC also implies data compression between raw formats (e.g., 14-bit or 16-bit) into display-ready data (8-bit).

In classical histogram equalization, the nonlinear mapping used for contrast enhancement is derived directly from the cumulative distribution function (*cdf*) of the raw intensity values. This approach allows one to achieve an approximate linear *cdf* on the compressed 8-bit data, yielding an image with intensity values spread across the full extent of the available 8-bit range. Figure 2 illustrates this AGC procedure for a raw 14-bit image converted into the 8-bit range.

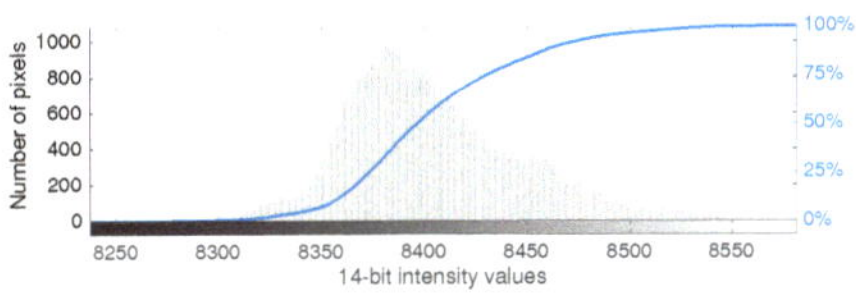

(a)

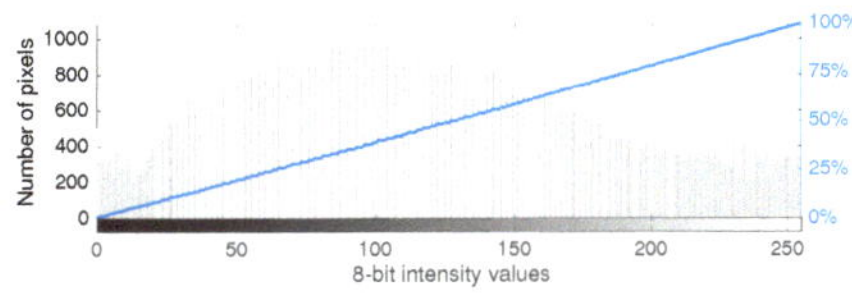

(b)

Figure 2. Automatic gain control with histogram equalization: (**a**) the raw data histogram is used to compute the cumulative distribution function (in blue) for the bounds of the 14-bit intensity values; (**b**) the data are compressed to 8-bit by flattening the distribution of the compressed data histogram, resulting in an approximate linear cumulative distribution function (in blue).

Note that although the bit resolution of the 14-bit sensor represents values up to 16,383, for environments with ambient temperatures around 20 °C, the raw data captured are represented in a narrow band of the full range, as can be observed in Figure 2. Therefore, compression and contrast enhancement play a pivotal role in the encoding of thermal images. However, note also that enhancement operations in thermal images artificially distort the data, meaning that the physical correlation that relates the radiant flux from infrared radiation and pixel intensity is lost.

Alternatively, for cases where it is important to preserve the correspondence between the pixel intensity and temperature of objects, for instance, a "linear" mapping function is better suited. The linear approach also relies on the *cdf* to define the image transformation table (ITT) by defining the slope and clipping points of the resulting nonlinear mapping function. Figure 3 depicts the application of this AGC algorithm to the previous example.

In addition to the *cdf*, the linear transformation, given by $T(x) = mx + b$, requires setting the midpoint of the image transfer table, $\mathrm{ITT}_{\mathrm{mid}}$, which is normally the average value of the range of the 8-bit range (128), and the tail rejection percentage $r_{\%}$. These parameters are used to determine the points on the *cdf* employed to define the linear equation and respective slope, m, and intercept, b, as:

$$b = \mathrm{ITT}_{\mathrm{mid}} - \mathrm{mean}(x_{100-r_{\%}}, x_{r_{\%}}) \cdot m \tag{1}$$

$$m = 255/(x_{100-r_{\%}} - x_{r_{\%}}), \tag{2}$$

with x representing the raw values indexed in the x-axis of the histogram. The image transformation function, $T'(x)$, is defined for the range of values of the histogram by clipping the values to the lower and upper bounds of the output domain $[0, 255]$, as can be observed in Figure 3. Note that the clipping operation increases the absolute frequency of 0 and 255 intensities in the 8-bit histogram. Subsequently, the data are converted through the ITT look-up table.

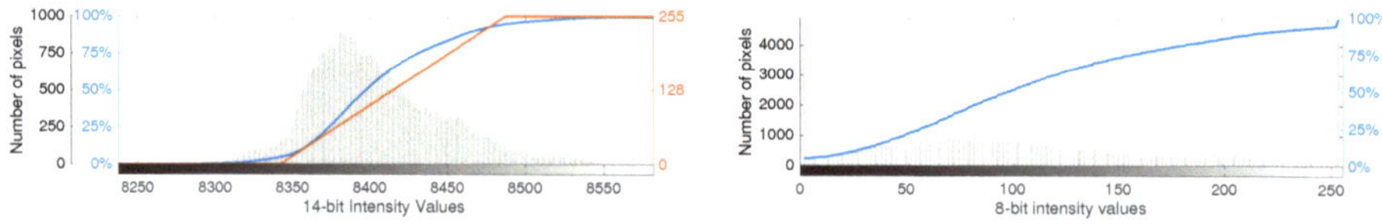

(a) (b)

Figure 3. Automatic gain control with the linear algorithm: (**a**) the raw data histogram is used to compute the cumulative distribution function (in blue), for which a percentage of the values are clipped at the lower and upper bounds to obtain the resulting nonlinear transformation (in red); (**b**) the data are compressed through the previous transformation, resulting in the histogram for 8-bit intensity values with the corresponding cumulative distribution function (in blue).

In practice, the ITT midpoint influences the brightness of the image in the sense that increasing or lowering the midpoint shifts the equation horizontally to the left or right, respectively. As a result, this clips the data to zero at a corresponding lower or higher raw value, and vice-versa for the 255 upper bound. This aspect is especially relevant because, for fire detection scenarios that present high temperature gradients, both low and high raw values are important for situation awareness. Likewise, the definition of the tail rejection percentage follows the same principle. Therefore, parameter tuning should be approached with caution so as not to discard relevant data.

Nonetheless, the linear algorithm is not the most used AGC method because this compression technique implies a considerable loss of detail. To avoid this limitation, thermal cameras usually employ the plateau equalization algorithm, which aims to balance the distribution of all intensity levels in the image scene, thereby enhancing image contrast and highlighting differences in temperature.

The plateau equalization algorithm [29] implements a nonlinear mapping that compromises between histogram projection and histogram equalization. The concept is to bound the representation of the different intensity levels to a defined threshold, termed as the plateau value, P, while limiting the slope of the transformation function through the maximum gain value, G, which is set to 1 by default. However, for images with low dynamic range, where a small interval of values has high bin counts, the plateau equalization algorithm may yield fewer intensity values than the 256 available in 8-bit space. To ensure that the entire contrast depth is leveraged, the maximum gain sets the upper limit of gain that can be used to stretch the data to the full extent of the 8-bit range. Considering that fire detection applications most likely exhibit high dynamic range situations, this parameter will not be tuned in the development of this work.

First, the image histogram is clipped according to the plateau value and represented through the effective count, c_x, for each raw intensity level x, which conditions the absolute frequencies not to exceed the prescribed plateau value, P, as:

$$c_x \equiv \min(C_x, P), \tag{3}$$

where C_x represents the original count of pixels with raw intensity value x. The plateau equalization mapping function is based on the cumulative distribution function of the clipped histogram values, denoted by cdf', which is computed according to:

$$cdf'(x) = \sum_{j=0}^{x} c_x, \qquad 0 \leq k < 2^N, \tag{4}$$

with N representing the exponent corresponding to the original bit resolution of the sensor. Then, the transformation function based on plateau equalization for an 8-bit compression is defined as:

$$T_{PE}(x) = \left\lfloor \frac{255 \cdot cdf'(x)}{cdf'(2^N)} \right\rfloor, \quad (5)$$

where $\lfloor\rfloor$ represents the truncation operator for the next lower integer. Note that if the plateau value equals the maximum absolute frequency in the original histogram, this algorithm is equivalent to histogram equalization. In turn, if the threshold is defined as 1, this algorithm behaves as histogram projection. The plateau value is established as a percentage of the maximum value of the bin count of the histogram by default, e.g., 7%, but varies depending on the camera model and specifications. Figure 4 illustrates the application of the plateau equalization algorithm for the previous example.

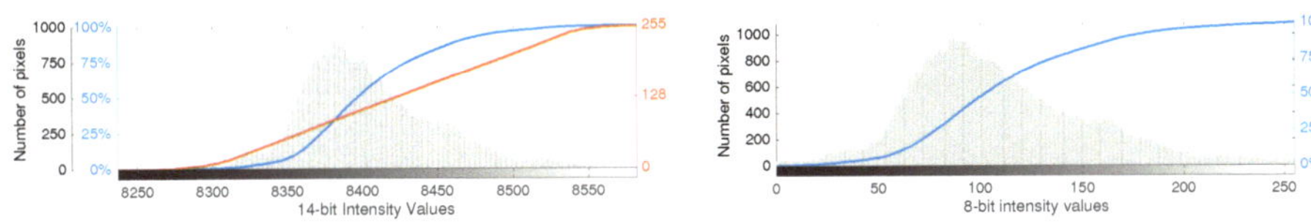

(a) (b)

Figure 4. Automatic gain control with the plateau equalization algorithm: (**a**) the raw data histogram, with the respective cumulative distribution function (in blue), is clipped according to a prescribed plateau value to generate the transformation function (in red); (**b**) the data are compressed through the previous transformation, resulting in the histogram for 8-bit intensity values with the corresponding cumulative distribution function (in blue).

Besides these algorithms, some cameras also have the information-based equalization variant that combines the plateau equalization algorithm with basic image enhancement techniques to yield more detail from scene context, irrespective of the data distribution. This method is not implemented in FLIR ResearchIR but was implemented herein for illustration purposes.

Information-based equalization aims to allocate a proportional amount of the dynamic range to different parts of the scene to capture the most information from the scene context, irrespective of these being represented with a large part of the image, such as the background, or by small areas with slight variations in temperature in the foreground. This method performs this by using a high-pass filter to highlight detail in the image, which is subtracted from the original image before the application of plateau equalization. The low-pass histogram is subsequently modified by increasing the bin count of the pixels in the high-pass image, hence increasing image contrast whilst also including greater detail. For instance, in a fire scenario, this may be useful to distinguish the fire while also being able to discern if there are people in the scene. Since this is an extreme case, with high dynamic range in the raw data, the effect of the compression into 8-bit data implies a significant loss of detail, therefore making more important these image enhancement techniques to highlight even subtle variations in temperature.

While in controlled environments, adequate tuning of this type of algorithm can yield better results, in outdoor contexts, adjusting these methods for a robust and consistent performance becomes very complex because the uncertainty and measurement errors associated with these open and unknown environments are greater due to the distinct emissivity properties of the multitude of heterogeneous materials that can be encountered [26]. In this way, by distorting the actual measurements, these contrast enhancement techniques also introduce greater inaccuracy in the image representation, which can hinder the development of robust robotic perception methods.

2.2.2. Thermal Imaging Metadata

In addition to the image data, thermal imaging files also store different types of metadata encoded under standard metadata formats, e.g., TIFF, Exif and XMP. By accessing this information through adequate interfaces, several valuable parameters can be retrieved. In addition to encoding the standard properties of the digital camera (e.g., focal length and focal plane resolution) and GPS coordinates, the manufacturers can also store camera calibration parameters relevant for conversion between raw intensity values and temperature values.

Herein, we explore relevant metadata tags that help to shed light on the working mechanisms of these sensors, providing important insight into useful parameters for robotic integration. More specifically, since we tackle the relation between raw data and RGB-encoded data, only the parameters that influence this transformation will be addressed.

The metadata were retrieved from raw video files and image files encoded in "SEQ" or "TIFF" and a proprietary format from FLIR that encodes both JPEG and metadata, as outlined in Figure 5.

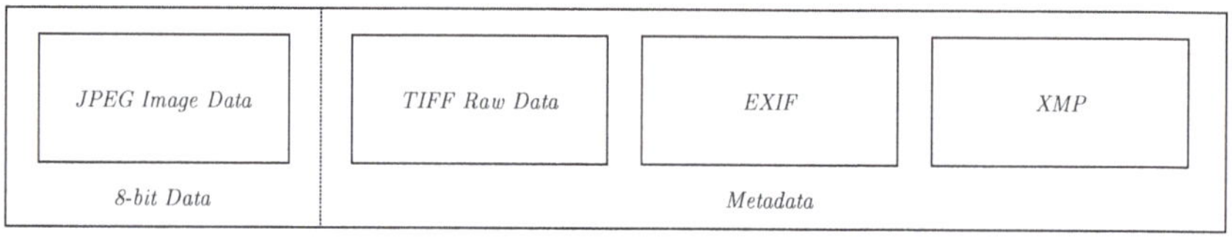

Figure 5. Schematic of file structure of FLIR image format.

By analyzing the metadata of the image and video files, two Exif tags were identified that are closely related to the histogram-based algorithms presented in Section 2.2.1, namely the raw value average and raw value range. The range of raw values and the average values are encoded separately in the metadata, but may not correspond to the values that would be computed from the raw frames for every instance. The reason for this is probably the proprietary in-camera processing to deal with noise and other types of outliers. For the FLIR SC660, the calculated values do not match the encoded values in the files, whereas for the FLIR Vue Pro, both match exactly for data recorded in image mode. This indicates that the update of these values also varies depending on the device and the mode of capture, as it is related to the firmware. With the average and the range of raw values from the metadata tags, the maximum and minimum values of the color scale are computed as follows:

$$\text{Raw Value Max} = \text{Raw Value Average} + \frac{\text{Raw Value Range}}{2} \tag{6}$$

$$\text{Raw Value Min} = \text{Raw Value Average} - \frac{\text{Raw Value Range}}{2}. \tag{7}$$

However, note that if these metadata tags are not available, the maximum and minimum values of the raw data can be computed directly on-line, as observed in the histogram-based algorithms presented in Section 2.2.1, by extracting maximum and minimum values directly from the raw frames.

Given that the histogram-based mapping functions and the respective color encoding adapt according to the scene context, these parameters are essential for understanding how the color scale adjusts over time. Thus, these will be explored next in the data-driven analysis approach proposed in Section 3.

Considering that fire surveillance applications are an extreme case with a high dynamic range, it is important to evaluate how these techniques behave in such scenarios. Thus, to delve into this issue, in the following, the color encoding used to enhance the interpretation of thermal data is presented, along with the comparison of an example for a controlled burn performed in a real-world context.

2.2.3. Color Mapping

Following the conversion into an 8-bit image format, the data are represented in the 0 to 255 range. To encode these values in an RGB color space representation, the cameras offer several color palettes with distinct characteristics, which are adequate for different applications. In cases where the cameras provide the raw data in addition to the RGB-encoded images, these data can be post-processed with different color palettes for further analysis.

The color palette is a discrete set of color samples composed of values for each color channel, and it is defined in a look-up table (LUT) in the camera firmware or external software. In practice, this discrete sequence of values provides a continuous representation of the mapping of values in the image. Figure 6 illustrates this for some widely available color palettes, presenting the set of discrete colors that form the color mapping applied to the 8-bit data. The color sequences are depicted according to an increasing order of bit values, meaning that these are attributed from lower values to higher values of sensed radiant flux. Notwithstanding, recall that the color scale is adaptive; thus, the color assignment also depends on the AGC algorithm. For this reason, the full spectrum of colors is not necessarily used in each image.

The different types of colormaps depicted can be advantageous in different applications. For instance, WhiteHot or Ironbow employ a sequential colormap with a uniform distribution between two main colors. In turn, Lava uses the contrast between several colors to enhance subtle differences between temperatures of objects in the scene, whereas the GrayRed alternative employs a divergent color distribution to highlight large temperature gradients in a scene.

(a) WhiteHot (b) Ironbow (c) Lava (d) GrayRed

Figure 6. Examples of thermal imaging color palettes.

Since the principal interest herein is to adapt thermal imaging sensors for wildfire detection and monitoring, the GrayRed color palette was selected. As depicted in Figure 6d, this palette applies high-contrast colors with a divergent color scheme, which is useful to draw attention to the hottest objects in the scene. Furthermore, an extensive color-based data analysis also employing this colormap in the detection of fire situations was presented in previous work [22]. Thus, in the interest of extending the scope of this investigation to the analysis of both color-encoded images and the raw data, the same color palette will be used in this article.

For comparison of alternative automatic gain control techniques, Figure 7 demonstrates the resulting images obtained with the different histogram-based methods for a controlled burn in a real context. The linear and plateau equalization (PE) methods follow the principles explained previously, whereas two additional proprietary FLIR methods that are available through FLIR ResearchIR are also presented, namely digital detail enhancement (DDE) and advanced plateau equalization (APE).

(a) Linear (b) PE (c) APE (d) DDE

Figure 7. Comparison of results of automatic gain control (AGC) algorithms: (a) linear, (b) plateau equalization (PE), (c) advanced plateau equalization (APE), and (d) digital detail enhancement (DDE).

From observing these examples, it becomes clear that advanced contrast enhancement techniques degrade the ability to distinguish the fire from the surroundings. Note that these algorithms were applied with the default configurations, i.e., in their less intense modes, so further tuning of the parameters would deteriorate the image interpretation even further.

In addition to these core AGC methods, the image processing pipeline can also include, for some cameras, an array of proprietary features, such as active contrast enhancement (ACE) and smart scene optimization (SSO). These algorithms also tune the image output, and their parameters can be adjusted manually using manufacturer software applications, but in this study, the default values were used.

3. Data-Driven Thermal Imaging Analysis

Since fires are sources of extreme heat with a strong emission of infrared radiation, thermal imaging sensors can potentially bring significant advantages for fire surveillance applications. However, to implement autonomous systems for early detection and monitoring, i.e., independent of human-in-the-loop approaches, the robotic perception algorithms have to be data-driven.

To provide a preliminary intuition on the behavior of this type of sensor in the advent of a fire ignition, this section starts by introducing an experimental example of a fire detection instance in a controlled environment with static image capture conditions. Subsequently, a feature engineering approach is designed for data-driven situational awareness in fire scenarios.

3.1. Thermal Imaging in Fire Scenarios

The following example presents an illustrative demonstration relating image response, the temperature profile of the scene, and raw measurements to offer some insight into the sensor behavior. The experiment was performed in a laboratory environment with a direct line of sight between the fire and the thermal camera, in this case, the FLIR SC660. Figure 8 presents selected samples from the ignition of a straw fuel bed at different time instances. The corresponding temperature profile, which is represented in Figure 9, was extracted from the radiometric data provided through FLIR ResearchIR software. At this juncture, this example is presented as a preliminary analysis. For further information concerning the experimental setup, refer to Section 4, where the experiments are covered in full detail.

Observing the samples depicted in Figure 8, it should be noted that although the strongest source of thermal radiation is depicted in red, this does not mean it is necessarily fire. Hence, it would be a naive approach to focus only on the detection of the hottest objects in the scene. This is caused by the adaptive nature of the histogram-based algorithms in the image processing pipeline.

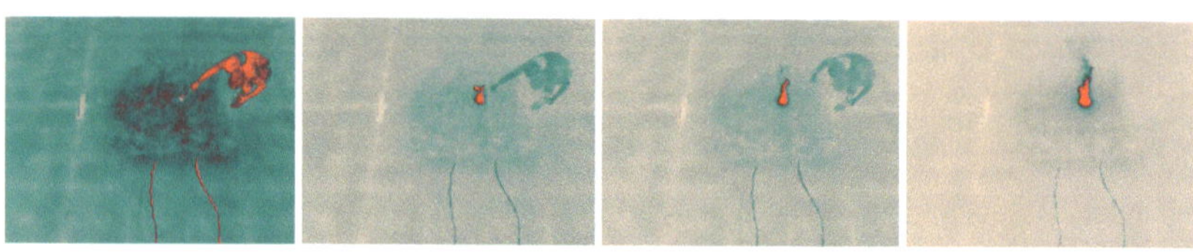

(**a**) Frame 80 (**b**) Frame 112 (**c**) Frame 121 (**d**) Frame 167

Figure 8. Fire ignition detection with a radiometric thermal camera.

In the first frame, right before the ignition, the majority of the frame has green-colored pixels, whereas from the second frame onwards, as the fire starts, the contrast progressively shifts. As a result, the objects with temperatures close to the ambient temperature are subsequently represented by pixels with a whiter shade. It is important to notice that, in this context, the color scale that translates the differences in temperature adjusts with respect to the object in the scene with the maximum temperature. This effect can be observed by comparing the first and second frames. Whereas in the first, the person can be observed in red, in the second, the person is depicted in dark green. This happens because radiation being emitted by the greatest heat source is measured by the sensor with a much higher intensity than the radiation emitted by the person. While the temperature of the person remained unaltered, the range of raw measurements expanded, prompting a significant adjustment in the color scale. The underlying reason for the color evolution is that, at this stage, the body temperature is much closer to the ambient temperature than to the temperature of the fire. Comparing the first and last frames highlights that as the temperature differences increase, the contrast in the image adjusts accordingly.

The temperature profile of the experiment is illustrated in Figure 9, along with the raw statistics computed from the data frames. Note that the temperature profile depicts the maximum and minimum values of temperature extracted from the pixels in the image, as well as the average temperature computed with the maximum and minimum values. Note that the first increase in maximum temperature around frame 80 concerns an ignition that did not fully develop, so a re-ignition was promptly conducted, which can be observed by the increase in maximum temperature from around frame 110 onwards.

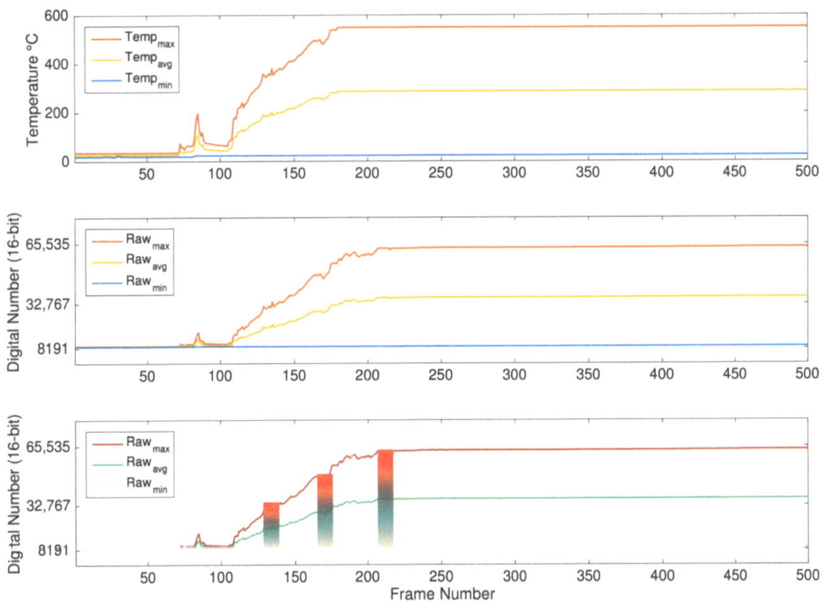

Figure 9. Comparison of temperature (**top**), raw data (**center**), and the evolution of the adaptive color scale (**bottom**).

In juxtaposition, Figure 9 also represents the corresponding evolution of the raw values, including the maximum and minimum raw values extracted from the image, as well as the average between these two metrics, to illustrate how the color scale adjusts as the dynamic range extends. Note that this depiction does not take into account the application of histogram-based algorithms, in the sense that it only translates a linear application of the GrayRed colormap to the raw values. Nevertheless, it provides a broad intuition regarding the adaptive color scale while relating the raw measurements and the temperature data with the pseudocolor images presented in Figure 8.

Since the camera was configured to measure up to 500 °C for this trial, it is noticeable by observing both graphs that the sensor saturates when that temperature is reached.

The analysis of these variables highlights the importance of testing the thermal imaging cameras in real operation scenarios and characterizing the behavior of these sensors in extreme conditions, as they may exhibit differences depending on the camera model. In the case of the FLIR Vue Pro in image mode, the values in the metadata match the maximum and minimum values of the raw frames.

Therefore, in order to employ thermal imaging sensors for situation awareness in fire detection and monitoring scenarios, it is important to translate the insights from this analysis into meaningful features for robotic perception, as will be explored in the following.

3.2. Feature Engineering

Autonomous robotics require real-time processing of sensor data for navigation and guidance, e.g., to enable sensor-driven planning, as it is crucial for the robot to have precise estimates of its position and attitude, as well as its surroundings, in order to plan its trajectory on-line. Even though emerging on-board computing platforms have higher processing capabilities, to incorporate powerful exteroceptive sensors capable of generating high-resolution samples of the surrounding environment, it is essential to handle the burden of data dimensionality. Furthermore, considering the required high sampling frequencies suitable for these tasks, mobile robotics applications are increasingly data-intensive; thus, feature engineering plays a prominent role in deriving efficient algorithms.

In light of the insights exposed in the previous sections, the following presents a feature engineering approach for situation awareness using thermal imaging, which combines feature extraction from raw data and pseudocolor images to yield relevant features for robotic perception in fire scenarios.

3.2.1. Raw-Based Features

The feature extraction approach for raw data is based on the construction of features that aim to preserve the intuition of the actual working mechanism of thermal imaging sensors and the histogram algorithms of the image processing pipeline. Recalling the examples presented in Section 2.2.1, the raw-based features will explore the maximum, minimum and average values of the range of the raw data frames. As explained in the metadata description in Section 2.2.2, these variables can be computed according to Equations (6) and (7), but since these do not always match the metadata tags for the cameras tested, it is advisable to extract these variables on-line from the raw data frames.

3.2.2. Color-Based Features

To extract features that translate the evolution of the color statistics of the image in fire scenarios, as observed in Figure 8, this work employs the color segmentation heuristic proposed in [22]. The approach divides the GrayRed color palette into three parts according to the three main colors that compose this palette: gray, green and red. The complete color scale that comprises 120 distinct color representations defined in the RGB color space is partitioned with the division illustrated in Figure 10.

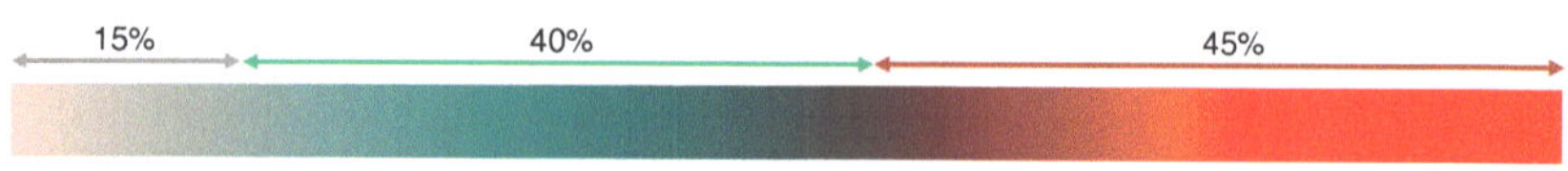

Figure 10. Division of the color scale [22].

The first segment, represented in gray, corresponds to lower raw values, i.e., lower temperatures, and is defined by 18 color levels, which represent 15% of the color scale. The green segment is defined by 48 color levels, which represent the mid-range of raw values, corresponding to 40% of the color palette. Since this color palette is designed to draw attention to the hottest elements in a scene, the largest portion, corresponding to 45% of the color scale, is dedicated to 54 red color levels. Although these colors exhibit high contrast between them, color gradients between each color are low. The segmentation is performed according to the parameters in RGB color space defined in Table 2.

In contrast with strictly data-based approaches, the proposed feature engineering method allows the incorporation of expert knowledge about the sensor by deriving features from the information retrieved from raw data in combination with the data-driven color segmentation heuristic that describes the behavior of the sensor over time. Furthermore, these features are designed with data interpretability in mind, which is especially important for safety-critical applications, such as wildfire detection and monitoring systems.

Table 2. RGB segmentation thresholds for feature construction [22].

Features	Gray			Green			Red		
channels	R	G	B	R	G	B	R	G	B
upper limit	253	199	185	143	169	157	255	73	71
lower limit	149	171	160	98	90	86	103	89	85

In order to study the response of thermal imaging sensors, a comprehensive set of experiments was conducted for analysis of the sensor image processing pipeline and its application in real scenarios, which are presented in the following.

4. Controlled Fire Experiments

This section presents a set of controlled fire experiments performed in laboratory and field contexts. The trials comprise the burning of wild fuels, such as straw, common heather (*Calluna vulgaris*) or Baccharis (*Baccharis trimera*), and artificial materials present in a caravan that was also experimentally burned. In the following, each section describes the experimental setup and conditions used in the trials, as well as the means employed for data acquisition.

4.1. Laboratory Test

The laboratory experiment enables testing of the response of the camera to a fire ignition in a controlled environment, allowing a high-resolution sampling of the phenomena under static environmental conditions. Since the test was performed indoors, there was no influence of wind. The horizontal fuel bed was composed of straw with a fuel load of 600 g/m^2 and a moisture content of 13%. The environmental temperature and relative humidity were 20 °C and 78%, respectively. By using these testing conditions, the behavior of the sensor and the subsequent image processing is not affected by external factors.

For data acquisition purposes, the FLIR SC660 was positioned on top of an elevated platform with a direct line of sight to the fire, as depicted in Figure 11. The line of sight made an angle of 45° with the horizontal plane. In this setup, the data were processed and recorded in real-time using FLIR ResearchIR software through a cable connection to a desktop computer.

(**a**) Experimental setup (**b**) Top view of the fire in the straw fuel bed

Figure 11. Laboratory trials with the FLIR SC660 mounted on an elevated platform.

The objective of this type of test centers on capturing the transition to a fire scenario; thus, it requires a high frame rate. In this case, the device was configured to a 15 Hz frame rate.

4.2. Caravan Burning Test

The caravan burning test comprised the burning of man-made fuels, unlike the other trials, which only included natural fuels. This scenario is important to take into account, especially for airborne fire surveillance in wildland-urban-interface regions. In addition, due to the presence of some materials like plastic, the temperatures expected are much higher than those registered for wild fuel burning. The test was performed with average values of relative humidity of 75% and temperature of 23 °C. The wind velocity was below 2 m/s.

Like in the previous trial, the device used for data acquisition was the FLIR SC660 operated in connection with FLIR ResearchIR software with a 15 Hz frame rate. In this case, the camera was mounted on a tripod at ground level at approximately 10 m from the caravan, which is depicted in Figure 12.

(**a**) Before flashover (**b**) After flashover

Figure 12. Recording of the caravan burning trial.

To study this case, as fire-related temperatures reached higher levels, the automatic gain control configuration was changed manually mid-trial to expand the measurement range to cover higher temperatures.

In this trial, unlike in the previous example, there was no direct line of sight to the fire at the point of ignition, which occurred in the interior of the caravan, as can be observed in the visible range image presented in Figure 12a. Indeed, only after the flashover, i.e., the point where the heat and pressure build-up reached the maximum immediately before breaking the caravan wall, the fire started to be visible in the visible range.

4.3. Summer Festival Trials

To design algorithms with robustness in real-world contexts, it is crucial to analyze data acquired in highly dynamic scenarios. In that sense, a set of tests was performed at the venue of a festival outdoors in midsummer to emulate conditions of real operation of fire safety systems. The trials were conducted in a region with dry weather in the summer, with strong winds and high temperatures, when conditions of sustainable ignition propensity were consequently high as well. More specifically, during the period of the trials, the following meteorological conditions were registered: minimum humidity of 17%, maximum wind speed of 21 km/h, and temperatures up to 36 °C.

The tests performed encompassed the aerial surveying of the festival area at several different locations. Figure 13 illustrates part of the venue surroundings, and provides some detail on the nature of the camping areas. As can be observed, the festival occurred in the fire-prone area, where the vegetation of trees and shrubs is very combustible. In addition to the vegetation, the inclusion of more combustible materials, such as tents and caravans, increases fire risk, which is exacerbated by the human presence, raising fire safety concerns.

To provide aerial monitoring for several hours to detect possible fire ignitions, the thermal camera was installed on a tethered helium balloon, as depicted in Figure 14. With this setup, the experiments were conducted at various locations of the venue to test distinct conditions. There were no fire events in these field tests, so for the purposes of this study, a fire in the rocket stoves of the community kitchens was used instead.

Concerning image acquisition, due to the limited payload budget of the balloon, only the FLIR Vue Pro has suitable weight to be installed onboard. The camera was programmed to capture images with a 5 s time-lapse, recording both the raw data and the pseudocolor images encoded with the GrayRed color palette in the camera storage.

In contrast to previous trials, where the acquisition conditions were static, since these experiments employed an aerial platform without active means of actuation for stabilization, the dynamics of the balloon influenced the image acquisition by varying the image field of view as a result of wind disturbances.

(a) Festival venue (b) Camping areas

Figure 13. Festival venue and camping areas during the event where the trials were conducted.

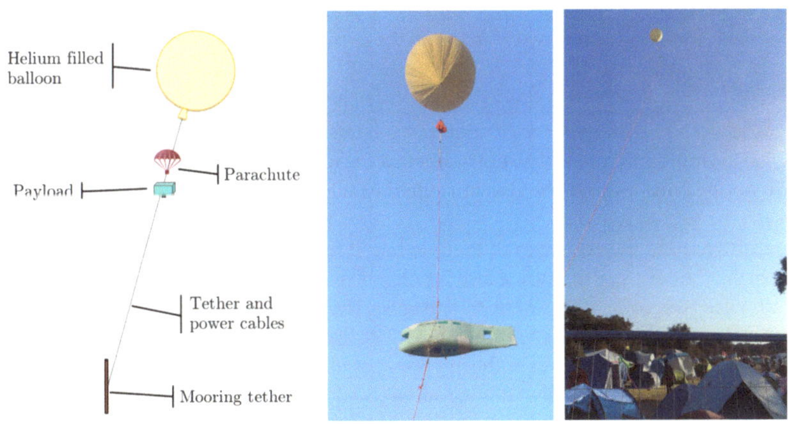

Figure 14. Thermal camera setup on tethered helium balloon.

4.4. Mountain Range Field Trials

In addition to the previous experiments, several tests were performed in a real-world scenario in mountain range field trials to test the capabilities of the sensors in a long-distance monitoring scenario. For these tests, the main interest was to evaluate the performance of a compact thermal camera, namely the FLIR Vue Pro, for future airborne monitoring operations. The trials were undertaken at an altitude of over 1000 m with the following meteorological conditions: wind speed in a range of 13–16 km/h from the northwest, average relative humidity of 45%, and average temperature of 21 °C.

Figure 15 illustrates the experimental conditions of the trials. The field experimental burns were conducted on a hill with an average slope of 25%, and comprised the burning of shrubland vegetation, including common heather (*Calluna vulgaris*) and Baccharis (*Baccharis trimera*).

Concerning the image acquisition, in the trials, both thermal cameras were employed, but given the aforementioned objective of these experiments in particular, this work only explores the data from the FLIR Vue Pro. The images were captured at ground level from across the valley at around the same altitude as the field burnings, with a direct line of sight between the camera and the fire at a distance of approximately 600 m. The data acquisition was configured to record the data in the camera storage, saving both the raw data and RGB-encoded images with the GrayRed color palette at a rate of one frame per second.

(a) Perspective view

(b) Top view

Figure 15. Experiments in mountain range field trials.

5. Results and Discussion

This section presents the results of the application of the data-driven thermal imaging analysis to the controlled fire experiments performed in laboratory and real contexts according to the specifications presented in Section 4. As previously introduced, the proposed approach explores raw-based features, as well as the color-based features derived from the segmentation heuristic proposed in Section 3.

5.1. Laboratory Test

In the laboratory trial, under static conditions of image acquisition, it was possible to record with FLIR SC660 the transition from a situation without a fire to a fire scenario through a high-rate sampling that includes the point of ignition. This procedure enables capturing the influence this phenomenon produces in the data acquired.

Figure 16 depicts the evolution of the raw-based features, illustrating how the dynamic range expands as the fire develops. In turn, it also intuitively demonstrates how the variation of the dynamic range influences the color statistics of the images. The sample frames concern the transient stage immediately before and after the start of the fire. Note that the first small peak around frame 80 occurs in a first attempt at the ignition, but as the fire does not fully develop, the straw was reignited. The fast response of the sensor indicates that it is sensitive enough for this type of application.

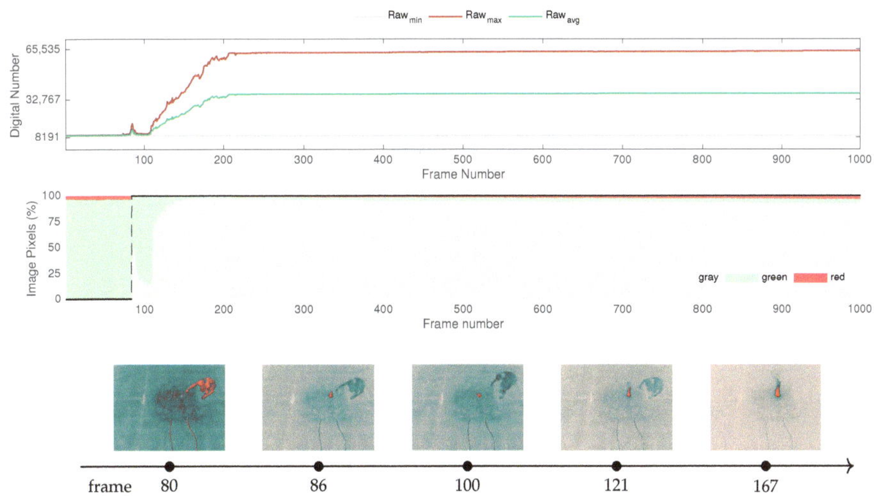

Figure 16. Feature-based representation of the sensor response of the thermal camera to a fire ignition: raw features (**top**); color features (**center**); sample frames (**bottom**).

In Figure 16, it is possible to identify that the sensor undergoes three distinct states: (i) before the point of ignition; (ii) transient state; (iii) steady-state. As indicated in the figure, at first, the environment is in equilibrium, with generally green levels corresponding to the mid-range. Note that the dynamic range in this state is narrow, since the incoming radiation originates predominantly from objects at ambient temperature.

In the second stage, which is prompted by the fire ignition, the camera detects different strengths of incoming radiation corresponding to a severe temperature gradient and, as a result, the dynamic range starts increasing. This drives the adaptation of the color representation according to the histogram-based algorithms presented in Section 2.2. Consequently, the increase in the percentage of gray pixels is prompted by the fire ignition, represented in red, since it is the hottest object in the scene and has a strong emission of infrared radiation, whereas objects at ambient temperature remain at the previous raw level.

In the third stage, a new equilibrium state emerges after the sensor stabilizes shortly after the beginning of the fire, at approximately frame 167. However, in these circumstances, due to the extended dynamic range, the color statistics have a considerably different representation with a high percentage of gray shades in the image context.

For this case, it is important to take into account that this experiment captures the image sequences with a frame rate of 15 Hz; thus, it is possible to record the transition between the equilibrium states. For cases with lower sampling frequency, the data may not capture this behavior, but only similar conditions resembling the first and third states.

Note also that although the dynamic range adapts over time, its measurement range is not altered; as such, the complete bit resolution is used in this configuration. Accordingly, since the measurement range is set between −40 and +550 °C, the raw measurements saturate at a digital number that would correspond to these temperatures; in this case, these match the full bit resolution from 0 to 65,535.

5.2. Caravan Burning Test

In the caravan burning test performed in an outdoor environment, as described in the experimental setup, the image acquisition conditions were also kept static, with the FLIR SC660 configured to capture image sequences at 15 frames per second. While this setup also allows for a high sampling of the phenomenon, in this case, the fire ignition is not in direct line of sight because it occurs inside the caravan. For this reason, it is also necessary to characterize how these circumstances impact the behavior of the thermal cameras in such scenarios. Moreover, the measurement range of the camera was adjusted mid-trial to encompass an extended range of temperatures. Consequently, this allows the evaluation of how the scaling of the bit resolution affects the raw digital data values.

However, before delving into the trial results, the effect of automatic temperature-dependent non-uniformity correction (NUC) should be addressed. This method consists of an internal process employed in thermal cameras to deal with spatial non-uniformity (NU) noise, which produces a fixed pattern over the thermal images that varies in intensity due to the internal temperature instability of such cameras [24]. To that end, on occasion, the camera performs a NUC reset, which results in a disturbance in the data acquisition process. Figure 17 exemplifies an instance illustrating this behavior along with relevant sample frames.

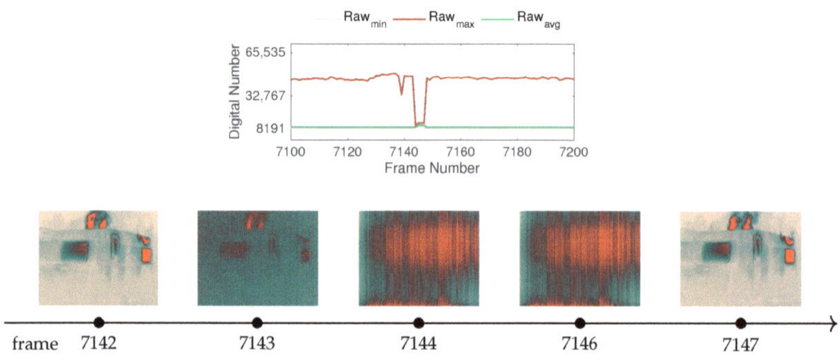

Figure 17. Effect of temperature-dependent non-uniformity correction (NUC), represented by selected samples of the raw data (**top**), along with the respective color-encoded frames (**bottom**).

In the selected window of the raw-based features depicted in Figure 17, it can be observed that when the camera triggers the NUC reset in frame 7143, the raw values exhibit a sudden drop, which continues for several instants until frame 7146. Given that this behavior is sporadic, but it is expected to occur, as was verified for the case in the caravan burning test, appropriate filtering of this type of frame should be included in the image processing pipeline in automatic systems, particularly if integrated in autonomous navigation of robotic platforms for early detection and monitoring. In this work, a median filter was used with a window of 15 frames, which introduces a one-second delay in the case of the FLIR SC660, since this camera operates at 15 Hz.

Figure 18 illustrates the variation of the features extracted from the raw data in comparison with the color features derived with the color segmentation heuristic for images encoded with a linear histogram approach (L) and the plateau equalization algorithm (PE) in the second and third graphs, respectively. The data presented in the graphs were pre-processed to filter out the effect of the NUC frames. In this example, we present the sensor response for two automatic gain control methods: linear (L) and the plateau equalization (PE).

Observing Figure 18, the adjustment of the measurement range from −40 to +550 °C to the upper temperature limit of +1500 °C that occurs at frame 3406 has a prominent impact on the raw data, which can be identified by the steep drop in the raw features at frame 3406. Notably, the scaling of the bit resolution amplifies the measurement range, but at first, this results in a narrow distribution of the raw features at lower bit values.

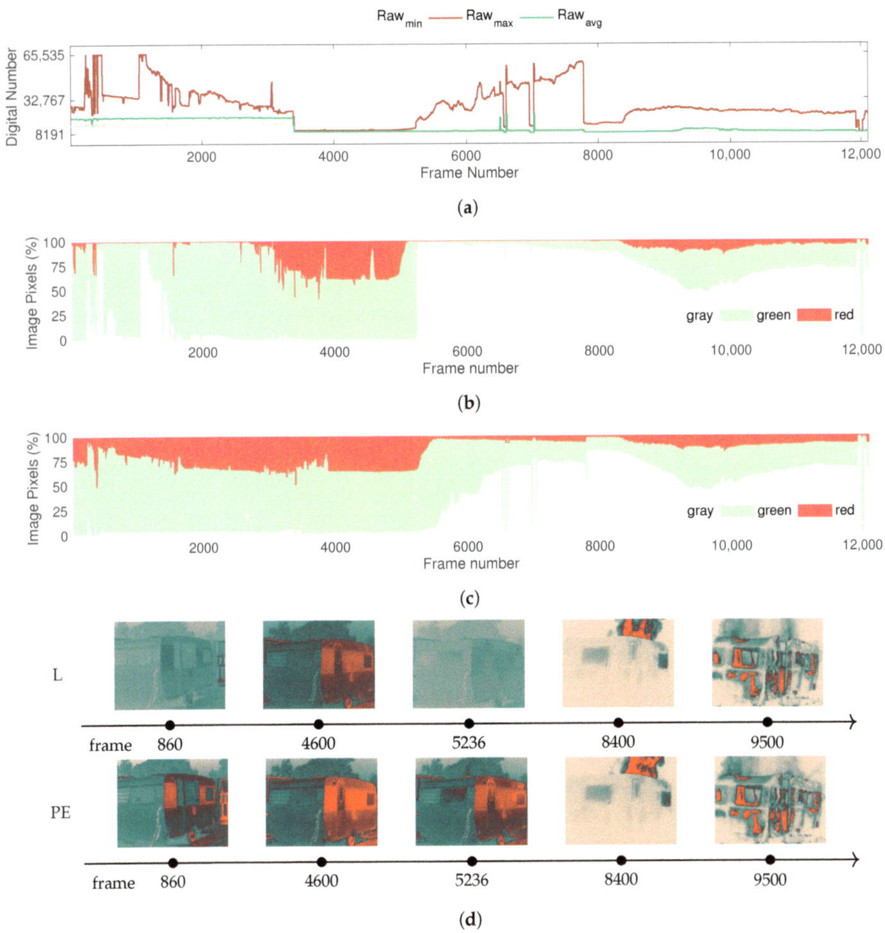

Figure 18. Representation of the filtered sensor response of the thermal camera to a fire ignition inside the caravan: (**a**) raw variables; (**b**) color-based features with the linear AGC (L); (**c**) color-based features with the plateau equalization AGC (PE); (**d**) comparison of frames between both approaches.

However, as the fire starts around frame 5100, the dynamic range begins to expand. With the evolution of the intensity of the fire, the dynamic range extends to comprehend higher values within the available resolution, reaching a maximum of 52,343 at frame 7736, corresponding to a maximum raw value of 60,978 and minimum raw value of 8635.

Subsequently, the sudden drop at frame 7782 indicates a second adjustment of the measurement range. The flashover point occurs at frame 8310, and as a result, the fire starts burning the exterior of the caravan with greater intensity, as can be seen in the corresponding frames (8400 and 9500). In addition, it can also be verified that with the fire visible in plain sight, the sensor response tends to stabilize.

Concerning the color-based features, the effect of the fire starts to be detected by the sensor around frame 5236, as indicated by the increase in the gray pixel percentage and as had been verified in the previous test. This occurs precisely when the fire burns a small hole in the upper right side of the exterior of the caravan, as depicted in frame 5236 (see the frame from the linear approach). In frame 8400, the effect after the flashover can be observed, where the fire can be seen in plain sight, causing the contrast in the scene to considerably increase. Even with mid-trial adjustments in the measurement range, the behavior of the color segmentation heuristic is generally stable.

Comparing the response in terms of color features for the linear (L) and plateau equalization (PE) algorithms, it becomes clear that the linear approach can detect the beginning of the fire in a prompt manner, as can be observed by the abrupt increase of the percentage of gray pixels. Conversely, for the plateau equalization algorithm, the response occurs more gradually, but it is still possible to detect the fire with this color segmentation heuristic. However, this difference in terms of image-based features is an important aspect to consider when developing intelligent fire detection systems to operate in real-time.

Following the analyses presented regarding tests performed in controlled environments, to validate the insights exposed, the next sections concern field trials undertaken in real contexts.

5.3. Summer Festival Trials

In the tests performed at the summer festival, the FLIR Vue Pro was installed onboard a tethered helium balloon, as illustrated previously in Figure 14. Although the movement of the balloon enabled surveying a wider area of the venue, for several periods of up to 2 h long, given that this unactuated aerial platform moves according to the direction of the wind, the image capture is considerably less stable since the field of view of the camera drifts due to wind disturbances. Due to payload restrictions, there was also no active means of camera stabilization; therefore, the sensor response for this set of data will inherently be more stochastic.

Notwithstanding, the proposed feature engineering approach is theoretically viable for images acquired from mobile platforms because it employs features extracted from each frame, but does not rely on temporal dependencies. Hence, the analysis of these experiments is paramount for validating the applicability of this technique for integration in robotic perception pipelines for mobile platforms in the context of wildfire surveillance tasks.

To that end, from several tests carried out in different parts of the festival venue, two examples were selected to showcase representative situations encountered in real contexts. The first example, presented in Figure 19, comprises a test in which fire situations were not detected, whereas the second case, depicted in Figure 20, covers areas without fire and also parts of the venue where fire was detected, namely in the community kitchens where rocket stoves were installed. The images were recorded with a 5 s time lapse between frames at an altitude from the ground of approximately 80 m.

Observing Figure 19 with respect to the evolution of the raw-based features, it becomes clear that under normal conditions, the sensor of the camera does not exhibit significant variations. Considering that the measurement range would allow the sensing of incoming infrared radiation up to a dynamic range of 16,383, i.e., the upper bound saturation level, the data demonstrate that in scenarios without fire, the maximum and minimum raw values do not exceed 8841 or go below 7812, respectively. Note also that in these trials, ambient temperatures reached up to 36 °C, so it is expected that the surfaces of objects captured in the field of view of the camera may have exceeded that value due to the different absorption properties of materials in such a heterogeneous context. Hence, including the raw-based features in detection and monitoring algorithms is rather promising for avoiding false alarms.

Conversely, since the festival took place in the summer, it is also normal to capture objects for cooling purposes that are at temperatures below the average ambient temperature. While this does not seem to have a considerable effect in the magnitude of the raw values, this effect can be observed according to the representation of the color-based features over time. More specifically, as different areas of the venue were surveyed, the color statistics vary considerably, predominantly influencing the red and green pixel percentages. For instance, in frame 365, it is possible to observe that a small set of items at cool temperatures caused the adaptation of the color representation, making the surroundings become depicted in red, even though there was not a sudden increase in temperature. In this way, since the histogram-based algorithms promote highly nonlinear effects in image data, color-based features require careful interpretation.

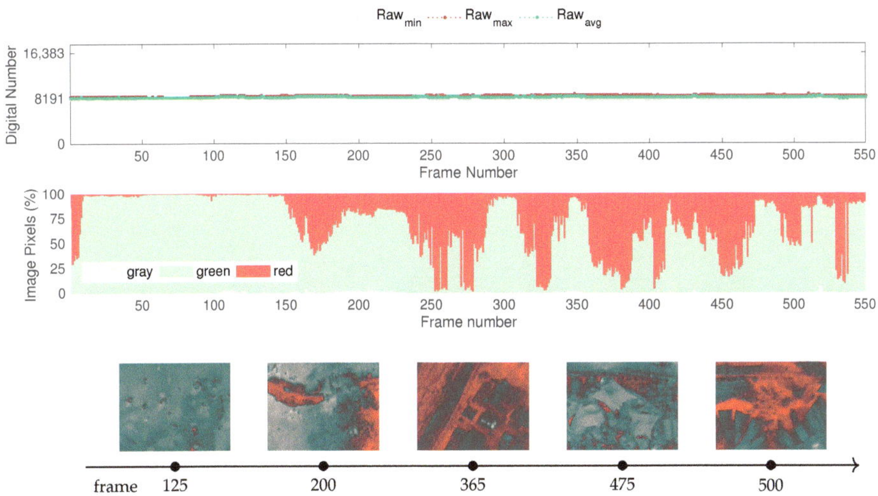

Figure 19. Results for data acquired when surveying the festival venue: raw features (**top**); color features (**center**); sample frames (**bottom**).

In turn, the presented sample frames also demonstrate that the appearance of red hotspots can be illusive, as it does not necessarily mean that fire sources exist in the scene, but rather that these areas are slightly above the average temperature. The representation of the color features takes this into account, since it considers in the color segmentation that the upper 45% of the scale is represented by red-shaded tones, as illustrated in Figure 10. The uneven distribution of colors in the GrayRed color palette also justifies the behavior of the sensor mentioned in the previous comment, particularly regarding how the presence of cool objects in the scene easily makes red the most dominant color in the image.

Concerning the second case represented in Figure 20, which includes situations at the venue where fire was detected, the behavior of the sensor was consistent with that in the controlled fire scenarios discussed previously. In normal situations, the raw features followed the same behavior of the baseline case presented in Figure 19, but when covering the community kitchens area, the dynamic range expanded due to the presence of fire sources. More specifically, regarding the complete sequence, the maximum raw value reached 12,477, the minimum raw value was 7918, and the maximum amplitude of the dynamic range, recorded in a fire detection instance, was 7290, which is well above the variations registered in the baseline case.

Accordingly, the color representation also suffers an abrupt adjustment, with the surroundings becoming depicted mostly by gray pixels. This effect can be attested in the accompanying sample frames illustrating the first detection instance, which were captured while covering the community kitchens area. Since fire was strictly forbidden in the festival premises due to fire safety concerns, with the exception of specific fires lit and controlled by the organization for scenic purposes and the community kitchens area, the remaining detection points in this sequence also concern the same area illustrated in Figure 20.

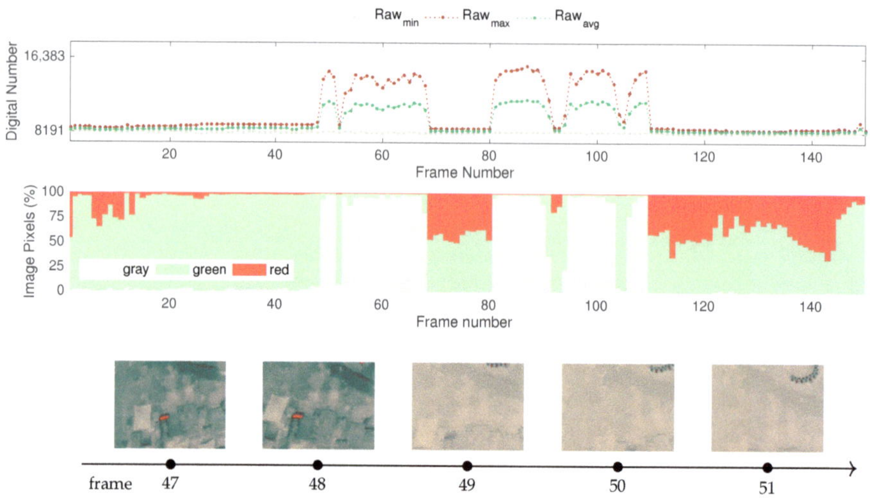

Figure 20. Results for data acquired when surveying the community kitchens area: raw features (**top**); color features (**center**); sample frames (**bottom**).

Considering that in these field trials, the images were captured from a considerably longer distance than in the laboratory tests, the data demonstrate that the color scale adapts even when hot objects have low spatial resolutions, which is essential for allowing early fire detection in aerial surveillance scenarios, therefore confirming the validity of the proposed approach for fire detection using mobile platforms.

Furthermore, recalling that the FLIR Vue Pro has a more limited measurement range, with an upper saturation level at 150 °C, this may raise some concerns regarding false alarms. However, under real operation conditions, no false detections were encountered.

5.4. Mountain Range Field Trials

In the field trials conducted in a mountain range, the images were captured from a longer distance than in the previous tests. The purpose of these tests was to evaluate a key issue regarding the use of compact thermal cameras, particularly concerning their capabilities when operating outside the nominal conditions specified in the official hardware datasheets. The field trials performed allow the evaluation of the behavior of the camera in a long-distance image capture setup to assess if the signal-to-noise ratio in this scenario is sufficient for fire detection applications. In this case, the tests were performed with the FLIR Vue Pro, which was configured to record images with a 1 s time-lapse at a distance of approximately 600 m, as specified in the description of the experimental setup (Section 4.4).

From the several tests performed in the mountain range, two cases were selected for analysis, encompassing a situation with an active fire, which is presented in Figure 21, and another in the aftermath of a fire, depicted in Figure 22.

Regarding the monitoring of an active fire scenario, it can be observed in Figure 21 that the statistics of the raw variables do not exhibit particular variation, remaining with an extended dynamic range for the several minutes featured in this trial. Notably, despite the long distance, the sensed infrared radiation does not suggest a significant signal attenuation, as the sensor bit resolution is used to a great extent. Indeed, the maximum raw value registered is the sensor upper saturation level at 16,383, while the minimum raw value recorded is 7235, which corresponds to pixels representing the sky in this case, as can be confirmed in the sample frames.

For this case, the evolution of the color-based features does not reveal particular changes and is consistent with the insights exposed in the previous trials. The sensor response stays generally constant, with slight variations attributed to the change of field of view of the camera. The gray and green levels dominate the image context, and a reduced percentage of red pixels depict the area of the fire, as can be observed in the sequence of frames selected.

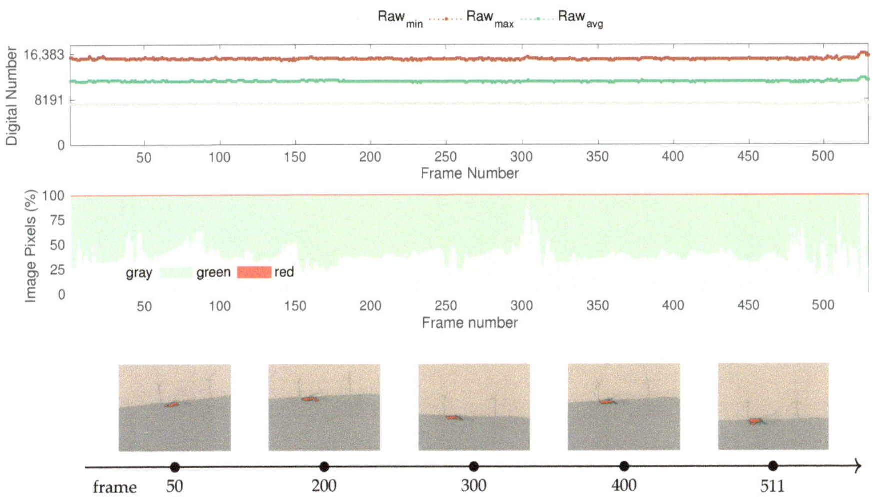

Figure 21. Results for fire detection at long distance in a mountain range: raw features (**top**); color features (**center**); sample frames (**bottom**).

Although the FLIR Vue Pro was not specifically designed for this type of task, these results validate the capability of detecting fire sources at long distances. In turn, an important aspect to take into account when considering extended distances is that due to the limited resolution of the focal plane array and, consequently, the output image resolution, with the increase in distance, the minimum size of the fire that can be detected naturally increases as well.

Since wildfires often extend over large periods and can be inactive after firefighting operations take place, but re-ignite before the fire event is completely extinct, the monitoring of areas in the fire perimeter is important for preventing rekindles. In this context, the ground is usually at higher temperatures, adding difficulty in pinpointing potential re-ignition spots.

The following test, performed in the field trials and targeting the aftermath of a controlled fire, which is depicted in Figure 22, aims to uncover how these circumstances differ from the previously studied cases.

Observing the time-series of the raw-based features, it is immediately noticeable that the dynamic range is reduced once again, but with visible gaps between the minimum, average and maximum values, unlike in the baseline case encountered in the festival trials. In contrast, in this case, the maximum raw value extends as high as 9782, the minimum raw value remains around 8141, and when the dynamic range is greater at 2426, the average raw value is 8569.

Concerning the adaptation of the color-based features, it should be noted that the field of view of the camera was changed considerably to capture distinct parts of the monitoring area, as illustrated by the set of sample frames presented in Figure 22. In this case, the green and gray percentages are predominant when capturing the burned area, which is depicted in red in frames 13 and 60 because it is at a higher temperature than the surroundings. The sky area remains depicted in gray; however, in contrast with the situation with an active fire, the ground is represented by a considerably darker tone of green. In that sense, the color-based features alone do not evidence a difference from the previous case, which highlights that the color segmentation heuristic has to be fine-tuned depending on the image capture setup or defined with a more restrictive range for the green class. Interestingly, when the burned area is not in the field of view of the camera, the ground is generally depicted in red, as can be observed in frames 51 or 94.

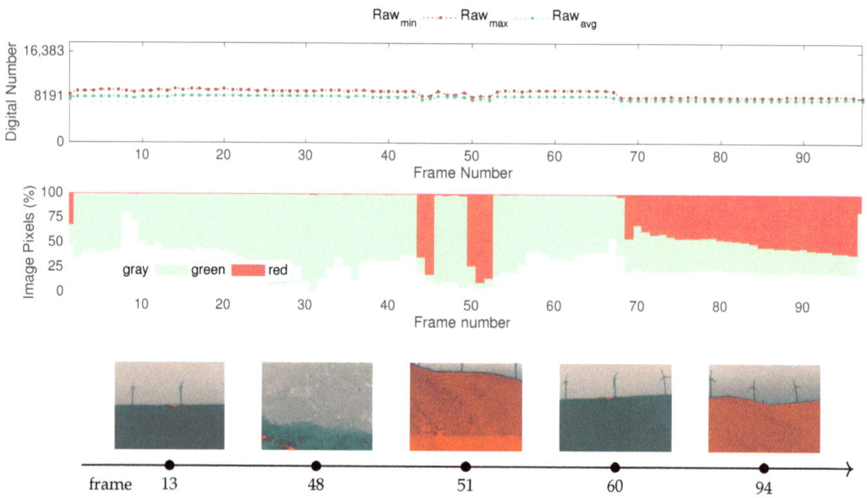

Figure 22. Results for the fire aftermath captured at long distance in a mountain range: raw features (**top**); color features (**center**); sample frames (**bottom**).

In turn, when facing downward on the valley, as illustrated in frame 48, the distance influences the color representation, with closer objects represented by the bottom part of the image, and the gray part representing the mountain on the other side of the valley. This result shows that in addition to being capable of detecting fire at long distances, the thermal sensitivity of this sensor is also able to convey depth perception. While this does not play a significant role in the detection in itself, it has interesting implications for autonomous navigation by robotic platforms.

5.5. Discussion

From the analyses presented throughout this study, it becomes evident that thermal cameras can potentially bring improved reliability for fire detection and monitoring systems in several aspects in which visible range systems have limitations. On the one hand, the proposed data-driven feature engineering approach that combines expert knowledge on the sensor working principles with color-based features derived from processed image data was demonstrated to be robust to false alarms, even when tested in highly dynamic environments. On the other hand, it was possible to validate the applicability of commercial off-the-shelf camera models in long-distance scenarios for fire detection applications, as verified with the onboard compact thermal camera.

However, adapting thermal imaging cameras for wildfire detection and monitoring systems entails an adequate understanding of the application requirements and capabilities of each device, which change depending on manufacturers and camera models. Furthermore, considering UAV-based applications, the integration of this type of device in robotic perception frameworks is not straightforward, requiring knowledge of the underlying image processing algorithms, which are highly nonlinear. For this reason, based on insights gained from the comprehensive analyses presented, several remarks deserve to be highlighted.

This work explored a series of automatic gain control methods available in the firmware and software of thermal cameras, which are based on histogram-based algorithms for data compression and contrast enhancement. In particular, testing the applicability of several alternatives in fire scenarios, the results revealed that advanced contrast enhancement techniques degrade the ability to distinguish the fire from the surroundings, preventing an accurate identification of the burning area. Indeed, the linear approach registered the best results, as identified in Figure 7.

Since histogram-based algorithms promote highly nonlinear effects in image data, color-based features require careful interpretation because tracking the hottest areas can be misleading depending on the scene context. In that sense, extending state-of-the-art color-based approaches [22] by also including raw-based data provides enhanced situation awareness, thus potentially enabling improvements in robotic perception.

By testing the proposed method in distinct scenarios, under static image acquisition conditions, and also from onboard mobile platforms (e.g., a tethered helium balloon), it was possible to validate the approach in both contexts. Nonetheless, for integration in robotic platforms, it is advisable to employ means of camera stabilization to minimize the effect of the external disturbances due to the movement of the platform and consequent changes in the field of view of the camera.

In turn, the operating distance conditions of these sensors play an important role when applying heuristic methods and, as such, the performance shall be fine-tuned depending on the image capture setup. Moreover, the characterization of the device in terms of sensor response is fundamental for this type of application because although manufacturers define nominal operating conditions, since fire is a source of extreme temperatures, it is possible to successfully detect fire at long distances as well, since a high signal-to-noise ratio is preserved.

Regarding the integration in robotic perception pipelines for autonomous navigation of UAVs, it was also verified that the thermal sensitivity of these sensors is also able to convey depth perception. However, in order to apply the methodology proposed in this work, the camera model has to provide access to the raw data on-line so that it can be incorporated and used in the image processing pipeline of the robotic perception framework.

6. Conclusions

The study conducted in this work provides a comprehensive introduction to adapting thermal imaging sensors for wildfire detecting and monitoring systems. Since these phenomena are increasing in intensity and frequency, there is increased urgency in the early detection, monitoring and surveillance of these events. For these reasons, leveraging mobile robotics in this type of extreme environment requires devising of application-specific techniques.

For that purpose, this paper explored the foundations of image processing of thermal imaging, covering the most used automatic gain control methods and the color mapping schemas responsible for generating thermal images encoded in pseudocolor.

The proposed data-driven feature engineering approach, combining expert knowledge about thermal imaging sensors as well as color-based segmentation heuristics, offers improvements in situation awareness in comparison to existing methods focusing only on one data type.

The study of the response of thermal imaging sensors in fire scenarios as well as situations without fire considering raw sensor data and color-based features was revealed to be fundamental for providing a better intuition of the working mechanisms of these sensors and the expected behavior under such conditions. By evaluating how the features evolve over time, these analyses provided a clear understanding of the nonlinear adaptive color scale, providing important insights relative to the sensing capabilities of commercial off-the-shelf thermal cameras, and thus establishing a solid foundation for the development of robotic perception algorithms.

Considering that wildfire detection and monitoring are safety-critical applications, there is a great benefit in exploring highly interpretable feature engineering approaches in order to subsequently combine these methods with intelligent soft sensor approaches, enabling better model understanding and interpretation.

Author Contributions: Conceptualization, M.J.S., A.M., and M.A.; methodology, M.J.S.; software, M.J.S.; validation, M.J.S., A.M., and M.A.; formal analysis, M.J.S.; investigation, M.J.S., A.M., and M.A.; resources, M.J.S., A.M., and M.A.; data curation, M.J.S.; writing—original draft preparation, M.J.S.; writing—review and editing, M.J.S., A.M., and M.A.; visualization, M.J.S.; supervision, A.M. and M.A.; project administration, A.M.

and M.A.; funding acquisition, A.M. and M.A. All authors have read and agreed to the published version of the manuscript.

Funding: This work was financed by national funds through the FCT – Fundação para a Ciência e Tecnologia, I.P., through the IDMEC under the project Eye in the Sky (PCIF/SSI/0103/2018), and through the IDMEC under LAETA, project UIDB/50022/2020. M.J. Sousa acknowledges the support from the FCT through the Ph.D. scholarship SFRH/BD/145559/2019.

Acknowledgments: We thank Good Mood—Eco & Art Org for facilitating part of this study in the Boom Festival venue (https://boomfestival.org/).

Conflicts of Interest: The authors declare no conflict of interest.

References

1. San-Miguel-Ayanz, J.; Durrant, T.; Boca, R.; Libertà, G.; Branco, A.; de Rigo, D.; Ferrari, D.; Maianti, P.; Vivancos, T.A.; Tomas, A.; et al. *Forest Fires in Europe, Middle East and North Africa 2018*; Publications Office of the European Union: Luxembourg, 2019.
2. Spreading like wildfire. *Nat. Clim. Chang.* **2017**, *7*, 755. [CrossRef]
3. Molina, J.R.; González-Cabán, A.; y Silva, F.R. Potential Effects of Climate Change on Fire Behavior, Economic Susceptibility and Suppression Costs in Mediterranean Ecosystems: Córdoba Province, Spain. *Forests* **2019**, *10*, 679. [CrossRef]
4. Christensen, B.R. Use of UAV or remotely piloted aircraft and forward-looking infrared in forest, rural and wildland fire management: Evaluation using simple economic analysis. *N. Z. J. For. Sci.* **2015**, *45*. [CrossRef]
5. de Dios, J.R.M.; Merino, L.; Ollero, A.; Ribeiro, L.M.; Viegas, X. Multi-UAV Experiments: Application to Forest Fires. In *Springer Tracts in Advanced Robotics*; Springer: Berlin/Heidelberg, Germany, 2007; pp. 207–228. [CrossRef]
6. Merino, L.; Caballero, F.; de Dios, J.R.M.; Maza, I.; Ollero, A. An Unmanned Aircraft System for Automatic Forest Fire Monitoring and Measurement. *J. Intell. Robot. Syst.* **2011**, *65*, 533–548. [CrossRef]
7. Ambrosia, V.G.; Hinkley, E. NASA Science Serving Society: Improving Capabilities for Fire Characterization to Effect Reduction in Disaster Losses. In Proceedings of the IGARSS 2008—2008 IEEE International Geoscience and Remote Sensing Symposium, Boston, MA, USA, 7–11 July 2008.
8. Ambrosia, V.G.; Wegener, S.; Zajkowski, T.; Sullivan, D.V.; Buechel, S.; Enomoto, F.; Lobitz, B.; Johan, S.; Brass, J.; Hinkley, E. The Ikhana unmanned airborne system (UAS) western states fire imaging missions: From concept to reality (2006–2010). *Geocarto Int.* **2011**, *26*, 85–101. [CrossRef]
9. Pastor, E.; Barrado, C.; Royo, P.; Santamaria, E.; Lopez, J.; Salami, E. Architecture for a helicopter-based unmanned aerial systems wildfire surveillance system. *Geocarto Int.* **2011**, *26*, 113–131. [CrossRef]
10. Yang, G.Z.; Bellingham, J.; Dupont, P.E.; Fischer, P.; Floridi, L.; Full, R.; Jacobstein, N.; Kumar, V.; McNutt, M.; Merrifield, R.; et al. The grand challenges of Science Robotics. *Sci. Robot.* **2018**, *3*, eaar7650. [CrossRef] [PubMed]
11. Çetin, A.E.; Dimitropoulos, K.; Gouverneur, B.; Grammalidis, N.; Günay, O.; Habiboğlu, Y.H.; Töreyin, B.U.; Verstockt, S. Video fire detection—Review. *Digit. Signal Process.* **2013**, *23*, 1827–1843. [CrossRef]
12. Allison, R.; Johnston, J.; Craig, G.; Jennings, S. Airborne Optical and Thermal Remote Sensing for Wildfire Detection and Monitoring. *Sensors* **2016**, *16*, 1310. [CrossRef] [PubMed]
13. Lippitt, C.D.; Stow, D.A.; Coulter, L.L. *Time-Sensitive Remote Sensing*; Springer: Berlin/Heidelberg, Germany, 2015.
14. Yuan, C.; Zhang, Y.; Liu, Z. A survey on technologies for automatic forest fire monitoring, detection, and fighting using unmanned aerial vehicles and remote sensing techniques. *Can. J. For. Res.* **2015**, *45*, 783–792. [CrossRef]
15. Bosch, I.; Gomez, S.; Molina, R.; Miralles, R. Object Discrimination by Infrared Image Processing. In Proceedings of the Bioinspired Applications in Artificial and Natural Computation, IWINAC 2009, Santiago de Compostela, Spain, 22–26 June 2009.
16. Töreyin, B.U.; Cinbis, R.G.; Dedeoglu, Y.; Cetin, A.E. Fire detection in infrared video using wavelet analysis. *Opt. Eng.* **2007**, *46*, 067204. [CrossRef]

17. Verstockt, S.; Vanoosthuyse, A.; Van Hoecke, S.; Lambert, P.; Van de Walle, R. Multi-sensor Fire Detection by Fusing Visual and Non-visual Flame Features. In *Image and Signal Processing*; Elmoataz, A., Lezoray, O., Nouboud, F., Mammass, D., Meunier, J., Eds.; Springer: Berlin/Heidelberg, Germany, 2010; pp. 333–341.
18. de Dios, J.M.; Arrue, B.; Ollero, A.; Merino, L.; Gómez-Rodríguez, F. Computer vision techniques for forest fire perception. *Image Vis. Comput.* **2008**, *26*, 550–562. [CrossRef]
19. Bosch, I.; Serrano, A.; Vergara, L. Multisensor Network System for Wildfire Detection Using Infrared Image Processing. *Sci. World J.* **2013**, *2013*, 1–10. [CrossRef] [PubMed]
20. Valero, M.M.; Rios, O.; Pastor, E.; Planas, E. Automated location of active fire perimeters in aerial infrared imaging using unsupervised edge detectors. *Int. J. Wildland Fire* **2018**, *27*, 241–256. [CrossRef]
21. Mengod, P.C.; Bravo, J.A.T.; Sardá, L.L. The influence of external factors on false alarms in an infrared fire detection system. *Int. J. Wildland Fire* **2015**, *24*, 261–266. [CrossRef]
22. Sousa, M.J.; Moutinho, A.; Almeida, M. Classification of potential fire outbreaks: A fuzzy modeling approach based on thermal images. *Expert Syst. Appl.* **2019**, *129*, 216–232. [CrossRef]
23. Gade, R.; Moeslund, T.B. Thermal cameras and applications: A survey. *Mach. Vis. Appl.* **2013**, *25*, 245–262. [CrossRef]
24. Wolf, A.; Pezoa, J.; Figueroa, M. Modeling and Compensating Temperature-Dependent Non-Uniformity Noise in IR Microbolometer Cameras. *Sensors* **2016**, *16*, 1121. [CrossRef] [PubMed]
25. Rogalski, A. *Infrared Detectors*; CRC Press: Boca Raton, FL, USA, 2010.
26. Minkina, W.; Dudzik, S. *Infrared Thermography: Errors and Uncertainties*; John Wiley & Sons: Hoboken, NJ, USA, 2009.
27. Sagan, V.; Maimaitijiang, M.; Sidike, P.; Eblimit, K.; Peterson, K.; Hartling, S.; Esposito, F.; Khanal, K.; Newcomb, M.; Pauli, D.; et al. UAV-Based High Resolution Thermal Imaging for Vegetation Monitoring, and Plant Phenotyping Using ICI 8640 P, FLIR Vue Pro R 640, and thermoMap Cameras. *Remote Sens.* **2019**, *11*, 330. [CrossRef]
28. Solomon, C.; Breckon, T. *Fundamentals of Digital Image Processing: A Practical Approach with Examples in Matlab*; John Wiley & Sons: Hoboken, NJ, USA, 2011.
29. Vickers, V.E. Plateau equalization algorithm for real-time display of high-quality infrared imagery. *Opt. Eng.* **1996**, *35*, 1921. [CrossRef]

Publisher's Note: MDPI stays neutral with regard to jurisdictional claims in published maps and institutional affiliations.

© 2020 by the authors. Licensee MDPI, Basel, Switzerland. This article is an open access article distributed under the terms and conditions of the Creative Commons Attribution (CC BY) license (http://creativecommons.org/licenses/by/4.0/).

sensors

MDPI

Review

A Review on Early Forest Fire Detection Systems Using Optical Remote Sensing

Panagiotis Barmpoutis, Periklis Papaioannou, Kosmas Dimitropoulos and Nikos Grammalidis *

Centre for Research and Technology Hellas, Information Technologies Institute, 57001 Thessaloniki, Greece; panbar@iti.gr (P.B.); ppapaioa@iti.gr (P.P.); dimitrop@iti.gr (K.D.)
* Correspondence: ngramm@iti.gr

Received: 15 October 2020; Accepted: 10 November 2020; Published: 11 November 2020

check for updates

Abstract: The environmental challenges the world faces nowadays have never been greater or more complex. Global areas covered by forests and urban woodlands are threatened by natural disasters that have increased dramatically during the last decades, in terms of both frequency and magnitude. Large-scale forest fires are one of the most harmful natural hazards affecting climate change and life around the world. Thus, to minimize their impacts on people and nature, the adoption of well-planned and closely coordinated effective prevention, early warning, and response approaches are necessary. This paper presents an overview of the optical remote sensing technologies used in early fire warning systems and provides an extensive survey on both flame and smoke detection algorithms employed by each technology. Three types of systems are identified, namely terrestrial, airborne, and spaceborne-based systems, while various models aiming to detect fire occurrences with high accuracy in challenging environments are studied. Finally, the strengths and weaknesses of fire detection systems based on optical remote sensing are discussed aiming to contribute to future research projects for the development of early warning fire systems.

Keywords: early fire detection; multispectral imaging systems; terrestrial; aerial; satellite; artificial intelligence

1. Introduction

Over the last few years, climate change and human-caused factors have a significant impact on the environment. Some of these events include heat waves, droughts, dust storms, floods, hurricanes, and wildfires. Wildfires have extreme consequences on local and global ecosystems and cause serious damages to infrastructure, injuries, and losses in human lives; therefore, fire detection and the accurate monitoring of the disturbance type, size, and impact over large areas is becoming increasingly important [1]. To this end, strong efforts have been made to avoid or mitigate such consequences by early fire detection or fire risk mapping [2]. Traditionally, forest fires were mainly detected by human observation from fire lookout towers and involved only primitive tools, such as the Osborne fire Finder [3]; however, this approach is inefficient, as it is prone to human error and fatigue. On the other hand, conventional sensors for the detection of heat, smoke, flame, and gas typically take time for the particles to reach the point of sensors and activate them. In addition, the range of such sensors is relatively small, hence, a large number of sensors need to be installed to cover large areas [4].

Recent advances in computer vision, machine learning, and remote sensing technologies offer new tools for detecting and monitoring forest fires, while the development of new materials and microelectronics have allowed sensors to be more efficient in identifying active forest fires. Unlike other fire detection review papers that have focused on various sensing technologies [5], on video flame or/and smoke methodologies in visible or/and InfraRed (IR) range [6–9], on various environments [10], and airborne systems [11,12], in this paper, we provide a comprehensive study of the most representative forest fire detection systems, focusing on those that use optical remote sensing, as well as digital image processing [13] and classification techniques [14]. Depending on the acquisition level, three broad categories of widely used systems that can detect or monitor active fire or smoke incidents in real/near-real-time are identified and discussed, namely terrestrial, aerial, and satellite. These systems are usually equipped with visible, IR, or multispectral sensors whose data are processed by machine learning methods. These methods rely either on the extraction of handcrafted features or on powerful deep learning networks (Figure 1) for early detection of forest fires (Figure 2) as well as for modeling fire or smoke behavior. Finally, we present the strengths and weaknesses of the aforementioned methods and sensors, as well as future trends in the field of early fire detection.

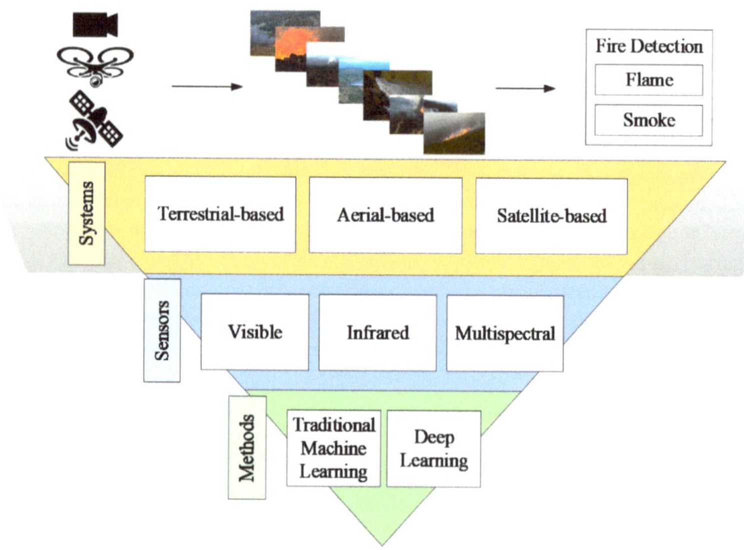

Figure 1. Generalized multispectral imaging systems for early fire detection.

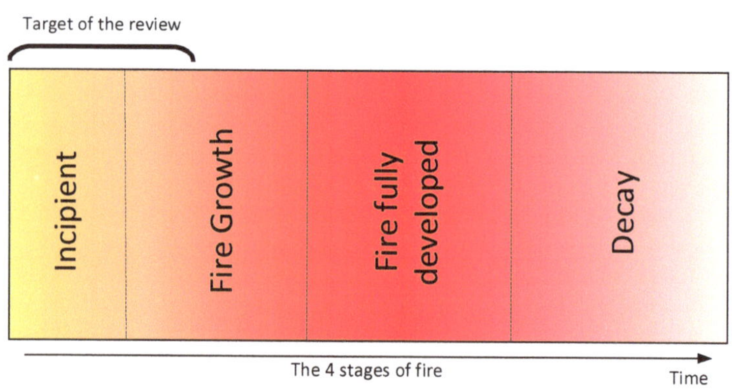

Figure 2. Systems discussed in this review target the detection of fire in the early stages of the fire cycle.

This paper is organized as follows: Section 2 covers different optical remote sensing systems for early fire detection, organized into three subsections—for Terrestrial, Aerial, and Satellite systems respectively—Section 3 includes discussion and the future scope of research.

2. Early Fire Detection Systems

2.1. Terrestrial Systems

Terrestrial-based early detection systems consist of either individual sensors (fixed, PTZ, or 360° cameras) or networks of ground sensors. These sensors need to be carefully placed to ensure adequate visibility. Thus, they are usually located in watchtowers, which are structures located on high vantage points for monitoring high-risk situations and can be used not only for detection but also for verification and localization of reported fires. There are two types of cameras used for early fire detection, namely optical cameras and IR cameras that can capture data ranging from low resolution to ultra-high resolution for different fire detection scenarios [15]. Optical cameras provide color information, whereas IR imaging sensors can provide a measure of the thermal radiation emitted by objects in the scene [16]. More recently, early detection systems that combine both types have also been introduced. The computer-based methods can process a high number of data aiming to achieve a consistent level of accuracy maintaining a low false alarm rate. In the following, we first present traditional approaches that are based on handcrafted features followed by more recent methods using deep learning for automated feature extraction.

2.1.1. Traditional Methods

Detection methods that use optical sensors or RGB cameras combine features that are related to the physical properties of flame and smoke, such as color, motion, spectral, spatial, temporal, and texture characteristics. The following color spaces have been used for the task of early fire detection: RGB [17–19], YCbCr [20], CIELAB [21], YUV [22,23], and HSV [24]; however, a drawback of color-based fire detection models is the high false alarm rates, since single-color information is insufficient in most cases for the early and robust fire detection. Thus, many of the developed methodologies combine color and motion information in images and videos [25]. Zhang et al. [26] used a probabilistic color-based model for the detection of fire regions and motion features for the final fire existence occurrence decision. Avgerinakis et al. [27] identified the smoke candidate blocks and then constructed the histograms of oriented gradients (HOGs) and histograms of optical flow (HOFs), thus taking into account both appearance and motion information. Likewise, Mueller et al. [28] used two optical flow schemes, namely optimal mass transport models and data-driven optical flow models.

Other researchers focused on the flickering effect of fire. This is observed in flame contours at a frequency of around 10 Hz, independently of the burning material and the burner [29]. To this end, Gunay et al. [30] distinguished flame flicker from the motion of fire-colored moving objects in videos using hidden Markov models (HMMs). Training HMMs leads to the reduction of data redundancy and improvement of reliability, while real-time detection is also achieved [31]. Furthermore, HMM-based methods for fire detection in the compressed domain have been proposed in MJPEG2000 [32] and H. 264 [33] compressed video. The use of multi-feature fire-based detection can offer more accurate results. Chen et al. [34] combined motion detection using a Gaussian mixture model, color analysis using an RGB color filtering, and flickering temporal analysis. The algorithm was applied to a video dataset consisting of different daytime and nighttime environments; however, at night, color analysis is less useful and night smoke is less visible. Thus, nighttime wildfire detection typically relies on motion analysis. Also, Töreyin, et al. [35] proposed a system equipped with an optical camera and a methodology that combines feature extraction (moving pixel/region extraction, color-based segmentation, and wavelet analysis in temporal and spatial domains), followed by a voting-based classifier. In [36], Barmpoutis et al. extracted an additional feature aiming to estimate a spatio–temporal consistency energy. Thereafter, they used a support vector machine (SVM) classifier to increase the robustness of fire detection.

Many other researchers have used infrared cameras aiming to reduce the false alarm rates of the optical-based terrestrial systems. The IR cameras measure the thermal radiation emitted by objects within the spectral range from either 3 to 5 μm (middle wavelength infrared, MWIR) or 8 to 14 μm (long wavelength infrared, LWIR). The MWIR band detectors, although they are optimal for fire detection, are expensive due to the cooling system required, so typically LWIR cameras are used. In IR videos the existence of rapid time-varying contours is an important sign of the presence of fire in the scene. Arrue et al. [37] observed that fire detection systems fail for distant fires and proposed a system for false alarm reduction in infrared forest-fire detection. More specifically, they used an adaptive infrared threshold, a segmentation method, and a neural network for early fire detection. Furthermore, Töreyin, et al. [16] used IR video to overcome the limitations of optical cameras in the detection of fires with little radiance in the visible spectrum. Specifically, they first estimated the boundaries of moving bright regions in each frame and then used spatio–temporal analysis in the wavelet domain using HMMs.

In contrast to single-sensor systems, multisensor systems typically cover wider areas and can achieve higher accuracies by fusing data from different sensors. The idea of integrated early fire detection and monitoring by combining data from optical and infrared cameras as well as a wireless sensor network (WSN) of temperature/humidity sensors was proposed by Grammalidis et al. [38]. Sensor data were processed and transmitted to a monitoring center employing computer vision and pattern recognition algorithms for automated fire detection and localization. The algorithm took into account color, spatial, and temporal information for flame detection, while for smoke detection, an online adaptive decision fusion (ADF) framework was developed. This framework consisted of several algorithms aiming to detect slow-moving objects, smoke-colored regions, and smoke region smoothness. Furthermore, improved early wildfire detection was achieved by fusing smoke detection from visual cameras and flame detection from infrared (LWIR) cameras. Similarly, Bosch et al. [39] proposed a system consisting of a wireless sensor network with a central monitoring station. In this, each sensor consists of an optical and a thermal camera and an integrated system for the processing of data and communication.

More recently, Barmpoutis et al. [40] and Dimitropoulos et al. [41] introduced fire detection systems based on smoke and flame dynamic texture analysis through linear dynamical systems (LDSs). Their modeling, combining color, motion, and spatio–temporal features led to higher detection rates and a significant reduction of false alarms. Temporal and spatial dynamic texture analysis of flame for forest fire detection was performed in [42]. Dynamic texture features were derived using two-dimensional (2D) spatial wavelet decomposition in the temporal domain and three-dimensional (3D) volumetric wavelet decomposition. In [43], the authors improved the smoke modeling of the fire incidents through dynamic textures solving higher order LDS (h-LDS). Finally, in [44], the authors took the advantage of the geometric properties of stabilized h-LDS (sh-LDS) space, and they proposed a novel descriptor, namely, histograms of Grassmannian points (HoGP) to improve the classification of both flame and smoke sequences.

2.1.2. Deep Learning Methods

In contrast to previously discussed methods that rely on handcrafted features, deep learning (DL) methods [45] can automatically extract and learn complex feature representations. Since the seminal work of Krizhevsky et al. [46], which achieved exceptional image classification performance by training a convolutional neural network (CNN) with efficient computation resources, deep learning is one of the most rapidly evolving fields and has been successfully applied to numerous computer vision problems. To this end, Luo et al. [47] developed a smoke detection algorithm based on the motion characteristics of smoke and a CNN. Firstly, they identified the candidate regions based on the background dynamic update and dark channel a priori method [48]. Then, the features of the candidate region were extracted automatically by a CNN consisting of five convolutional layers and three fully connected layers. In [49], the authors combined deep learning and handcrafted features to recognize the fire and smoke areas. For static features, the AlexNet architecture was adapted, while for dynamic features an adaptive weighted direction algorithm was used. Moreover, Sharma et al. [50] used optical images and re-tuned two pre-trained CNNs, based on VGG16 and ResNet50 backbones to distinguish images that contain fire and images that do not. It is worth mentioning that for the training, they created an unbalanced dataset including more non-fire images. Zhang et al. [51] proposed deep CNNs for forest fire detection in a cascaded fashion. Firstly, they trained a full image fire classifier to decide whether the image contains the fire or not and then applied a fine-grained patch classifier to localize the fire patches within this image. The full image classifier is a deep CNN that has been fine-tuned from AlexNet and the fine-grained patch classifier is a two-layer fully connected neural networks trained with the upsampled Pool-5 features. Muhammad et al. [52], inspired by GoogleNet architecture and proposed a fine-tuned fire detection CNN model for surveillance videos, while Shen et al. [53] used an optimized YOLO model for flame detection from video frames. Frizzi et al. [54] built a simple CNN consisting of nine layers. The architecture of this model was similar to LeNet-5 including dropout layers and used a leaky rectified linear unit (ReLU) activation function. Muhamad et al. [55] proposed a fine-tuned CNNs architecture based on the SqueezeNet model and developed a feature map selection algorithm for fire segmentation and background analysis. In [56], the authors combined AlexNet as a baseline architecture and the internet of multimedia things (IoMT) for fire detection and disaster management. The developed system introduced an adaptive prioritization mechanism for cameras in the surveillance system allowing high-resolution cameras to be activated to confirm the fire and analyze the data in real-time. Furthermore, Dunnings and Breckon [57] used low-complexity CNN architectural variants and applied a superpixel localization approach aiming to reduce the computational performance offering up to 17 fps processing time.

Since the number of publicly available wildfire datasets is still limited, Sousa et al. [58] proposed a fire detection method based on data augmentation and transfer learning. A method that had been pre-trained on the ImageNet Inception-v3 model was retrained and evaluated using ten-fold cross-validation on the Corsican Fire Database [59]. Extending deep learning approaches, Barmpoutis et al. [60] combined the power of faster region-based convolutional neural network (R-CNN) with multidimensional dynamic texture analysis based on higher-order LDSs aiming the early forest fire detection. A modified faster R-CNN and a 3D CNN were combined in [61]. To that end, the faster R-CNN with non-maximum annexation was utilized to realize the smoke target location based on static spatial information and then a 3D CNN was used for smoke recognition by combining dynamic spatial–temporal information. Jadon et al. [62] developed the FireNet convolution neural network using a standard fire dataset and a self-proposed dataset, achieving an encouraging performance in terms of a series of evaluation metrics. Zhang et al. [63] trained a faster R-CNN for forest smoke detection by creating synthetic smoke images and achieved great performance when tested on real smoke images. Moreover, in [64] the authors extracted spatial features through a faster R-CNN for the detection of the suspected regions of fire (SroFs) and non-fire. Then, the features of the detected SroFs in successive frames were used by a long short-term memory (LSTM) for them to identify whether there is a fire or not in a short-term period. Finally, a majority voting method and the exploitation of fire dynamics were used for the final decision. Shi et al. [65] inspired by the idea of R-CNN and combined image saliency detection and convolutional neural networks. More specifically, they utilized the pixel-wise image saliency aggregating (PISA) method [66] to identify the candidate regions and then classified them into fire or non-fire regions.

Instead of extracting bounding boxes, Yuan et al. [67] used a two-path encoder-decoder fully convolutional network (FCN) for visual smoke segmentation. FCNs can achieve end-to-end pixel-wise segmentation so the precise location of smoke can be identified in images. They also created synthetic smoke images instead of labeling the real smoke images manually for training and then tested the network on both synthetic and real videos. Cheng et al. [68] proposed a smoke detection model using Deeplabv3+ and a generative adversarial network (GAN). The smoke pixels were first identified by fusing the result Deeplabv3+ and the heatmap of smoke based on HSV features. Then, a GAN was employed for predicting the smoke trend heatmap based on the space–time analysis of the smoke videos. Finally, in [69] the authors used a two-stage training of deep convolutional GANs for smoke detection. This procedure included a regular training step of a deep convolutional (DC)-GAN with real images and noise vectors and a training step of the discriminator separately using the smoke images without the generator.

2.2. Unmanned Aerial Vehicles

Terrestrial imaging systems can detect both flame and smoke, but in many cases, it is almost impossible to view, in a timely manner, the flames of a wildfire from a ground-based camera or a mounted camera on a forest watchtower. To this end, autonomous unmanned aerial vehicles (UAVs) can provide a broader and more accurate perception of the fire from above, even in areas that are inaccessible or considered too dangerous for operations by firefighting crews. Either fixed or rotary-wing UAVs cover wider areas and are flexible, allowing the change of monitoring area, but they are affected by weather conditions and have limited flight time. UAVs mostly use ultra-high-resolution optical or infrared charge-coupled device (CCD) cameras to capture images as well as other sensors for navigation, such as global positioning system (GPS) receivers or inertial measurement units (IMUs).

2.2.1. Traditional Methods

The first attempts for aerial fire detection began around 1920 when planes were used for forest fire detection as a result of their unsuccessful deployment for the extinguishing of fires [70]. In 1986, Stearns et al. [71] captured IR images by the Flying Infrared Signatures Technology Aircraft of the U.S. Air Force Geophysics Laboratory and described a spatial and spectral analysis methodology to provide wildfire detection and monitoring for fire control. Similarly, Den Breejen et al. [72] used a single manned aerial vehicle for forest fire detection and tracking; however, although the operation of manned aerial vehicles is safer in busy airspace around fire [12], these vehicles are typically large and require increased operational costs making them a less useful tool for fire detection.

In recent times, the deployment of UAVs is considered to be a better option for the task of forest fire detection. More specifically, to achieve forest fire detection and tracking, Yuan et al. [73] used median filtering for noise reduction, color analysis based on CIELAB color space, Otsu threshold for fire segmentation, and blob counting. More recently, in [74] they used visual images captured by an optical camera of an unmanned aerial vehicle. Then, two color spaces, namely, RGB and HIS were chosen as inputs of a fuzzy logic rule and an extended Kalman filter was employed to adapt environmental condition variations and to perform smoke detection. Extending the color-based methods in [75,76] fire flame and smoke pixels are segmented using both color and motion characteristics. For the estimation of color features, they utilized three color spaces RGB, YCbCr, and HIS, whereas for the extraction of motion characteristics they noted that flames have turbulent movement or disordered characteristics. Thus, an optical flow algorithm was used to examine the motion characteristic of forest fires and extract fire motion pixels using dynamic background analysis.

Aiming to identify the fire location in terms of latitude, longitude, and altitude in [77], the authors used a DJI F550 hexacopter and applied two coordinate system transformations between the body-fixed frame, namely north-east-down frame (NED) and Earth-centered Earth-fixed (ECEF) frame. Then, a rule-based color model combining RGB and YCbCr color spaces was used for the identification of fire pixels. A fire simulation platform based on the Unity game engine and robot operating systems (ROS) was developed by Esfahlani [78]. The video data were collected through a monocular camera and navigation relied on a simultaneous localization and mapping (SLAM) system. A fire detection algorithm based on color, movement attributes, temporal variation of fire intensity, and its accumulation around a point was deployed. Finally, a mixed reality (MR) system incorporating physical and virtual elements was adopted to visualize and test the developed system.

Sudhakar et al. [79] proposed a method for forest fire detection through UAVs equipped with an optical and an infrared camera. They used a LAB color model and a motion-based algorithm followed by a maximally stable extremal regions (MSERs) extraction module. For improved presentation, the extracted forest fire detections are joined with landscape information and meteorological data. In [80] two types of UAVs, a fixed-wing drone and a rotary-wing drone equipped with optical and thermal cameras were used. As soon as the fixed-wing drone detects a fire, the rotary-wing drone will fly at a much lower altitude (10 to 350 m) compared to a fixed-wing UAV (350 to 5500 m), thus having better and more detailed visibility of the area and reducing false alarms through a neural network. Chen et al. [81] used optical and infrared sensors and data to train a CNN first for smoke detection and then for flame detection. In [82], the authors developed a system consisting of a central station and several aerial vehicles equipped with infrared or visual cameras, aiming to increase the coverage area. For fire detection, they applied a threshold for fire segmentation and then performed color and fire contour analysis.

2.2.2. Deep Learning Methods

Zhao et al. [83] used a UAV equipped with GPS and deployed a saliency detection algorithm for localization and segmentation of the fire area in aerial images. Then, a 15 layered deep convolutional neural network architecture was employed for both low and high-level fire feature extraction and classification. Tang et al. [84] captured 4K data using a ZenMuse XT2 dual vision sensor and applied an adaptive sub-region select block to detect fire candidate areas in 4K resolution images. Then, a YOLOv3 backbone architecture was used for fire detection. Jiao et al. [85] used a UAV to capture and transmit images to the ground station in real-time deploying a YOLOv3 network for fire detection. Furthermore, they deployed a UAV equipped with a visible and an infrared camera for image acquisition [86]. The onboard computer carried by UAV can perform local image processing and mission planning through a YOLOv3-tiny architecture. In addition, a ground station receives images and location information of fire spots, contributing to the detection of forest fires. Furthermore, it provides operational commands to the UAV for path planning and re-planning. Integrating fog computing and CNNs, Srinivas and Dua [87] employed a UAV to reduce the false alarm rates towards the task of forest fire detection at an early stage. More recently, Barmpoutis, et al. [88,89] used an optical 360-degree complementary metal-oxide-semiconductor (CMOS) camera mounted on a UAV to capture an unlimited field of view images. More specifically, they converted the equirectangular raw data to cubemap and stereographic projections, respectively. Then, they used deep neural networks and exploited fire dynamic textures aiming to reduce false alarms that are caused due to clouds, sunlight reflections, and fire/smoke-colored objects. Experimental results demonstrate the great potential of the proposed system for both flame and smoke detection.

2.3. *Spaceborne (Satellite) Systems*

Recently, mainly due to the large number of satellites launched and the decrease of associated costs, there are many research efforts to detect forest wildfires from satellite images. Specifically, a set of satellites were designed for Earth observation (EO, e.g., environmental monitoring or meteorology). Depending on their orbit, satellites can be broadly classified into various categories, each having their advantages and disadvantages. The most important categories include: (a) the geostationary orbit (GEO), which is a circular orbit with an altitude of 35,786 kilometers and zero inclination, so that the satellite does not move at all relative to the ground, providing a constant view of the same surface area, (b) the low Earth orbit (LEO), which has an altitude of 2000 km or less, requires the lowest amount of energy for satellite placement and provides high bandwidth and low communication latency, (c) the polar sun-synchronous orbit (SSO), which is a nearly polar orbit that passes the equator at the same local time on every pass. Most EO satellites are in specific low Earth polar SSO orbits, whose altitude and inclination are precisely calculated so that the satellite will always observe the same scene with the same angle of illumination coming from the Sun, so shadows appear the same on every pass.

Data from sun-synchronous satellites have high spatial resolution but low temporal resolution, whereas geostationary satellites have high temporal resolution but low spatial resolution. Some satellites of the first category, like Landsat or Sentinel satellites, have a large revisit time (eight-day repeat cycle for LandSat-7/8 and approximately 2–3 days at mid-latitudes for Sentinel 2A/2B). Hence, they are unsuitable for real-time active forest fire detection, but only for less time-sensitive tasks, e.g., burnt area estimation, and therefore their studies fall within the scope of this paper.

2.3.1. Fire and Smoke Detection from Sun-Synchronous Satellites

Imaging sensors in sun-synchronous satellites include three multispectral imaging sensors, namely advanced very-high-resolution radiometer (AVHRR) [90], moderate resolution imaging spectroradiometer (MODIS) [91], and visible infrared imaging radiometer suite (VIIRS) [92,93], whose data have also been used for wildfire detection. The advanced very-high-resolution radiometer (AVHRR/3) is a multipurpose imaging instrument that measures the reflectance of the Earth and has been used for global monitoring of cloud cover, sea surface temperature, ice, snow, and vegetation cover characteristics [90]. AVHRR instruments are or have been carried by the National Oceanic and Atmospheric Administration (NOAA) family of polar-orbiting platforms (polar-orbiting operational environmental satellite—POES) and European Meteorological Operational (MetOp) satellites. The instrument provides six channels, three in the visible/near-infrared region and three thermal infrared channels, with 1 km spatial resolution. The moderate resolution imaging spectroradiometer (MODIS), onboard the National Aeronautics Space Administration (NASA) EO Terra and Aqua satellites that have a revisit time of 1–2 days, capture data in 36 spectral bands ranging in wavelengths from 0.4 to 14.4 μm and at varying spatial resolutions (2 bands at 250 m, 5 bands at 500 m, and 29 bands at 1 km). MODIS was succeeded by the visible infrared imaging radiometer suite (VIIRS) instrument onboard the Suomi National Polar-orbiting Partnership (NPP) and NOAA-20 weather satellites. The instrument provides 22 different spectral bands, i.e., 16 moderate-resolution bands (M-bands, 750 m), 5 imaging resolution bands (I-bands, 375 m), and 1 day/night panchromatic band (750 m).

Traditional Methods

These imaging sensors have also been extensively applied for near-real-time wildfire detection. For instance in [94], Sayad et al. combined big data, remote sensing, and data mining algorithms (artificial neural network and SVM) to process big data collected from MODIS images and extract insights from them to predict the occurrence of wildfires. More specifically, they used pre-processed MODIS data to create a dataset based on three parameters related to the state of the crops: namely normalized difference vegetation index (NDVI), land surface temperature (LST), and thermal anomalies. For wildfire prediction, they used two different supervised classification approaches based on neural networks and SVM, achieving good prediction accuracies, i.e., 98.32% and 97.48%, respectively. Results were assessed using several validation strategies (e.g., classification metrics, cross-validation, and regularization) and comparisons with other wildfire prediction models, demonstrating the efficiency of the model in predicting the occurrence of wildfires.

Several papers deal with the problem of smoke detection based on MODIS data, which is a very challenging problem given its strong similarity with clouds, haze, and other similar phenomena. Shukla et al. [95] proposed an algorithm for automatic detection of smoke using MODIS data, which was based on a multiband thresholding technique for discriminating between smoke plumes and clouds. Results suggested that the algorithm was able to isolate smoke pixels in the presence of other scene types, such as clouds, although it performed better in identifying fresh dense smoke as compared to highly diffused smoke. Similarly, Li et al. [96] proposed an approach to automatically separate smoke plumes from clouds and background by analyzing MODIS data. Specifically, a previous approach proposed by Li et al. [97] for the AVHRR sensor was improved based on spectral analysis among the smoke, cloud, and underlying surface using MODIS data. Specifically, a multi-threshold method was used for extracting training sample sets to train a back-propagation neural network (BPNN) to discriminate between three classes: (smoke, cloud, and underlying surface). Results using MODIS data of several forest fires occurred in different places and different dates were satisfactory. Advantages include the ability of the algorithm to detect smoke plumes in different seasons using seasonal training data sets, as well as that it provides quantitative and continuous outputs of smoke and other objects.

Many researchers used active fire products derived from these sensors to assess various other proposed algorithms. Hally et al. [98] examined the performance of a threshold algorithm against commonly used products such as the VIIRS active fire product, to determine the completeness of anomaly capture. Specifically, the study considers two commonly used active fire products: the MODIS Collection 6 (MOD/MYD14) 1 km active fire product as outlined in Giglio et al. [99] and the VIIRS 375 m (VNP14IMG) active fire product described in Schroeder et al. [93]. In both cases, the geographic position of the detected hotspots, as well as the time of satellite overpass, were used. Also, Wickramasinghe et al. [100] compared the Advanced Himawari Imager—Fire Surveillance Algorithm (AHI-FSA) across the Northern Territory of Australia (1.4 million km^2) over ten days with the well-established active fire products from satellites: MODIS and VIIRS.

Finally, the Chinese HuanJing sun-synchronous satellites ("HuanJing" means "environment" in Chinese) are satellites for disaster and environmental monitoring that are capable of visible, infrared, multi-spectral, and synthetic aperture radar imaging. Lin et al. [101] presented a spatio–temporal model (STM) based forest fire detection method that uses multiple images of the inspected scene based on Huanjing-1B satellite images. A comparison of detection results demonstrated that the proposed algorithm is useful to represent the spatio–temporal information contained in multi-temporal remotely sensed data.

Deep Learning Methods

Deep Learning methods have also been recently applied for fire and smoke detection from multispectral satellite images. Ba et al. [102] presented a new large-scale satellite imagery dataset based on MODIS data, namely USTC_SmokeRS, consisting of 6225 satellite images from six classes (i.e., cloud, dust, haze, land, seaside, and smoke) and covering various areas/regions over the world. Using this dataset, they evaluated several state-of-the-art deep learning-based image classification models for smoke detection and proposed *SmokeNet*, a new CNN model that incorporated spatial- and channel-wise attention in CNN to enhance feature representation for scene classification. Also, Priya et al. [103] used a dataset of 534 RGB satellite images from different sources, including MODIS images from the NASA Worldview platform and Google. An effective approach using an Inception-v3 CNN framework and transfer learning was used for fire and non-fire image classification. Then, the fire regions were extracted based on thresholding and local binary patterns.

2.3.2. Fire and Smoke Detection from Geostationary Satellites

Regarding satellite imagery from geostationary satellites, important work for fire and smoke detection has already been performed using the advanced Himawari imager (AHI) sensor of the Himawari-8 weather satellite. Himawari 8 is a new generation of Japanese geostationary weather satellites operated by the Japan Meteorological Agency. AHI-8 has significantly higher radiometric, spectral, and spatial resolution than its predecessor.

Regarding Europe and the US, two additional sensors that are installed in geostationary satellites are the European Space Agency (ESA) Meteosat second generation (MSG, a satellite series)-spinning enhanced visible and infrared imager (SEVIRI) sensor and the NASA geostationary operational environmental satellite (GOES)-16 advanced baseline imager (ABI) sensor. The MSG-SEVIRI geostationary sensor is a 50 cm diameter aperture, line-by-line scanning radiometer, which provides image data in 12 spectral channels (four visible and near-infrared (NIR), including a broadband high resolution (1 km) visible channel, and eight thermal IR with a resolution of 3 km) with a baseline repeat cycle of 15 min. GOES-16 is the first of the GOES-R series of the geostationary operational environmental satellite (GOES) operated by NASA and the NOAA. The advanced baseline imager (ABI) is its primary instrument, providing high spatial and temporal resolution imagery of the Earth through 16 spectral bands at visible and infrared wavelengths.

Traditional Methods

Hally et al. [98] extended previous work by the same authors Hally et al. [104] using AHI sensor data from the Himawari geostationary satellite in the application of a multi-temporal method of background temperature estimation, known as the broad area training (BAT). This method involves a two-step process for geostationary data: a preprocessing step, where AHI Band 7 images are aggregated and then a fitting step, where this spatially aggregated data are used in individual pixel fitting using a single value decomposition (SVD) process. These fittings at the pixel level can then be compared to the raw brightness temperature data as measured by the satellite sensor, to identify thermal anomalies such as those caused by an active fire. Results are seen to compare favorably to active fire products produced by low Earth orbit satellite data during the period of study. Fatkhuroyan et al. [105] perform a study of data from fires in Sumatera and Kalimantan regions in August, September, and October 2015 and concluded that smoke detection and monitoring is feasible using pseudo-RGB images consisting of one visual channel and two near-infrared channels. A limitation revealed by this study is that Himawari-8/AHI is a passive sensor that very dependent on the reflection of solar radiance, so it can only monitor the forest fire during the day-time. Xu et al. [106] investigated the feasibility of extracting real-time information about the spatial extents of wildfires using the Himawari-8 satellite. The algorithm is based on previous work using the MODIS sensor: it first identifies possible hotspots and then eliminates false alarms by applying certain thresholds, similar to Giglio et al. [99]. False alarms are then rejected by cloud, water, and coast tests based on the additional bands and comparison with neighboring pixels. Results demonstrated that fire detection is robust to smoke and moderate cloud obscuration and sensitive enough for early detection of wildfires.

Typically, only temporal-based fire detection algorithms are used for geostationary orbital sensors, detecting the fire by analyzing multi-temporal changes of brightness temperature (BT). On the other hand, polar-orbiting platforms, use spatial-based fire detection algorithms, which are commonly classified either "fixed-threshold" or "contextual". Aiming to combine the merits of both approaches, Xie et al. [107] presented a spatio–temporal contextual model (STCM) that fully exploits geostationary data's spatial and temporal dimensions using data from Himawari-8 Satellite. They applied an improved robust fitting algorithm to model each pixel's diurnal temperature cycles (DTC) in the middle and long infrared bands. For each pixel, a Kalman filter was used to blend the DTC to estimate the true background brightness temperature.

Significant research results have also been produced using data from the GOES ABI and MSG SEVIRI instruments. A multi-temporal change-detection technique, namely robust satellite techniques for fires detection and monitoring (RST-FIRES) using data from the MSG-SEVIRI sensor. Filizzola et al. [108] was seen to be very efficient for the timely detection of even small/short-living fire incidents. Furthermore, Di Biase et al. [109] updated the satellite fire detection (SFIDE) algorithm, previously proposed by the same authors Laneve et al. [110] to reduce the false alarm rate. Specifically, they improved the estimation of the reference temperature used to define a fire pixel, the cloud mask accuracy, and the exploitation of the high refresh rate of the images to implement several tests for more accurate detection of forest fires. In Hall et al. [111], both satellite imaging sensors (GOES ABI and MSG SEVIRI) were found to be very efficient in detecting active fire incidents. Additional sensing systems providing such broad spatial coverage and favoring improved geostationary satellite fire data consistency across regions could further improve performance. Reference fire data were derived from Landsat-8 Operational Land Imager (OLI) 30 m imagery using the Schroeder et al. [112] Landsat-8 OLI automated fire detection algorithm.

Deep Learning Methods

Very recently, Larsen et al. [113] presented a deep FCN for predicting fire smoke in satellite imagery in near-real-time (NRT) using data from the Himawari-8/AHI sensor. Also, Phan et al. [114] proposed a novel remote wildfire detection framework based on GOES-16 sensor imagery, which contains three distinct stages: (i) a streaming data processing component to purify and examine raw image data for ROI's, (ii) an early wildfire prediction component using deep learning architectures to grab spatial and spectral designs for more accurate and robust detection, and (iii) a streaming data visualization dashboard for potential wildfire incidents.

2.3.3. Fire and Smoke Detection Using CubeSats

A recent trend in remote sensing from satellites is using "Cubsats", i.e., miniaturized satellites that typically weigh between 1 and 10 kg and follow the popular 'CubeSat' standard [115], which defines the outer dimensions of the satellite within multiple cubic units of 10 × 10 × 10 cm and can accommodate small technology payloads to perform different scientific research or commercial functions and explore new space technologies. Technically, it is easier for CubeSats to function in the LEO-zone due to their unique characteristics. More than 1100 CubeSats have been successfully launched by several universities and companies worldwide, as of January 2020.

The latest advances in satellite imagery had allowed CubeSats to rapidly discover wildfires. Barschke et al. (2017) [116] described a nanosatellite called TUBIN (Technische Universität Berlin Infrared Nanosatellite), which was designed to validate an infrared microbolometer for wildfire remote sensing. Another constellation of four 6U Cubesats for monitoring forest fires and other natural disasters was proposed in Africa and Asia [117]. The payload is an optical sensor with three spectral bands (green, red, and near-infrared) with a revisit time of 72 h. Shah et al. [118] proposed a system consisting of a constellation of nanosatellites equipped with multi-spectral visible to Infrared (IR) cameras and a ground station, which will allow all surface points on the planet to be revisited at least once in an hour. Capturing a surface location with high resolution in MWIR and LWIR allows for the precise estimation of the thermal output of the surface. Simulations indicated that a fire of about four hundred square meters can be detected using a payload of a multispectral IR camera measuring incident power in two thermal infrared bands (mid-wave and long-wave). The system will use onboard data processing, enabling an early wildfire warning within 30 min and minimizing bandwidth requirements. Additionally, compressed raw images can be transmitted to the ground station to provide global thermal data updated every 90 min. The first satellite is planned for launch in late 2020 with the data available for research purposes.

3. Discussion and Conclusions

Three categories of early fire and smoke detection systems have been analyzed and compared (Table 1) thoroughly in this literature review paper, namely terrestrial, unmanned aerial vehicles, and satellite-based systems. In general, terrestrial systems tend to be more efficient in terms of accuracy and response time to wildfire incidents. Furthermore, these systems offer high spatial resolution depending on the camera resolution and the viewing angle/distance; however, their coverage is limited when compared to the other two solutions, as the cameras are placed in fixed positions and additional limitations may apply (e.g., occlusions).

On the other hand, aerial-based systems gained recently a lot of attention due to the rapid development of UAV technology. Such systems provide a broader and more accurate perception of the fire, even in regions that are inaccessible or considered too dangerous for fire-fighting crews. In addition, UAVs can cover wider areas and are flexible, in the sense that they monitor different areas, as needed. Recent technological achievements have led to better camera resolutions offering high spatial resolution, wider field of view, and better battery autonomy. The latest UAVs are equipped with both visible and infrared cameras, improving the detection accuracy and allowing night operation; however, they are affected by weather conditions and, in many cases, their flight time is limited.

Table 1. Multispectral imaging systems and their characteristics.

	(Satellite)-Sensor	Spectral Bands	Access to the Data	Specs/Advantages/Limitations	Spatial Scale	Spatial Resolution	Data Coverage	Accuracy Range
Terrestrial	**Optical**	Visible spectrum	Both web cameras and image and video datasets are available	Easy to operate, limited field of view, need to be carefully placed in order to ensure adequate visibility.	Local	Very high spatial resolution (centimeters) depending on camera resolution and distance between the camera and the event	Limited coverage depending the specific task of each system	85%–100% [35,40,58,60]
	IR	Infrared spectrum						
	Multimodal	Multispectral						
Aerial	**Optical**	Visible spectrum	Limited number of accessible published data	Broader and more accurate perception of the fire, cover wider areas, flexible, affected by weather conditions, limited flight time.	Local—Regional	High spatial resolution depending on flight altitude, camera resolution and distance between the camera and the event	Coverage of hundred hectares depending on battery capacity.	70%–94.6% [75,86,89]
	IR	Infrared spectrum						
	Multimodal	Multispectral						
Satellite	**Terra/Aqua-MODIS** [118]	36 (0.4–14.4 µm)	Registration Required (NASA)	Easily accessible, limited spatial resolution, revisit time: 1–2 days	Global	0.25 km (bands 1–2) 0.5 km (bands 3–7) 1 km (bands 8–36)	Earth	92.75%–98.32% [94–96,99,102]
	Himawari-8/9—AHI-8 [119]	16 (0.4–13.4 µm)	Registration Required/ (Himawari Cloud)	Imaging sensors with high radiometric, spectral, and temporal resolution. 10 min (Full disk), revisit time: 5 min for areas in Japan/Australia)	Regional	0.5 km or 1 km for visible and near-infrared bands and 2 km for infrared bands	East Asia and Western Pacific	75%–99.5% [98,100,104–107,113]
	MSG—SEVIRI [120]	12 (0.4–13.4 µm)	Registration Required (EUMETSAT	Low noise in the long-wave IR channels, tracking of dust storms in near-real-time, susceptibility of the larger field of view to contamination by cloud and lack of dual-view capability, revisit time: 5–15 min	Regional	1 km for the high-resolution visible channel 3 km for the infrared and the 3 other visible channels	Atlantic Ocean, Europe and Africa	71.1%–98% [108–111]
	GOES-16ABI [121]	16 (0.4–13.4 µm)	Registration Required (NOAA)	Infrared resolutions allow the detection of much smaller wildland fires with high temporal resolution but relatively low spatial resolution, and delays in data delivery, revisit time: 5–15 min	Regional	0.5 km for the 0.64 µm visible channel 1 km for other visible/near-IR 2 km for bands > 2 µm	Western Hemisphere (North and South America)	94%–98% [111,114]

Table 1. *Cont.*

(Satellite)-Sensor	Spectral Bands	Access to the Data	Specs/Advantages/Limitations	Spatial Scale	Spatial Resolution	Data Coverage	Accuracy Range
HuanJing (HJ)-1B—WVC (Wide View CCD Camera)/**IRMSS** (Infrared Multispectral Scanner) [122]	WVC: 4 (0.43–0.9 μm) IRMSS: 4 (0.75–12.5 μm)	Registration Required	Lack of an onboard calibration system to track HJ-1 sensors' on-orbit behavior throughout the life of the mission, revisit time: 4 days	Regional	WVC: 30 m IRMSS: 150–300 m	Asian and Pacific Region	94.45% [101]
POES/MetOp—AVHRR [123]	6 (0.58–12.5 μm)	Registration Required (NOAA)	Coarse spatial resolution, revisit time: 6 h	Global	1.1 km by 4 km at nadir	Earth	99.6% [97]
S-NPP/ NOAA-20/NOAA—VIIRS-375 m [124,125]	16 M-bands (0.4–12.5 μm) 5 I-bands (0.6–12.4 μm) 1 DNB (0.5–0.9 μm)	Registration Required (NASA)	Increased spatial resolution, improved mapping of large fire perimeters, revisit time: 12 h	Global	0.75 km (M-bands) 0.375 km (I-bands) 0.75 km (DNB)	Earth	89%–98.8% [93]
CubeSats (data refer to a specific design from [126])	2: MWIR (3–5 μm) and LWIR (8–12 μm)	Commercial access planned	Small physical size, reduced cost, improved temporal resolution/response time, Revisit time: less than 1 h.	Global	0.2 km	Wide coverage in orbit	The first satellite is planned for launch in late 2020

Finally, Earth Observation satellite systems have been used successfully for wildfire detection, mainly due to their large-scale coverage. The majority of satellites providing earth imagery are either geostatic or in the near-polar sun-synchronous orbit and include multispectral imaging sensors. Sun-synchronous satellites provide data with high spatial resolution but low temporal resolution, while geostationary satellites have a high temporal resolution but low spatial resolution. More recently, advances in nanomaterials and micro-electronics technologies have allowed the use of tiny low-Earth-orbiting satellites, known as CubeSats. CubeSats have significant advantages in comparison with traditional satellites regarding smoke and fire detection, since they are more effective in terms of costs, temporal resolution/response time, and coverage. In addition, they are smaller in size than traditional satellites and need less time to be put into orbit; however, one issue that needs to be tackled is their poor ability to transmit large amounts of data to the ground. From the aforementioned analysis, it is clear that each category has its advantages and disadvantages. To this end, recent research efforts on wildfire detection [127–129] focus either on the combination of these technologies or on the use of additional input sources such as crowdsourcing, social media, and weather forecasting.

It is worth mentioning that most of the institutions and agencies aiming to support wildfire management at the national and regional level use either satellites or combine them with a small fleet of planes to detect and map the extent, spread, and impact of forest fires [130,131]. Furthermore, various organizations have installed network-connected optical cameras in or near wildland areas sharing live images on the web to assist early forest fire detection [132].

A detailed comparison between the three categories of early fire detection systems in terms of the performance (Accuracy), number of research papers (Volume of works), the potential for future improvement (Future potential), the minimum fire size that can be detected (Minimum fire size), the monitoring area covered by the system (Coverage area), and time needed for early fire detection (Response time) in the scale 0 (low) to 5 (high) is shown in Figure 3. Most of the literature shows that terrestrial systems have been extensively studied, achieving high accuracy rates and fast response times, despite the limited coverage area that they offer. Thus, large networks of ground sensors can be deployed to increase the coverage area; however, in this case, a trade-off between the number of sensors, cost, and complexity is required. On the other hand, aerial and satellite-based systems provide better coverage. These systems have already shown their great potential and accuracy rates and response times.

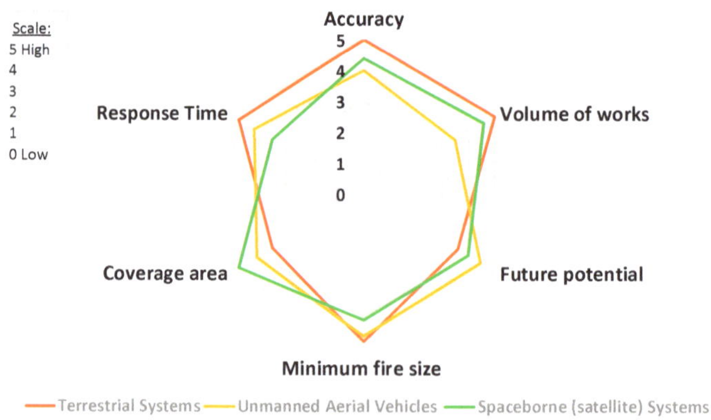

Figure 3. Radar chart showcasing the findings of this review for different early forest fire detection systems with regards to accuracy, response time, coverage area, future potential, and volume of works in the scale 0 (low) to 5 (high).

Also, the terrestrial and aerial-based systems can detect fires at a very early stage depending on their distance from the fire and their spatial resolution in parallel with short latency time [132]. In contrast, regarding satellite-based systems, time latency and minimum detectable fire size are expected to be improved in the following years. Currently, imaging sensors in sun-synchronous-orbiting satellites, such as MODIS, can detect after observation very small fires (up to 50 m^2) under near-ideal conditions and an average size of 30 × 30 = 900 m^2 under a variety of conditions [133]. Regarding the latency time, MODIS fire products are produced and delivered to fire managers partners in near-real-time (within 2–4 h of when MODIS collected the observations) [133]. On the other hand, geostationary sensors, like Himawari-8 AHI, can provide observations every 10–30 min, making them ideal for near-real-time fire surveillance, at the cost of a higher spatial resolution (2 km pixel size, which can be reduced to 500 m) [100]. Furthermore, the creation of new datasets for the training of novel deep learning algorithms (e.g., super-resolution), as well as advances in transmission technologies will further contribute towards this direction.

Similarly, Figure 4 demonstrates some attributes for each of the three sensor types: visible, infrared, and multispectral. Although the research community has thoroughly utilized optical-based systems, multispectral approaches seem to achieve better accuracy rates, due to the complementarity of information provided by different spectral bands; however, the use of multispectral technology increases significantly the overall cost of the system. This fact justifies the extensive use of low-cost optical sensors by many research works in the literature. Nevertheless, the wider use of multispectral sensors in different systems is expected to further improve the performance of early wildfire detection systems. To this end, extensive research is still needed on systems that integrate multimodal sensing technologies along with advanced deep-learning algorithms.

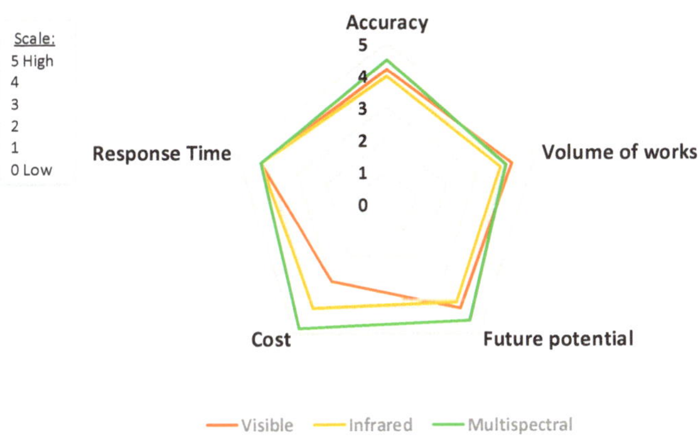

Figure 4. Radar chart showcasing the findings of this review for different sensor types with regards to accuracy, response time, cost, future potential, and volume of works in the scale 0 (low) to 5 (high).

Furthermore, to explore the evolution of the forest fire detection research domain, we carried out a bibliometric analysis. The initial search yielded over 2024 papers related to forest fire detection published in the Web of Science (WoS) database [134]. Figure 5 shows the trend in the number of articles published between 1 January 1990 and 31 October 2020. Narrowing the results to only the imaging research area, the search yielded 697 published articles (Figure 6). These results show that there is a growth in publications in the last 30 years. Of these, 378 are related to forest fire detection based on terrestrial systems, 59 based on aerial systems, and 260 based on satellite systems (Figure 7). The results of Figure 7 indicate that the field of forest fire detection in the image research area for terrestrial, aerial, and satellite-based systems is still evolving.

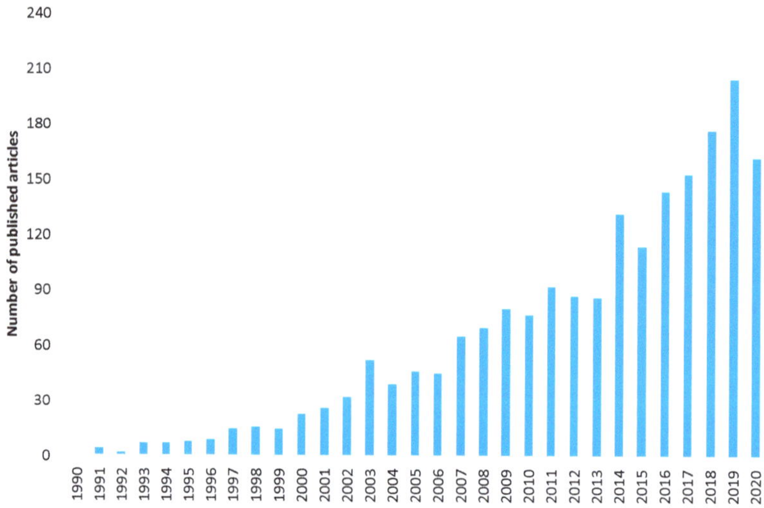

Figure 5. The number of published articles per year related to forest fire detection. Data retrieved from Web of Science [134] for dates between 1990 to October 2020.

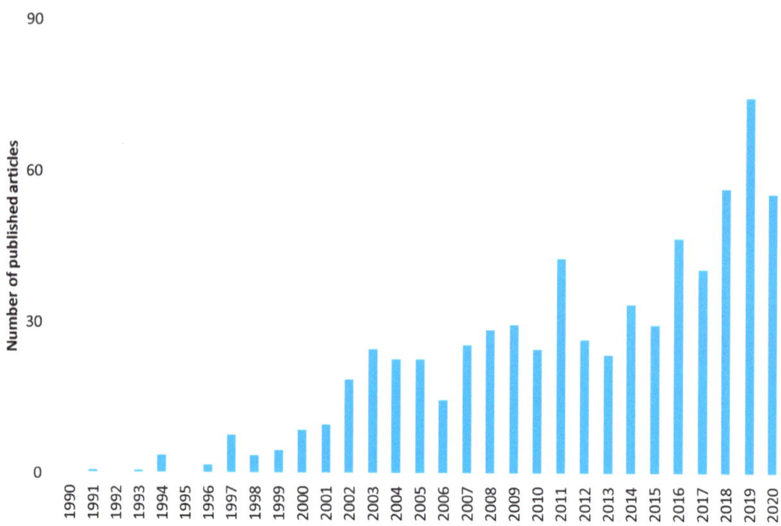

Figure 6. The number of published articles per year related to forest fire detection in the imaging research area. Data retrieved from Web of Science [134] for dates between 1990 to October 2020.

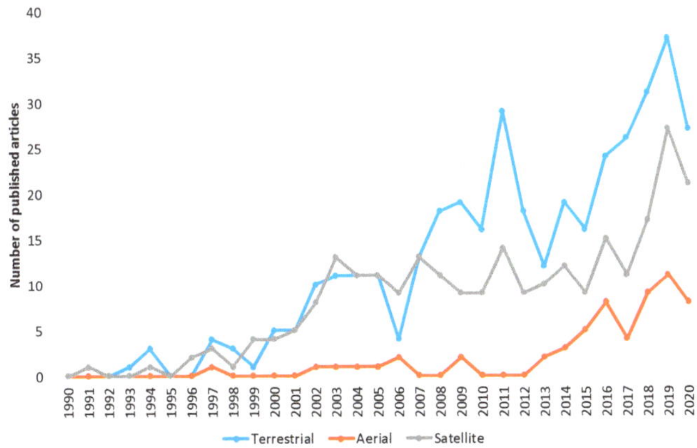

Figure 7. The number of published articles per year for terrestrial, aerial, and satellite-based systems. The analysis was performed for forest fire detection in the imaging research area. Data retrieved from Web of Science [134] for dates between 1990 to October 2020.

Subsequently, to provide information on the identity of papers and the corresponding systems that receive the most citations, we performed a quantitative citation analysis. A citation occurs when one paper refers to another paper. In this analysis (Figure 8), we identified that papers related to forest fire detection in the imaging satellite-based research area receive the most citations followed by terrestrial and aerial-based papers. Finally, to identify the study area of the aforementioned papers, a study analysis of the funding agencies was performed. In Figure 9, ten of the organizations and agencies that were the major sponsors of these papers are shown. More specifically, the National Natural Science Foundation of China (NSFC) has funded 42 papers related to forest fire detection in the imaging research area, while National Aeronautics Space Administration (NASA) has funded 37 papers. In addition, organizations and agencies from Canada, the European Union, France, the USA, and the UK have funded more than 79 papers related to the field of imaging-based forest fire detection. Similarly, in the second analysis, the authors' countries of affiliation were mapped (Figure 10). To this end, over 25 percent and 15 percent of the authors are affiliated with an organization that is based in the USA and China, respectively.

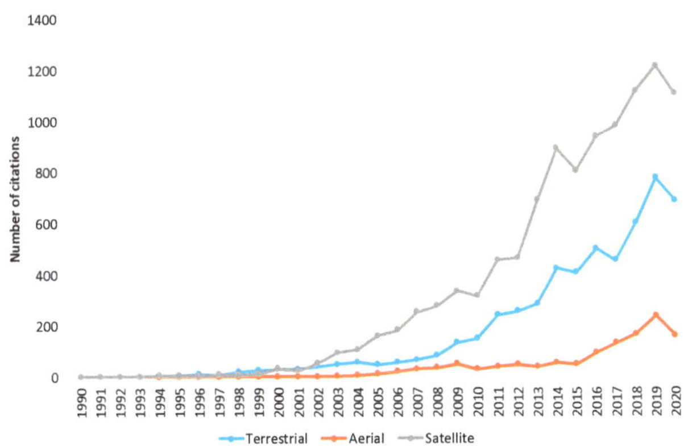

Figure 8. The number of times cited the published articles per year for terrestrial, aerial, and satellite-based systems. The analysis was performed for forest fire detection in the imaging research area. Data retrieved from Web of Science [134] for dates between 1990 to October 2020.

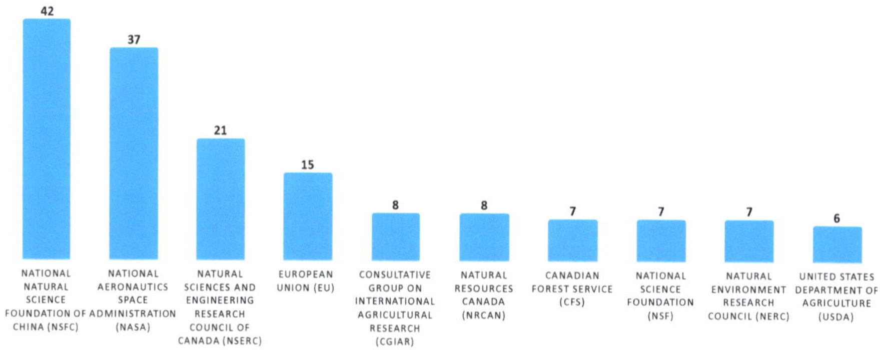

Figure 9. Organizations and agencies that funded most of the published articles for forest fire detection in the imaging research area. Data retrieved from Web of Science [134] for dates between 1990 to October 2020.

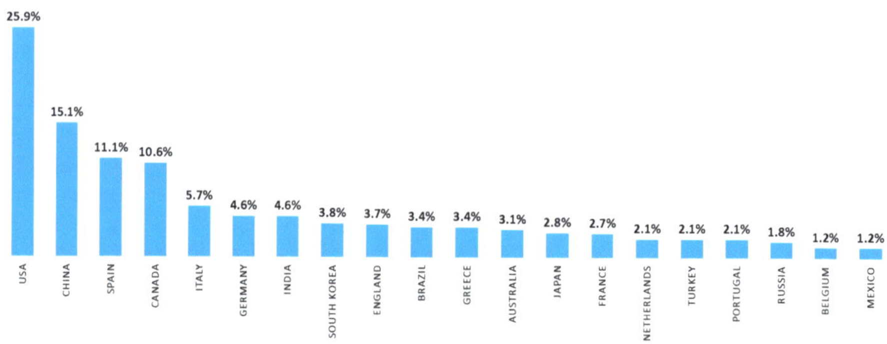

Figure 10. Authors' affiliation by country (%) for forest fire detection in the imaging research area. Data retrieved from Web of Science [134] for dates between 1990 to October 2020.

Author Contributions: Conceptualization, P.B., P.P., K.D., and N.G.; formal analysis, P.B., P.P., K.D., and N.G.; writing—original draft preparation, P.B. and P.P.; writing—review and editing, K.D. and N.G. All authors have read and agreed to the published version of the manuscript.

Funding: This research was funded by Greece and the European Union, projects i-FORESTER ("Reinforcement of Postdoctoral Researchers—2nd Cycle") and eOUTLAND ("INTERREG V-A COOPERATION PROGRAMME Greece-Bulgaria 2014–2020", grant number 1672).

Acknowledgments: This research is co-financed by Greece and the European Union (European Social Fund—ESF) through the Operational Programme "Human Resources Development, Education and Lifelong Learning" in the context of the call "Reinforcement of Postdoctoral Researchers—2nd Cycle" (MIS 5033021) for the project "i-FORESTER: Intelligent system for FOREST firE suRveillance", implemented by the State Scholarships Foundation (IKY). Periklis Papaioannou, Kosmas Dimitropoulos, and Nikos Grammalidis have received funding from INTERREG V-A COOPERATION PROGRAMME Greece-Bulgaria 2014–2020 project "eOUTLAND: Protecting biodiversity at NATURA 2000 sites and other protected areas from natural hazards through a certified framework for cross-border education, training, and support of civil protection volunteers based on innovation and new technologies".

Conflicts of Interest: The authors declare no conflict of interest.

Abbreviations

ABI	Advanced Baseline Imager
ADF	Adaptive Decision Fusion
AHI	Advanced Himawari Imager
AVHRR	Advanced Very-High-Resolution Radiometer
BAT	Broad Area Training
BPNN	Back-Propagation Neural Network
BT	Brightness Temperature
CCD	Charge-Coupled Device
CMOS	Complementary Metal-Oxide-Semiconductor
CNN	Convolutional Neural Network
DC	Deep Convolutional
DL	Deep Learning
DTC	Diurnal Temperature Cycles
ECEF	Earth-Centered Earth-Fixed
EO	Earth Observation
ESA	European Space Agency
FCN	Fully Convolutional Network
GAN	Generative Adversarial Network
GEO	Geostationary Orbit
GOES	Geostationary Operational Environmental Satellite
GPS	Global Positioning System
h-LDS	higher-order LDS
HMM	Hidden Markov Models
HOF	Histograms of Optical Flow
HOG	Histograms of Oriented Gradients
HoGP	Histograms of Grassmannian Points
IMU	Inertial Measurement Unit
IoMT	Internet of Multimedia Things
IR	InfraRed
LDS	Linear Dynamical Systems
LEO	Low Earth Orbit
LST	Land Surface Temperature
LSTM	Long Short-Term Memory
LWIR	Long Wavelength InfraRed
MetOp	Meteorological Operational
MODIS	Moderate Resolution Imaging Spectroradiometer
MR	Mixed Reality
MSER	Maximally Stable Extremal Regions
MSG	Meteosat Second Generation
MWIR	Middle Wavelength InfraRed
NDVI	Normalized Difference Vegetation Index
NED	North-East-Down
NIR	Near-InfraRed
NOAA	National Oceanic and Atmospheric Administration
NPP	National Polar-orbiting Partnership
NRT	Near-Real-Time
OLI	Operational Land Imager
PISA	Pixelwise Image Saliency Aggregating
POES	Polar-orbiting Operational Environmental Satellite
R-CNN	Region-Based Convolutional Neural Networks
ReLU	Rectified Linear Unit
RST	Robust Satellite Techniques
SEVIRI	Spinning Enhanced Visible and Infrared Imager

SFIDE	Satellite Fire Detection
sh-LDS	stabilized h-LDS
SLAM	Simultaneous Localization and Mapping
SroFs	Suspected Regions of Fire
SSO	Sun-Synchronous Orbit
STCM	Spatio–temporal Contextual Model
STM	Spatio–Temporal Model
SVD	Single Value Decomposition
SVM	Support Vector Machine
UAV	Unmanned Aerial Vehicles
VIIRS	Visible Infrared Imaging Radiometer Suite
WoS	Web of Science
WSN	Wireless Sensor Network

References

1. Tanase, M.A.; Aponte, C.; Mermoz, S.; Bouvet, A.; Le Toan, T.; Heurich, M. Detection of windthrows and insect outbreaks by L-band SAR: A case study in the Bavarian Forest National Park. *Remote Sens. Environ.* **2018**, *209*, 700–711. [CrossRef]
2. Pradhan, B.; Suliman, M.D.H.B.; Awang, M.A.B. Forest fire susceptibility and risk mapping using remote sensing and geographical information systems (GIS). *Disaster Prev. Manag. Int. J.* **2007**, *16*. [CrossRef]
3. Kresek, R. History of the Osborne Firefinder. 2007. Available online: http://nysforestrangers.com/archives/osborne%20firefinder%20by%20kresek.pdf (accessed on 7 September 2020).
4. Bouabdellah, K.; Noureddine, H.; Larbi, S. Using wireless sensor networks for reliable forest fires detection. *Procedia Comput. Sci.* **2013**, *19*, 794–801. [CrossRef]
5. Gaur, A.; Singh, A.; Kumar, A.; Kulkarni, K.S.; Lala, S.; Kapoor, K.; Srivastava, V.; Kumar, A.; Mukhopadhyay, S.C. Fire sensing technologies: A review. *IEEE Sens. J.* **2019**, *19*, 3191–3202. [CrossRef]
6. Gaur, A.; Singh, A.; Kumar, A.; Kumar, A.; Kapoor, K. Video Flame and Smoke Based Fire Detection Algorithms: A Literature Review. *Fire Technol.* **2020**, *56*, 1943–1980. [CrossRef]
7. Kaabi, R.; Frizzi, S.; Bouchouicha, M.; Fnaiech, F.; Moreau, E. Video smoke detection review: State of the art of smoke detection in visible and IR range. In Proceedings of the 2017 International Conference on Smart, Monitored and Controlled Cities (SM2C), Kerkennah-Sfax, Tunisia, 17 February 2017; pp. 81–86.
8. Garg, S.; Verma, A.A. Review Survey on Smoke Detection. *Imp. J. Interdiscip. Res.* **2016**, *2*, 935–939.
9. Memane, S.E.; Kulkarni, V.S. A review on flame and smoke detection techniques in video's. *Int. J. Adv. Res. Electr. Electr. Instrum. Eng.* **2015**, *4*, 885–889.
10. Bu, F.; Gharajeh, M.S. Intelligent and vision-based fire detection systems: A survey. *Image Vis. Comput.* **2019**, *91*, 103803. [CrossRef]
11. Yuan, C.; Zhang, Y.; Liu, Z. A survey on technologies for automatic forest fire monitoring, detection, and fighting using unmanned aerial vehicles and remote sensing techniques. *Can. J. For. Res.* **2015**, *45*, 783–792. [CrossRef]
12. Allison, R.S.; Johnston, J.M.; Craig, G.; Jennings, S. Airborne optical and thermal remote sensing for wildfire detection and monitoring. *Sensors* **2016**, *16*, 1310. [CrossRef]
13. Nixon, M.; Aguado, A. *Feature Extraction and Image Processing for Computer Vision*; Academic Press: Cambridge, MA, USA, 2019.
14. Mather, P.; Tso, B. *Classification Methods for Remotely Sensed Data*; CRC Press: Boca Raton, FL, USA, 2016.
15. Çetin, A.E.; Dimitropoulos, K.; Gouverneur, B.; Grammalidis, N.; Günay, O.; Habiboğlu, Y.H.; Töreyin, B.U.; Verstockt, S. Video fire detection–review. *Dig. Signal Process.* **2013**, *23*, 1827–1843. [CrossRef]
16. Töreyin, B.U.; Cinbis, R.G.; Dedeoglu, Y.; Cetin, A.E. Fire detection in infrared video using wavelet analysis. *Opt. Eng.* **2007**, *46*, 067204. [CrossRef]
17. Cappellini, Y.; Mattii, L.; Mecocci, A. An Intelligent System for Automatic Fire Detection in Forests. In *Recent Issues in Pattern Analysis and Recognition*; Springer: Berlin/Heidelberg, Germany, 1989; pp. 563–570.
18. Chen, T.H.; Wu, P.H.; Chiou, Y.C. An early fire-detection method based on image processing. In Proceedings of the 2004 International Conference on Image Processing (ICIP 04), Singapore, 24–27 October 2004; Volume 3, pp. 1707–1710.

19. Dimitropoulos, K.; Gunay, O.; Kose, K.; Erden, F.; Chaabene, F.; Tsalakanidou, F.; Grammalidis, N.; Çetin, E. Flame detection for video-based early fire warning for the protection of cultural heritage. In Proceedings of the Euro-Mediterranean Conference, Limassol, Cyprus, 29 October–3 November 2012; Springer: Berlin/Heidelberg, Germany; pp. 378–387.
20. Celik, T.; Demirel, H. Fire detection in video sequences using a generic color model. *Fire Saf. J.* **2009**, *44*, 147–158. [CrossRef]
21. Celik, T. Fast and efficient method for fire detection using image processing. *ETRI J.* **2010**, *32*, 881–890. [CrossRef]
22. Marbach, G.; Loepfe, M.; Brupbacher, T. An image processing technique for fire detection in video images. *Fire Saf. J.* **2006**, *41*, 285–289. [CrossRef]
23. Kim, D.; Wang, Y.F. Smoke detection in video. In Proceedings of the 2009 WRI World Congress on Computer Science and Information Engineering, Los Angeles, CA, USA, 31 March–2 April 2009; Volume 5, pp. 759–763.
24. Yamagishi, H.; Yamaguchi, J. Fire flame detection algorithm using a color camera. In Proceedings of the MHS'99, 1999 International Symposium on Micromechatronics and Human Science (Cat. No. 99TH8478), Nagoya, Japan, 23–26 November 1999; pp. 255–260.
25. Dimitropoulos, K.; Tsalakanidou, F.; Grammalidis, N. Flame detection for video-based early fire warning systems and 3D visualization of fire propagation. In Proceedings of the 13th IASTED International Conference on Computer Graphics and Imaging (CGIM 2012), Crete, Greece, 18–20 June 2012. Available online: https://zenodo.org/record/1218#.X6qSVmj7Sbg (accessed on 10 November 2020).
26. Zhang, Z.; Shen, T.; Zou, J. An improved probabilistic approach for fire detection in videos. *Fire Technol.* **2014**, *50*, 745–752. [CrossRef]
27. Avgerinakis, K.; Briassouli, A.; Kompatsiaris, I. Smoke detection using temporal HOGHOF descriptors and energy colour statistics from video. In Proceedings of the International Workshop on Multi-Sensor Systems and Networks for Fire Detection and Management, Antalya, Turkey, 8–9 November 2012.
28. Mueller, M.; Karasev, P.; Kolesov, I.; Tannenbaum, A. Optical flow estimation for flame detection in videos. *IEEE Trans. Image Process.* **2013**, *22*, 2786–2797. [CrossRef]
29. Chamberlin, D.S.; Rose, A. The First Symposium (International) on Combustion. *Combust. Inst. Pittsburgh* **1965**, *1965*, 27–32.
30. Günay, O.; Taşdemir, K.; Töreyin, B.U.; Çetin, A.E. Fire detection in video using LMS based active learning. *Fire Technol.* **2010**, *46*, 551–577. [CrossRef]
31. Teng, Z.; Kim, J.H.; Kang, D.J. Fire detection based on hidden Markov models. *Int. J. Control Autom. Syst.* **2010**, *8*, 822–830. [CrossRef]
32. Töreyin, B.U. Smoke detection in compressed video. In *Applications of Digital Image Processing XLI*; International Society for Optics and Photonics: Bellingham, WA, USA, 2018; Volume 10752, p. 1075232.
33. Savcı, M.M.; Yıldırım, Y.; Saygılı, G.; Töreyin, B.U. Fire detection in H. 264 compressed video. In Proceedings of the ICASSP 2019–2019 IEEE International Conference on Acoustics, Speech and Signal Processing (ICASSP), Brighton, UK, 12–17 May 2019; pp. 8310–8314.
34. Chen, J.; He, Y.; Wang, J. Multi-feature fusion based fast video flame detection. *Build. Environ.* **2010**, *45*, 1113–1122. [CrossRef]
35. Töreyin, B.U.; Dedeoğlu, Y.; Güdükbay, U.; Cetin, A.E. Computer vision based method for real-time fire and flame detection. *Pattern Recognit. Lett.* **2006**, *27*, 49–58. [CrossRef]
36. Barmpoutis, P.; Dimitropoulos, K.; Grammalidis, N. Real time video fire detection using spatio-temporal consistency energy. In Proceedings of the 10th IEEE International Conference on Advanced Video and Signal Based Surveillance, Krakow, Poland, 27–30 August 2013; pp. 365–370.
37. Arrue, B.C.; Ollero, A.; De Dios, J.M. An intelligent system for false alarm reduction in infrared forest-fire detection. *IEEE Intell. Syst. Their Appl.* **2000**, *15*, 64–73. [CrossRef]
38. Grammalidis, N.; Cetin, E.; Dimitropoulos, K.; Tsalakanidou, F.; Kose, K.; Gunay, O.; Gouverneur, B.; Torri, D.; Kuruoglu, E.; Tozzi, S.; et al. A Multi-Sensor Network for the Protection of Cultural Heritage. In Proceedings of the 19th European Signal Processing Conference, Barcelona, Spain, 29 August–2 September 2011.
39. Bosch, I.; Serrano, A.; Vergara, L. Multisensor network system for wildfire detection using infrared image processing. *Sci. World J.* **2013**, *2013*, 402196. [CrossRef]

40. Barmpoutis, P.; Dimitropoulos, K.; Grammalidis, N. Smoke detection using spatio-temporal analysis, motion modeling and dynamic texture recognition. In Proceedings of the 22nd European Signal Processing Conference, Lisbon, Portugal, 1–5 September 2014; pp. 1078–1082.
41. Dimitropoulos, K.; Barmpoutis, P.; Grammalidis, N. Spatio-temporal flame modeling and dynamic texture analysis for automatic video-based fire detection. *IEEE Trans. Circuits Syst. Video Technol.* **2015**, *25*, 339–351. [CrossRef]
42. Prema, C.E.; Vinsley, S.S.; Suresh, S. Efficient flame detection based on static and dynamic texture analysis in forest fire detection. *Fire Technol.* **2018**, *54*, 255–288. [CrossRef]
43. Dimitropoulos, K.; Barmpoutis, P.; Grammalidis, N. Higher order linear dynamical systems for smoke detection in video surveillance applications. *IEEE Trans. Circuits Syst. Video Technol.* **2016**, *27*, 1143–1154. [CrossRef]
44. Dimitropoulos, K.; Barmpoutis, P.; Kitsikidis, A.; Grammalidis, N. Classification of multidimensional time-evolving data using histograms of grassmannian points. *IEEE Trans. Circuits Syst. Video Technol.* **2016**, *28*, 892–905. [CrossRef]
45. Schmidhuber, J. Deep learning in neural networks: An overview. *Neural Netw.* **2015**, *61*, 85–117. [CrossRef]
46. Krizhevsky, A.; Sutskever, I.; Hinton, G.E. Imagenet classification with deep convolutional neural networks. In *Advances in Neural Information Processing Systems*; MIT Press: Cambridge, MA, USA, 2012; pp. 1097–1105.
47. Luo, Y.; Zhao, L.; Liu, P.; Huang, D. Fire smoke detection algorithm based on motion characteristic and convolutional neural networks. *Multimed. Tools Appl.* **2018**, *77*, 15075–15092. [CrossRef]
48. Zhao, L.; Luo, Y.M.; Luo, X.Y. Based on dynamic background update and dark channel prior of fire smoke detection algorithm. *Appl. Res. Comput.* **2017**, *34*, 957–960.
49. Wu, X.; Lu, X.; Leung, H. An adaptive threshold deep learning method for fire and smoke detection. In Proceedings of the 2017 IEEE International Conference on Systems, Man, and Cybernetics (SMC), Banff, AB, Canada, 5–8 October 2017; pp. 1954–1959.
50. Sharma, J.; Granmo, O.C.; Goodwin, M.; Fidje, J.T. Deep convolutional neural networks for fire detection in images. In Proceedings of the International Conference on Engineering Applications of Neural Networks, Athens, Greece, 25–27 August 2017; Springer: Berlin/Heidelberg, Germany, 2017; pp. 183–193.
51. Zhang, Q.; Xu, J.; Xu, L.; Guo, H. Deep convolutional neural networks for forest fire detection. In *2016 International Forum on Management, Education and Information Technology Application*; Atlantis Press: Paris, France, 2016.
52. Muhammad, K.; Ahmad, J.; Mehmood, I.; Rho, S.; Baik, S.W. Convolutional neural networks based fire detection in surveillance videos. *IEEE Access* **2018**, *6*, 18174–18183. [CrossRef]
53. Shen, D.; Chen, X.; Nguyen, M.; Yan, W.Q. Flame detection using deep learning. In Proceedings of the 2018 4th International Conference on Control, Automation and Robotics (ICCAR), Auckland, New Zealand, 20–23 April 2018; pp. 416–420.
54. Frizzi, S.; Kaabi, R.; Bouchouicha, M.; Ginoux, J.M.; Moreau, E.; Fnaiech, F. Convolutional neural network for video fire and smoke detection. In Proceedings of the IECON 2016-42nd Annual Conference of the IEEE Industrial Electronics Society, Florence, Italy, 23–26 October 2016; pp. 877–882.
55. Muhammad, K.; Ahmad, J.; Lv, Z.; Bellavista, P.; Yang, P.; Baik, S.W. Efficient deep CNN-based fire detection and localization in video surveillance applications. *IEEE Trans. Syst. Man Cybern. Syst.* **2018**, *49*, 1419–1434. [CrossRef]
56. Muhammad, K.; Ahmad, J.; Baik, S.W. Early fire detection using convolutional neural networks during surveillance for effective disaster management. *Neurocomputing* **2018**, *288*, 30–42. [CrossRef]
57. Dunnings, A.J.; Breckon, T.P. Experimentally defined convolutional neural network architecture variants for non-temporal real-time fire detection. In Proceedings of the 2018 25th IEEE International Conference on Image Processing (ICIP), Athens, Greece, 7–10 October 2018; pp. 1558–1562.
58. Sousa, M.J.; Moutinho, A.; Almeida, M. Wildfire detection using transfer learning on augmented datasets. *Expert Syst. Appl.* **2020**, *142*, 112975. [CrossRef]
59. Toulouse, T.; Rossi, L.; Campana, A.; Celik, T.; Akhloufi, M.A. Computer vision for wildfire research: An evolving image dataset for processing and analysis. *Fire Saf. J.* **2017**, *92*, 188–194. [CrossRef]
60. Barmpoutis, P.; Dimitropoulos, K.; Kaza, K.; Grammalidis, N. Fire Detection from Images Using Faster R-CNN and Multidimensional Texture Analysis. In Proceedings of the ICASSP 2019–2019 IEEE International Conference on Acoustics, Speech and Signal Processing, Brighton, UK, 12–17 May 2019; pp. 8301–8305.

61. Lin, G.; Zhang, Y.; Xu, G.; Zhang, Q. Smoke detection on video sequences using 3D convolutional neural networks. *Fire Technol.* **2019**, *55*, 1827–1847. [CrossRef]
62. Jadon, A.; Omama, M.; Varshney, A.; Ansari, M.S.; Sharma, R. Firenet: A specialized lightweight fire & smoke detection model for real-time iot applications. *arXiv* **2019**, arXiv:1905.11922.
63. Zhang, Q.X.; Lin, G.H.; Zhang, Y.M.; Xu, G.; Wang, J.J. Wildland forest fire smoke detection based on faster R-CNN using synthetic smoke images. *Procedia Eng.* **2018**, *211*, 441–446. [CrossRef]
64. Kim, B.; Lee, J. A video-based fire detection using deep learning models. *Appl. Sci.* **2019**, *9*, 2862. [CrossRef]
65. Shi, L.; Long, F.; Lin, C.; Zhao, Y. Video-based fire detection with saliency detection and convolutional neural networks. In *International Symposium on Neural Networks*; Springer: Cham, Switzerland, 2017; pp. 299–309.
66. Wang, K.; Lin, L.; Lu, J.; Li, C.; Shi, K. PISA: Pixelwise image saliency by aggregating complementary appearance contrast measures with edge-preserving coherence. *IEEE Trans. Image Process.* **2015**, *9*, 2115–2122. [CrossRef] [PubMed]
67. Yuan, F.; Zhang, L.; Xia, X.; Wan, B.; Huang, Q.; Li, X. Deep smoke segmentation. *Neurocomputing* **2019**, *357*, 248–260. [CrossRef]
68. Cheng, S.; Ma, J.; Zhang, S. Smoke detection and trend prediction method based on Deeplabv3+ and generative adversarial network. *J. Electron. Imaging* **2019**, *28*, 033006. [CrossRef]
69. Aslan, S.; Güdükbay, U.; Töreyin, B.U.; Çetin, A.E. Early wildfire smoke detection based on motion-based geometric image transformation and deep convolutional generative adversarial networks. In Proceedings of the ICASSP 2019–2019 IEEE International Conference on Acoustics, Speech and Signal Processing (ICASSP), Brighton, UK, 12–17 May 2019; pp. 8315–8319.
70. Hristov, G.; Raychev, J.; Kinaneva, D.; Zahariev, P. Emerging methods for early detection of forest fires using unmanned aerial vehicles and lorawan sensor networks. In Proceedings of the IEEE 28th EAEEIE Annual Conference, Hafnarfjordur, Iceland, 26–28 September 2018; pp. 1–9.
71. Stearns, J.R.; Zahniser, M.S.; Kolb, C.E.; Sandford, B.P. Airborne infrared observations and analyses of a large forest fire. *Appl. Opt.* **1986**, *25*, 2554–2562. [CrossRef] [PubMed]
72. Den Breejen, E.; Breuers, M.; Cremer, F.; Kemp, R.; Roos, M.; Schutte, K.; De Vries, J.S. *Autonomous Forest Fire Detection*; ADAI-Associacao para o Desenvolvimento da Aerodinamica Industrial: Coimbra, Portugal, 1998; pp. 2003–2012.
73. Yuan, C.; Liu, Z.; Zhang, Y. UAV-based forest fire detection and tracking using image processing techniques. In Proceedings of the 2015 International Conference on Unmanned Aircraft Systems (ICUAS), Denver, CO, USA, 9–12 June 2015; pp. 639–643.
74. Yuan, C.; Liu, Z.; Zhang, Y. Learning-based smoke detection for unmanned aerial vehicles applied to forest fire surveillance. *J. Intell. Robot. Syst.* **2019**, *93*, 337–349. [CrossRef]
75. Dang-Ngoc, H.; Nguyen-Trung, H. Aerial Forest Fire Surveillance-Evaluation of Forest Fire Detection Model using Aerial Videos. In Proceedings of the 2019 International Conference on Advanced Technologies for Communications (ATC), Hanoi, Vietnam, 17–19 October 2019; pp. 142–148.
76. Yuan, C.; Liu, Z.; Zhang, Y. Aerial images-based forest fire detection for firefighting using optical remote sensing techniques and unmanned aerial vehicles. *J. Intell. Robot. Syst.* **2017**, *88*, 635–654. [CrossRef]
77. De Sousa, J.V.R.; Gamboa, P.V. Aerial Forest Fire Detection and Monitoring Using a Small UAV. *KnE Eng.* **2020**, 242–256. [CrossRef]
78. Esfahlani, S.S. Mixed reality and remote sensing application of unmanned aerial vehicle in fire and smoke detection. *J. Ind. Inf. Integr.* **2019**, *15*, 42–49. [CrossRef]
79. Sudhakar, S.; Vijayakumar, V.; Kumar, C.S.; Priya, V.; Ravi, L.; Subramaniyaswamy, V. Unmanned Aerial Vehicle (UAV) based Forest Fire Detection and monitoring for reducing false alarms in forest-fires. *Comput. Commun.* **2020**, *149*, 1–16. [CrossRef]
80. Kinaneva, D.; Hristov, G.; Raychev, J.; Zahariev, P. Early forest fire detection using drones and artificial intelligence. In Proceedings of the 2019 42nd International Convention on Information and Communication Technology, Electronics and Microelectronics (MIPRO), Opatija, Croatia, 20–24 May 2019; pp. 1060–1065.
81. Chen, Y.; Zhang, Y.; Xin, J.; Yi, Y.; Liu, D.; Liu, H. A UAV-based Forest Fire Detection Algorithm Using Convolutional Neural Network. In Proceedings of the IEEE 37th Chinese Control Conference, Wuhan, China, 25–27 July 2018; pp. 10305–10310.
82. Merino, L.; Caballero, F.; Martínez-De-Dios, J.R.; Maza, I.; Ollero, A. An unmanned aircraft system for automatic forest fire monitoring and measurement. *J. Intell. Robot. Syst.* **2012**, *65*, 533–548. [CrossRef]

83. Zhao, Y.; Ma, J.; Li, X.; Zhang, J. Saliency detection and deep learning-based wildfire identification in UAV imagery. *Sensors* **2018**, *18*, 712. [CrossRef] [PubMed]
84. Tang, Z.; Liu, X.; Chen, H.; Hupy, J.; Yang, B. Deep Learning Based Wildfire Event Object Detection from 4K Aerial Images Acquired by UAS. *AI* **2020**, *1*, 166–179. [CrossRef]
85. Jiao, Z.; Zhang, Y.; Mu, L.; Xin, J.; Jiao, S.; Liu, H.; Liu, D. A YOLOv3-based Learning Strategy for Real-time UAV-based Forest Fire Detection. In Proceedings of the 2020 Chinese Control and Decision Conference (CCDC), Hefei, China, 22–24 August 2020; pp. 4963–4967.
86. Jiao, Z.; Zhang, Y.; Xin, J.; Mu, L.; Yi, Y.; Liu, H.; Liu, D. A Deep Learning Based Forest Fire Detection Approach Using UAV and YOLOv3. In Proceedings of the 2019 1st International Conference on Industrial Artificial Intelligence (IAI), Shenyang, China, 23–27 July 2019; pp. 1–5.
87. Srinivas, K.; Dua, M. Fog Computing and Deep CNN Based Efficient Approach to Early Forest Fire Detection with Unmanned Aerial Vehicles. In Proceedings of the International Conference on Inventive Computation Technologies, Coimbatore, India, 29–30 August 2019; Springer: Cham, Switzerland, 2020; pp. 646–652.
88. Barmpoutis, P.; Stathaki, T. A Novel Framework for Early Fire Detection Using Terrestrial and Aerial 360-Degree Images. In Proceedings of the International Conference on Advanced Concepts for Intelligent Vision Systems, Auckland, New Zealand, 10–14 February 2020; Springer: Cham, Switzerland, 2020; pp. 63–74.
89. Barmpoutis, P.; Stathaki, T.; Dimitropoulos, K.; Grammalidis, N. Early Fire Detection Based on Aerial 360-Degree Sensors, Deep Convolution Neural Networks and Exploitation of Fire Dynamic Textures. *Remote Sens.* **2020**, *12*, 3177. [CrossRef]
90. He, L.; Li, Z. Enhancement of a fire detection algorithm by eliminating solar reflection in the mid-IR band: Application to AVHRR data. *Int. J. Remote Sens.* **2012**, *33*, 7047–7059. [CrossRef]
91. He, L.; Li, Z. Enhancement of fire detection algorithm by eliminating solar contamination effect and atmospheric path radiance: Application to MODIS data. *Int. J. Remote Sens.* **2011**, *32*, 6273–6293. [CrossRef]
92. Csiszar, I.; Schroeder, W.; Giglio, L.; Ellicott, E.; Vadrevu, K.P.; Justice, C.O.; Wind, B. Active fires from the Suomi NPP Visible Infrared Imaging Radiometer, Suite: Product Status and first evaluation results. *J. Geophys. Res. Atmos.* **2014**, *119*, 803–816. [CrossRef]
93. Schroeder, W.; Oliva, P.; Giglio, L.; Csiszar, I.A. The New VIIRS 375m Active Fire Detection Data Product: Algorithm Description and Initial Assessment. *Remote Sens. Environ.* **2014**, *143*, 85–96. [CrossRef]
94. Sayad, Y.O.; Mousannif, H.; Al Moatassime, H. Predictive modeling of wildfires: A new dataset and machine learning approach. *Fire Saf. J.* **2019**, *104*, 130–146. [CrossRef]
95. Shukla, B.P.; Pal, P.K. Automatic smoke detection using satellite imagery: Preparatory to smoke detection from Insat-3D. *Int. J. Remote Sens.* **2009**, *30*, 9–22. [CrossRef]
96. Li, X.; Song, W.; Lian, L.; Wei, X. Forest fire smoke detection using back-propagation neural network based on MODIS data. *Remote Sens.* **2015**, *7*, 4473–4498. [CrossRef]
97. Li, Z.; Khananian, A.; Fraser, R.H.; Cihlar, J. Automatic detection of fire smoke using artificial neural networks and threshold approaches applied to AVHRR imagery. *IEEE Trans. Geosci. Remote Sens.* **2001**, *39*, 1859–1870.
98. Hally, B.; Wallace, L.; Reinke, K.; Jones, S.; Skidmore, A. Advances in active fire detection using a multi-temporal method for next-generation geostationary satellite data. *Int. J. Dig. Earth* **2019**, *12*, 1030–1045. [CrossRef]
99. Giglio, L.; Schroeder, W.; Justice, C.O. The Collection 6 MODIS Active Fire Detection Algorithm and Fire Products. *Remote Sens. Environ.* **2016**, *178*, 31–41. [CrossRef]
100. Wickramasinghe, C.; Wallace, L.; Reinke, K.; Jones, S. Intercomparison of Himawari-8 AHI-FSA with MODIS and VIIRS active fire products. *Int. J. Dig. Earth* **2018**. [CrossRef]
101. Lin, L.; Meng, Y.; Yue, A.; Yuan, Y.; Liu, X.; Chen, J.; Zhang, M.; Chen, J. A spatio-temporal model for forest fire detection using HJ-IRS satellite data. *Remote Sens.* **2016**, *8*, 403. [CrossRef]
102. Ba, R.; Chen, C.; Yuan, J.; Song, W.; Lo, S. SmokeNet: Satellite smoke scene detection using convolutional neural network with spatial and channel-wise attention. *Remote Sens.* **2019**, *11*, 1702. [CrossRef]
103. Vani, K. Deep Learning Based Forest Fire Classification and Detection in Satellite Images. In Proceedings of the 2019 11th International Conference on Advanced Computing (ICoAC), Chennai, India, 18–20 December 2019; pp. 61–65.
104. Hally, B.; Wallace, L.; Reinke, K.; Jones, S. A Broad-Area Method for the Diurnal Characterisation of Upwelling Medium Wave Infrared Radiation. *Remote Sens.* **2017**, *9*, 167. [CrossRef]

105. Fatkhuroyan, T.W.; Andersen, P. Forest fires detection in Indonesia using satellite Himawari-8 (case study: Sumatera and Kalimantan on august-october 2015). In *IOP Conference Series: Earth and Environmental Science*; IOP Publishing Ltd.: Bristol, UK, 2017; Volume 54, pp. 1315–1755.
106. Xu, G.; Zhong, X. Real-time wildfire detection and tracking in Australia using geostationary satellite: Himawari-8. *Remote Sens. Lett.* **2017**, *8*, 1052–1061. [CrossRef]
107. Xie, Z.; Song, W.; Ba, R.; Li, X.; Xia, L. A spatiotemporal contextual model for forest fire detection using Himawari-8 satellite data. *Remote Sens.* **2018**, *10*, 1992. [CrossRef]
108. Filizzola, C.; Corrado, R.; Marchese, F.; Mazzeo, G.; Paciello, R.; Pergola, N.; Tramutoli, V. RST-FIRES, an exportable algorithm for early-fire detection and monitoring: Description, implementation, and field validation in the case of the MSG-SEVIRI sensor. *Remote Sens. Environ.* **2017**, *192*, e2–e25. [CrossRef]
109. Di Biase, V.; Laneve, G. Geostationary sensor based forest fire detection and monitoring: An improved version of the SFIDE algorithm. *Remote Sens.* **2018**, *10*, 741. [CrossRef]
110. Laneve, G.; Castronuovo, M.M.; Cadau, E.G. Continuous monitoring of forest fires in the Mediterranean area using MSG. *IEEE Trans. Geosci. Remote Sens.* **2006**, *44*, 2761–2768. [CrossRef]
111. Hall, J.V.; Zhang, R.; Schroeder, W.; Huang, C.; Giglio, L. Validation of GOES-16 ABI and MSG SEVIRI active fire products. *Int. J. Appl. Earth Obs. Geoinf.* **2019**, *83*, 101928. [CrossRef]
112. Schroeder, W.; Oliva, P.; Giglio, L.; Quayle, B.; Lorenz, E.; Morelli, F. Active fire detection using Landsat-8/OLI data. *Remote Sens. Environ.* **2016**, *185*, 210–220. [CrossRef]
113. Larsen, A.; Hanigan, I.; Reich, B.J.; Qin, Y.; Cope, M.; Morgan, G.; Rappold, A.G. A deep learning approach to identify smoke plumes in satellite imagery in near-real time for health risk communication. *J. Expo. Sci. Environ. Epidemiol.* **2020**, 1–7. [CrossRef]
114. Phan, T.C.; Nguyen, T.T. Remote Sensing Meets Deep Learning: Exploiting Spatio-Temporal-Spectral Satellite Images for Early Wildfire Detection. No. REP_WORK. 2019. Available online: https://infoscience.epfl.ch/record/270339 (accessed on 7 September 2020).
115. Cal Poly, S.L.O. The CubeSat Program, CubeSat Design Specification Rev. 13. 2014. Available online: http://blogs.esa.int/philab/files/2019/11/RD-02_CubeSat_Design_Specification_Rev._13_The.pdf (accessed on 7 September 2020).
116. Barschke, M.F.; Bartholomäus, J.; Gordon, K.; Lehmenn, M.; Brie, K. The TUBIN nanosatellite mission for wildfire detection in thermal infrared. *CEAS Space* **2017**, *9*, 183–194. [CrossRef]
117. Kameche, M.; Benzeniar, H.; Benbouzid, A.B.; Amri, R.; Bouanani, N. Disaster monitoring constellation using nanosatellites. *J. Aerosp. Technol. Manag.* **2014**, *6*, 93–100. [CrossRef]
118. MODIS—Moderate Resolution Imaging Spectroradiometer, Specifications. Available online: https://modis.gsfc.nasa.gov/about/specifications.php (accessed on 15 September 2020).
119. Himawari-8 and 9, Specifications. Available online: https://earth.esa.int/web/eoportal/satellite-missions/h/himawari-8-9 (accessed on 15 September 2020).
120. The SEVIRI Instrument. Available online: https://www.eumetsat.int/website/wcm/idc/groups/ops/documents/document/mday/mde1/~{}edisp/pdf_ten_msg_seviri_instrument.pdf (accessed on 15 September 2020).
121. GOES-16ABI, Specifications. Available online: https://www.goes-r.gov/spacesegment/abi.html (accessed on 15 September 2020).
122. Huan Jing-1: Environmental Protection & Disaster Monitoring Constellation. Available online: https://earth.esa.int/web/eoportal/satellite-missions/h/hj-1 (accessed on 15 September 2020).
123. POES Series, Specifications. Available online: https://directory.eoportal.org/web/eoportal/satellite-missions/n/noaa-poes-series-5th-generation (accessed on 15 September 2020).
124. VIIRS-375m. Available online: https://earthdata.nasa.gov/earth-observation-data/near-real-time/firms/viirs-i-band-active-fire-data (accessed on 15 September 2020).
125. Visible Infrared Imaging Radiometer Suite (VIIRS) 375m Active Fire Detection and Characterization Algorithm Theoretical Basis Document 1.0. Available online: https://viirsland.gsfc.nasa.gov/PDF/VIIRS_activefire_375m_ATBD.pdf (accessed on 15 September 2020).
126. Shah, S.B.; Grübler, T.; Krempel, L.; Ernst, S.; Mauracher, F.; Contractor, S. Real-time wildfire detection from space—A trade-off between sensor quality, physical limitations and payload size. *Int. Arch. Photogramm. Remote Sens. Spat. Inf. Sci.* **2019**. [CrossRef]

127. Pérez-Lissi, F.; Aguado-Agelet, F.; Vázquez, A.; Yañez, P.; Izquierdo, P.; Lacroix, S.; Bailon-Ruiz, R.; Tasso, J.; Guerra, A.; Costa, M. FIRE-RS: Integrating land sensors, cubesat communications, unmanned aerial vehicles and a situation assessment software for wildland fire characterization and mapping. In Proceedings of the 69th International Astronautical Congress, Bremen, Germany, 1–5 October 2018.
128. Escrig, A.; Liz, J.L.; Català, J.; Verda, V.; Kanterakis, G.; Carvajal, F.; Pérez, I.; Lewinski, S.; Wozniak, E.; Aleksandrowicz, S.; et al. Advanced Forest Fire Fighting (AF3) European Project, preparedness for and management of large scale forest fires. In Proceedings of the XIV World Forestry Congress 2015, Durban, South Africa, 7–11 September 2015.
129. Bielski, C.; O'Brien, V.; Whitmore, C.; Ylinen, K.; Juga, I.; Nurmi, P.; Kilpinen, J.; Porras, I.; Sole, J.M.; Gamez, P.; et al. Coupling early warning services, crowdsourcing, and modelling for improved decision support and wildfire emergency management. In Proceedings of the IEEE International Conference on Big Data (Big Data), Boston, MA, USA, 11–14 December 2017; pp. 3705–3712.
130. European Forest Fire Information System (EFFIS). Available online: https://effis.jrc.ec.europa.eu/ (accessed on 2 November 2020).
131. NASA Tracks Wildfires From Above to Aid Firefighters Below. Available online: https://www.nasa.gov/feature/goddard/2019/nasa-tracks-wildfires-from-above-to-aid-firefighters-below (accessed on 2 November 2020).
132. Govil, K.; Welch, M.L.; Ball, J.T.; Pennypacker, C.R. Preliminary Results from a Wildfire Detection System Using Deep Learning on Remote Camera Images. *Remote Sens.* **2020**, *12*, 166. [CrossRef]
133. MODIS Data Product Non-Technical Description—MOD 14. Available online: https://modis.gsfc.nasa.gov/data/dataprod/nontech/MOD14.php (accessed on 1 November 2020).
134. Web of Science. Available online: http://apps.webofknowledge.com/ (accessed on 1 November 2020).

Publisher's Note: MDPI stays neutral with regard to jurisdictional claims in published maps and institutional affiliations.

© 2020 by the authors. Licensee MDPI, Basel, Switzerland. This article is an open access article distributed under the terms and conditions of the Creative Commons Attribution (CC BY) license (http://creativecommons.org/licenses/by/4.0/).

sensors

MDPI

Article

A Fast Segmentation Method for Fire Forest Images Based on Multiscale Transform and PCA

Lotfi Tlig [1,*], Moez Bouchouicha [2], Mohamed Tlig [1,2], Mounir Sayadi [1] and Eric Moreau [2]

1 Member of SIME Laboratory, ENSIT University of Tunis, Tunis 1008, Tunisia; mohamed.tlig@ensit.u-tunis.tn (M.T.); mounirsayadi@yahoo.fr (M.S.)

2 Aix Marseille Univ, Université de Toulon, CNRS, LIS, 83041 Toulon, France; moez.bouchouicha@univ-tln.fr (M.B.); eric.moreau@univ-tln.fr (E.M.)

* Correspondence: lotfi.tlig@isimg.tn

Received: 15 October 2020; Accepted: 5 November 2020; Published: 10 November 2020

Abstract: Forests provide various important things to human life. Fire is one of the main disasters in the world. Nowadays, the forest fire incidences endanger the ecosystem and destroy the native flora and fauna. This affects individual life, community and wildlife. Thus, it is essential to monitor and protect the forests and their assets. Nowadays, image processing outputs a lot of required information and measures for the implementation of advanced forest fire-fighting strategies. This work addresses a new color image segmentation method based on principal component analysis (PCA) and Gabor filter responses. Our method introduces a new superpixels extraction strategy that takes full account of two objectives: regional consistency and robustness to added noises. The novel approach is tested on various color images. Extensive experiments show that our method obviously outperforms existing segmentation variants on real and synthetic images of fire forest scenes, and also achieves outstanding performance on other popular benchmarked images (e.g., BSDS, MRSC). The merits of our proposed approach are that it is not sensitive to added noises and that the segmentation performance is higher with images of nonhomogeneous regions.

Keywords: Gabor filtering; PCA morphological transformations; fuzzy clustering; color image segmentation; fire forest

1. Introduction

The conventional detection systems of smoke and fire use sensors [1]. One of the major drawbacks, is that the systems do not issue the alarm unless the particles reach the sensors [2]. Recently, as an appropriate alternative to conventional techniques, vision-based fire and smoke detection methods have been adopted. Here, smoke and fire are regarded as a specific kind of texture. It is difficult to accurately detect the appearance of mentioned regions from images due to large variations of color intensities and texture. Although, many research works confirmed that texture features play a very important role in smoke and fire detection [3,4]. A wide recent work demonstrated that the multi-scale based techniques play an important role in smoke and texture classification [5,6]. Developed methods cover both areas; images and videos processing [4,7,8]. In this work, we aim to segment images into significant regions. This will be used to generate useful information for our project. In this paper, we propose a new segmentation approach based on Gabor filtering and Principal Component Analysis (PCA). The proposed method is based on the modification of the superpixels extraction methodology to increase the robustness to added noises and to improve the segmentation accuracy of the fire forests color images. For the extracted features clustering, we used the new version of the fuzzy classifier recently proposed in [9]. Our choice is done regarding the higher performance of the mentioned fuzzy method compared to a large variety of clustering methods; FCM, FGFCM, HMRF-FCM, FLICM, NWFCM, KWFLICM, NDFCM, FR- FCM, and Liu's algorithm. In [9], Lei et al. introduced a fast fuzzy clustering algorithm to address the problem of computational segmentation complexity of color image segmentation with higher resolution. This was conducted by the use of adaptive local spatial information provided by a pre-segmentation task. Despite its higher accuracy compared to a large number of used algorithms, this method remains limited due to many drawbacks. This is experimentally noted with blurred images and images with nonhomogeneous regions. In this work, our research is focused on the part where we have cited a lower efficiency. It is the task of superpixels extraction. As mentioned above, we have introduced the multiscale image transformation based on the Gabor filtering.

Many applications approved that Gabor feature sets are high-dimensional (typically defined by the number of orientations and frequencies). Concatenating all the feature images tends to exacerbate the dimensionality problems and computing complexity [4,6]. To overcome this issue, we introduce a dimensionality reducer before the superpixels extraction stage. In literature, there are many proposed dimensionality reduction methods; Independent Component Analysis (ICA), Principal Component Analysis (PCA), and Canonical Correlation Analysis (CCA) [10,11]. In this work, we find that PCA is sufficient.

Note that our intention was not to develop an end to end new segmentation approach of color image. Rather, we propose improving this task in general, by integrating several methodologies. As a first goal, these methodologies (multi-resolution filtering, superpixels computing and fuzzy clustering) work together to provide reliable segmentation results, characterized by a higher segmentation accuracy and robustness. The second is to reduce the computing complexity and speedup the segmentation process. By this work, we present two contributions as given below:

(1) A multiresolution image transformation based on 2-D Gabor filtering combined with a morphological gradient construction to generate a superpixel image with accurate boundaries. This proposition integrates a multiscale neighboring system to solve the problems of rotation, illumination, scale, and translation variance. This is very useful specially with images of high resolution.
(2) We introduce a Principal Component Analysis (PCA) to reduce the number of extracted Gabor features. From obtained regions we compute a simple color histogram to reduce the number of different intensities (pixels) and achieve a fast clustering for color image segmentation.

In summary, image segmentation methods can be roughly classified into two categories: supervised and unsupervised. In this paper, we mainly discuss a fuzzy unsupervised framework. No features learning is involved. This task remains one of the most challenging research topics because there is no unified approach to achieve fast, robust, and accurate segmentation.

In this work, a detailed study of existing color image segmentation approaches was carried out to investigate the most common stages in segmentation's techniques. In Section 2, we discuss the motivation for using the different implemented techniques. Furthermore, we thoroughly described each phase and introduced ideas for improvements. Next, we describe the development of the proposed method. Section 4 presents an evaluation study of the proposed improvement using a set of synthetic and real color images from the well-known dataset (BSD 500 and MSRC). As a validation stage, the developed method is applied on fire forest images and compared to the standard and recent methods.

2. Motivation

2.1. Motivation for Using Superpixels with Gabor Filtering

In color image segmentation, non-texture areas are relatively uniform, and it is easy to obtain the accurate boundaries. Color and spatial information are sufficient for the clustering task. In texture areas, the boundaries are the combination of micro and macro adjacent regions. Here, texture edges cannot be incorporated only by characteristics of single pixels: intensities and spatial coordinates. Hence, to obtain these boundaries requires a combination of multi scale characteristics. Many researchers have verified that multi-resolution features are able to get the main outline for various texture regions of the image [12,13]. In the last decade, the Gabor filters, firstly proposed by Dennis Gabor in 1946 in 1-D and extended, in 1985, to 2-D by Daugman, have received much attention. Their wide usage in multiple fields can be taken as proof of their success: image analysis, compression and restoration, object tracking and movement estimation, face recognition, smoke detection, texture retrieval, contour extraction, or image segmentation [14–17].

2.2. Motivation for Using Color Images Histograms

For C-means oriented algorithms, the clustering task has to compute the distance between each pixel and centers of different clusters. This task leads to a high computational complexity especially with images of higher resolution. Moreover, it is difficult to extend this idea of FCM for color image segmentation. This is due to the number of different colors which is usually close to the number of pixels in a color image. Compared to a grayscale image, the c-means clustering algorithms require a longer execution time to segment its corresponding color image. Because the histogram level is far less than the whole image pixels, the use of histogram-based features reduces the computational complexity of the clustering procedure. In [9], an enhanced FCM method for grayscale images was proposed. It is called the Spatial Fast Fuzzy C-means clustering algorithm (SFFCM). Authors demonstrate that it is faster to implement FCM on histogram-gray-levels than on pixel's intensities. This novel extension of fuzzy clustering algorithm is used in our segmentation pipeline (see Figure 1).

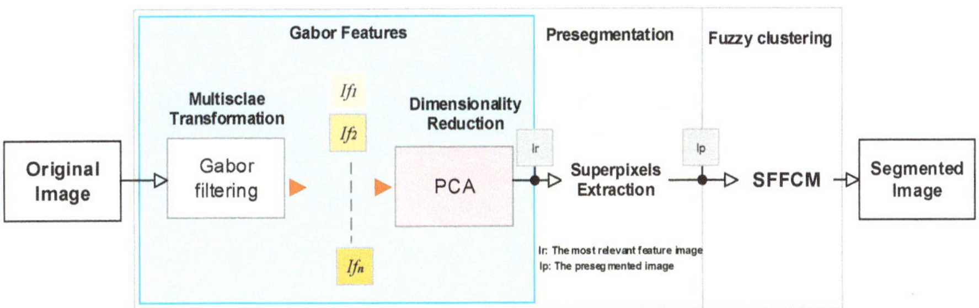

Figure 1. Overview of the proposed segmentation framework. Ifi is the ith image of Gabor features.

2.3. Fire Forest Image Application

Recently, wildfires devasted millions of hectares over the world. The lack of information about the current state and the dynamic evolution of fire plays a central role in the accidents. Nowadays the demand increases for remote monitoring of this natural disaster [2,18–20]. For that, artificial visual control is a new area that has gained interest. In literature, many techniques have been developed mainly for wildfire image processing [4,8,21,22]. In real applications, for smoke and fire, there is a different useful information: area, location, direction, etc. Because the forest environment suffers from many perception field drawbacks (uncontrollable and sudden changes in environmental conditions, calibration problems, non-rigid fire-model, etc.), this study involves many advanced computer vision techniques in 2D [13] and extends them to the 3D domain [23]. Our project is divided into different research interests: image segmentation, semantic fire and smoke detection, and flame direction estimation. In this work we developed a color image segmentation technique as a part of mentioned tasks. The goal of the proposed method is to improve the segmentation performance of wildfire noisy images and to reduce the clustering computational complexity.

3. Methodology

The developed method is based on two principal tasks:

- The Pre-segmentation, also called the Superpixels Extraction,
- The Clustering of firstly extracted superpixels

The framework of our proposed algorithm is shown in Figure 1.

3.1. *Superpixels Based on Gabor Filtering and Morphological Operations*

3.1.1. Superpixels Extraction: An Overview

Superpixels extraction, called also pre-segmentation, is the subdivision of the input image into a number of regions. Each region is a collection of pixels with homogenous characteristics. This procedure is always used for image classification and labeling. Compared to neighboring window-based methods, it is able to provide more representative local spatial information [9].

As given by [24], superpixel algorithms are classified into two principal categories:

Graph-based methods: each pixel is considered as a node in a graph. Similarities between neighboring pixels are defined as edge weights. Superpixels extraction minimizes a cost function defined over the graph. This category includes a large variety of developed methods: Normalized Cuts (NC), Homogeneous Superpixels (HS), Superpixels via Pseudo-Boolean Optimization (PB), and Entropy Rate Superpixels (ERS) [25,26].

Clustering-based methods: all image pixels are iteratively grouped until satisfying some convergence criteria. As given by [27], the most popular techniques are Simple Linear Iterative Clustering named (SLIC), Watersheds Transform (WT), Quick Shift (QS), and Turbo Pixel (TP). More details and evaluation of 15 superpixel algorithms are given in [24]. All mentioned approaches are usually considered as over-segmentation algorithms to improve the final segmentation. Referring to [9,27], in our work, we use the implementation of WT for the superpixels extraction. In the last part of experiments (Section 5.2), the SLIC is also implemented.

3.1.2. Gabor Filters and Their Characteristics

Image filtering based on Gabor filters is a procedure widely used for the extraction of spatially localized spectral features. The frequency and orientation representation of Gabor filters are similar to human visual system, and they have been found vital features that can be used for image segmentation [16,28]. In our project, the processed images of fire forest combine many complexities due to the higher intensity's variation and the texture geometrical diversity. To cope with complex image regions, we use a bank of filters as a multi-scale features extractor.

The Gabor filter is obtained by a Gaussian kernel function modulated by multiplying a sinusoidal plane wave. As shown in Figure 2, combining a 2D sinusoid with a Gaussian function results in a 2D Gabor filter.

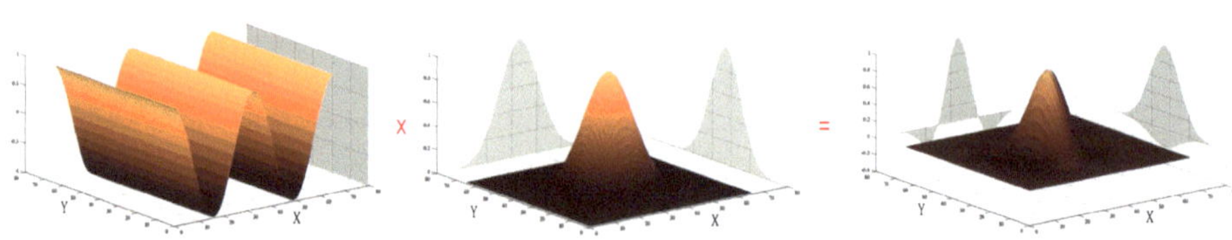

Figure 2. Spatial localization in 2D sinusoid (**left row**), Gaussian function (**middle row**), and corresponding 2D Gabor filter (**right row**).

Gabor features are extracted by the convolution of the original image $I(x,y)$ and the impulse response of the 2-D Gabor filter $g(x,y)$:

$$G(x,y) = I(x,y) \otimes g(x,y) \tag{1}$$

x and y are the spatial coordinates of the plane.

The Gabor kernel generating $g(x,y)$ is defined as follows:

As we have shown in [29], in the spatial domain, the 2-D Gabor function is formulated by:

$$g_{\lambda,\theta,\varphi}(x,y) = e^{-\frac{x'^2+\gamma^2 y'^2}{2\sigma^2}} \cos((2\pi x'/\lambda) + \varphi) \tag{2}$$

where

$$x' = x\cos\theta + y\sin\theta$$
$$y' = -x\sin\theta + y\cos\theta$$

σ is the standard deviation of the Gaussian factor that determines the size of the receptive field. The parameter λ is the wavelength and $F = 1/\lambda$ the spatial frequency of the cosine factor. They are, respectively, called the preferred wavelength and preferred spatial frequency of the Gabor function. The ratio σ/λ determines the spatial frequency bandwidth and the number of parallel excitatory and inhibitory stripe zones that can be observed in the receptive field (see Figure 3).

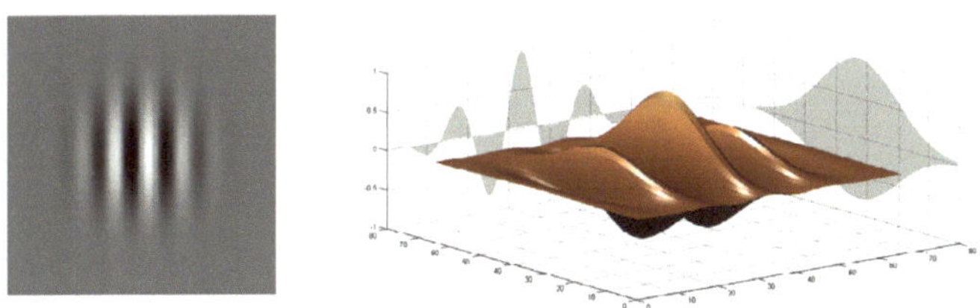

Figure 3. Example of the receptive field of the 2D-Gabor filter.

γ is a constant, called the spatial aspect ratio, that determines the ellipticity of the receptive field. θ represents the preferred orientation of the normal of the parallel stripes of a Gabor function, φ is the phase offset which defines the symmetry of Gabor filter.

As an example, with a different range of frequencies $f_k = \frac{\sqrt{2}}{k}\,(k = \{1,2,3\})$ and orientations $\theta_l = l\cdot\left(\frac{\pi}{8}\right)(l = \{0,1,\ldots,7\})$, the convolution generates a Gabor feature matrix given by:

$$G_{(f_l,\theta_k)} = \begin{bmatrix} r(x_0,y_0) & \cdots & r(x_N,y_0) \\ \vdots & \ddots & \vdots \\ r(x_0,y_M) & \cdots & r(x_M,y_N) \end{bmatrix} \tag{3}$$

The set of 3 spatial frequencies and 8 equidistant orientations is applied. Each Gabor kernel size is proportional to the wavelength value. The replication padding is used to reduce boundary artifacts. For each specific pair of frequency and orientation (f_k, θ_l), the feature image size is $(M \times N)$.

In our work, only the magnitude $r(x_i,\ y_j)$ is considered. $r(x_i,\ y_j)$ gives the intensity variations near the object boundaries (see Figure 4).

The Gabor features are processed with L2 normalization technique. The L2 norm is performed by:

$$g(x,y) = \frac{||G(x,y)||}{max\{||G(x,y)||\}} \tag{4}$$

g is the normalized Gabor feature image.

Figure 4. Example of object boundaries extraction using Gabor filters of (f_1, θ_3). First row: original images. Second row: image of boundaries.

3.1.3. Gabor Feature Reduction Based on PCA

High dimension data are extremely complex to process due to the inconsistences in the features which increase the computation time [30,31]. In our work, we only focus on the variation of frequency and orientation parameters of Gabor filters. In Figure 5, we present a convolution results of a synthetic image of sinusoids of different orientations, frequencies and magnitudes by Gabor filters of different orientations and frequencies.

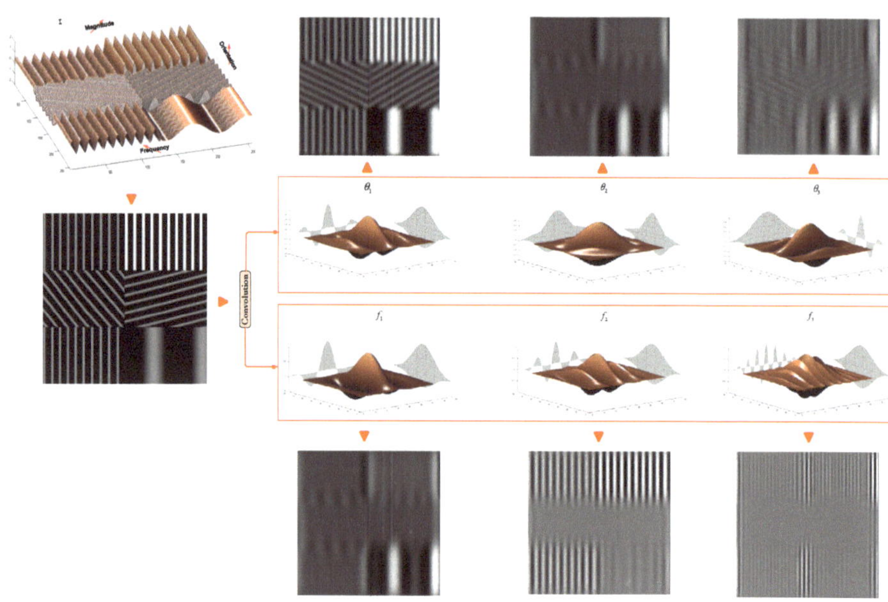

Figure 5. Convolution outputs of a synthetic image of sinusoids with various properties (orientations, frequencies and magnitudes) by Gabor filters of different frequencies and orientations (f_k, θ_l).

Both the mentioned parameters (frequencies and orientations) generate a large feature dimension $(K \times L)$. As mentioned above, a set of $K = 3$ different frequencies and $L = 8$ orientations are considered producing 24 features for each position of the filter. This is not performant because of the redundancy of features due to correlation of the overlapping filters. Moreover, as illustrated in Figure 5, by comparing the convolution results we notice a higher sensibility of filter parametrization. Many researchers propose the use of a small bank of filters [30–32]. In this work, the problem of redundancy was addressed. Because its performance compared to other dimensionality reducer methods [32], we have used the PCA retaining only the most representative response of 24 outputs. It will be considered as the input of the superpixesls extraction stage (see Figure 1).

3.1.4. Pre-segmentation Based on Gabor-WT

The WT produces a set of basins starting with a local minimal of a gradient image and searching lines between adjacent local minima that separate catchment watersheds. As given by [33], this is a relatively fast algorithm used for images with high resolution.

For noisy image segmentation, to fulfill both regional consistency and boundary keeping simultaneously, become more and more difficult. As shown by Figure 6, the MMGR-WT, introduced in [9], causes an over-segmentation or under-segmentation because it is sensitive to added noise.

Figure 6. MMGR-WT robustness test. First row: original images from the MSRC-dataset. Second row: Superpixels extraction results with original images. 3rd, 4th, and 5th rows: the obtained results for corrupted images by (5%, 10%, 15%) salt and pepper noise.

Moreover, these techniques greatly depend on the accurate extraction of region boundaries. The superpixels extraction performance of these methods deteriorates when the processed regions are textured or are of high varying intensities (see Figure 7).

Figure 7. MMGR-WT superpixels extraction: test on uniform and textured regions. First row: original images from the MSRC_dataset. Second row: obtained results.

As a summary of all the superpixels extraction given by Figures 6 and 7, the MMGR-WT results exhibit major limits, namely the poor boundary keeping and superpixels consistency. This is clearly noticed with noisy images and ones of textured regions (grass, trees, sand, etc.). In our work, it should be noted that Fire Forest images suffer from all the general drawbacks (noise, higher textured regions, environmental conditions, etc.). In literature, many algorithms have been introduced to avoid such issues. Major methods tend to modify the gradient output of original image. In this paper, a 2-D Gabor filtering stage is used for the enhancement of the boundaries of regions for better superpixels extraction.

3.2. Fuzzy Superpixels Clustering

3.2.1. Overview

A clustering divides data objects into homogeneous groups and performs a high similarity within a cluster (called compactness). Data partitioning is made according to a membership degree, in the range (0,1), which is proportional to the distance between the data and each cluster center. The partitioning result depends on the final centroid location [29]. The fuzzy oriented methods are based on the mentioned aspects and have been successfully used. For many an application, the traditional FCM clustering algorithm, firstly introduced by Bezdek, has depicted a higher performance. It is widely used for image segmentation. As an unsupervised clustering method, FCM does not need any prior knowledge about the image.

Let $X = \{x_1, x_2, \ldots, x_n\}$ be a color image and nc be the number of clusters. Each ith image pixel belongs to the jth cluster with a fuzzy membership degree denoted by u_{ij} according to its distance from the cluster center v_j. FCM can yield a good segmentation result by minimizing the following objective function:

$$J_{FCM} = \sum_{i=}^{n} \sum_{j=}^{nc} \mu_{ij}^{m} \| x_i - v_j \|^{2} \tag{5}$$

where u_{ij} and v_j are given as follows:

$$u_{ij} = \left(\sum_{k=1}^{nc} \left(\frac{\| x_i - v_j \|}{\| x_i - v_j \|} \right)^{\frac{2}{m-1}} \right)^{-1} \tag{6}$$

$$v_j = \frac{\sum_{j=1}^{n} u_{ij}^{m} x_j}{\sum_{j=1}^{n} u_{ij}^{m}} \tag{7}$$

and m is the degree of fuzziness.

The FCM algorithm is summarized in Algorithm 1.

Algorithm 1. Traditional FCM algorithm.

Input: X of n data to be clustered, the number of clusters nc, the convergence test $\varepsilon > 0$ (or the max number of iteration), randomly cluster centers $v^{(t=0)}$, the fuzzifier $m > 1$
Output: clustered data (pixel groups map)
Begin
  Step 1. Compute the membership matrix U by using Equation (6)
  Step 2. Update the cluster centers $v^{(t+1)}$ with Equation (7)
  Step 3. Test if $||v^{(t+1)} - v^{(t)}|| < \varepsilon$, execute **step 4**; otherwise, $t = t + 1$ and go to **step 1**
  Step 4. Output the pixels group map
End

In literature, a large variety of modified versions of the FCM clustering algorithm have been proposed. In 2003, Zhang developed a new extension called KFCM by introducing the "Kernel method". Later, in 2012, Zanaty et al. included a spatial information in the objective function of KFCM [34]. From 2005 to 2013 Pal et al. developed a possibilistic fuzzy clustering method called PFCM [35]. Another modification of FCM was proposed in 2015 by Zheng et al. [36] named the Generalized and Hierarchical FCM (GFCM), (HFCM). In 2017, in order to remove the information redundancy, Gu et al. proposed a novel version of FCM called SL-FCM [37]. Most of the mentioned methods are still time-consuming and unable to provide the desired segmentation accuracy. As mentioned above, Lei et al. developed the SFFCM algorithm. The modification is also based on the integration of the spatial information [9].

3.2.2. The Proposed Clustering Method

Further to the time-consuming, FCM describes an image in terms of fuzzy classes. It only depends on global features. As given by Figure 8, we developed a Gabor-PCA superpixels-based method to extract the most representative local spatial information. By this, the input data to be clustered include only the subsegment levels. In our work, the proposed segmentation method has three goals. The first is to reach a higher robustness to added noise with the multiscale processing based on Gabor Filters. The second goal is the improvement of the segmentation accuracy by incorporating local features. The third, is reducing the computational complexity and time consuming by minimizing the size of data to be clustered.

In this paper, the SFFCM algorithm, firstly proposed by [9], is adopted. Adding the spatial information, the problem of fuzzily partitioning into nc clusters becomes formulated as the minimization of the objective function given by:

$$J_m = \sum_{i=1}^{ns} \sum_{j=1}^{nc} S_i u_{ij}^m ||Med_i - v_j||^2 \tag{8}$$

With Med_i the mean value of color pixels within the corresponding region R_i of ith superpixel image given by:

$$Med_i = \frac{1}{S_i} \sum_{p \in Ri} x_p \tag{9}$$

where:

ns is the number of superpixels, $1 \leq i \leq ns$ the color level, and S_i the number of pixels with color x_p in R_i. The new objective function incorporates the histogram information by the level's frequencies given by S_i. Thereby, each color pixel in original image is replaced by the mean value Med_i of the region for which was assigned. The "*Med-image*" is called the pre-segmented image (see Figure 8).

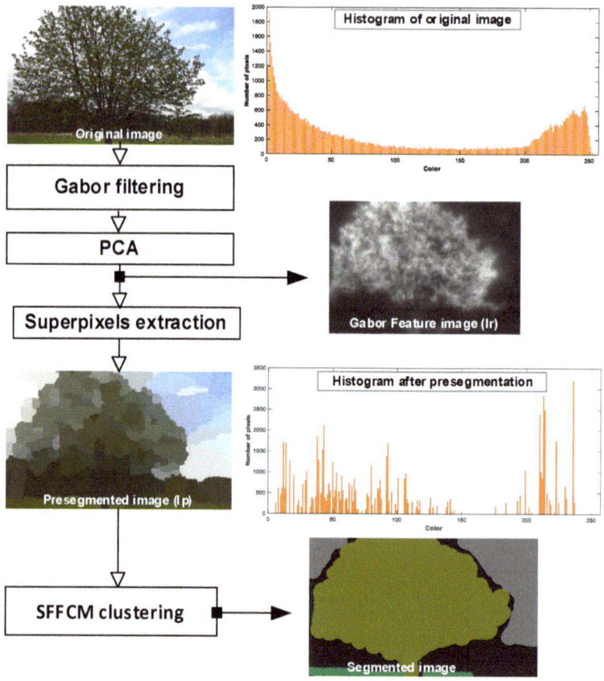

Figure 8. Example of the "End to End" segmentation pipeline with our proposed method.

New SFFCM objective function generates two novel formulation memberships (u_{ij}) and centroid functions (v_j) as follows:

$$u_{ij} = \frac{\|Med_i - v_j\|^{\frac{-2}{m-1}}}{\sum\limits_{k=1}^{nc} \|Med_i - v_k\|^{\frac{-2}{m-1}}} \quad (10)$$

$$v_j = \frac{\sum_{i=1}^{ns} u_{ij}^m \sum_{p \in R_i} x_p}{\sum_{i=1}^{ns} S_i u_{ij}^m} \quad (11)$$

In Algorithm 2, we show the pseudo-code of the Spatial Fast Fuzzy C-means clustering method (SFFCM).

Algorithm 2. SFFCM Algorithm.

Input: $S = \{S_1, \ldots, S_{ns}\}$ number of pixels with color corresponding to the presegmented regions $R = \{R_1, \ldots, R_{ns}\}$, $Med = \{Med_1, \ldots, Med_{ns}\}$ the mean values of superpixels levels (equation …), the number of clusters nc, the convergence test $\varepsilon > 0$ (or the max number of iteration), randomly cluster centers $v^{(t=0)}$, the fuzzifier $m > 1$

Output: clustered data (pixel groups map)

Begin

  Step 1. Compute the membership matrix U by using Equation (10)

  Step 2. Update the cluster centers $v^{(t+1)}$ with Equation (11)

  Step 3. Test if $\|v^{(t+1)} - v^{(t)}\| < \varepsilon$, execute **step 4**; otherwise, $t = t + 1$ and go to **step 1**

  Step 4. Output the pixels group map

End

3.3. Evaluation Criteria

In the last decade, several metrics have been applied to evaluate the segmentation methods [38]. The major ones focus on segmentation accuracy, superpixels compactness, regularity, coherence, and efficiency. In [24], Wang et al. divided the set of metrics into three groups: segmentation quality evaluation, superpixels quality, and the efficiency measure based on runtime. In this work, two metrics categories are considered.

a. Segmentation accuracy

To test the clustering performance, we use two metrics given in [9]. The first measures the Equality Degree (ED) between Clustered Pixels (CP) and Ground truth Prediction (GP). The second measures the Segmentation Accuracy (SA) based on the sum of correctly classified pixels. Both metrics are, respectively, given by:

$$ED = \sum_{k=1}^{nc} \frac{CP_k \cap GP_k}{CP_k \cup GP_k} \quad (12)$$

$$SA = \sum_{k=1}^{nc} \frac{CP_k \cap GP_k}{\sum_{j=1}^{c} GP_j} \quad (13)$$

where, CP_k is the set of pixels assigned to k^{th} cluster and GP_k the set of pixels belonging to the same class k of Ground Truth (GT). nc denotes the number of clusters.

$CP_k \cap GP_k$: the comprised of the labeled pixels *AND* the ground truth of the kth cluster.

$CP_k \cup GP_k$: the comprised of all pixels found in either the prediction *OR* the ground truth of the kth cluster.

b. Sensitivity and Specificity

These measures are based on region overlapping. Here, two aspects are considered: the matching direction and the corresponding criteria. For the sensitivity measure, the matching direction is defined as a ground truth to segmentation result directional correspondence and vice versa for the recall measure. Sensitivity (SEN) and Specificity (SPE) are formulated as follows:

$$SEN(CP,\ GP) = \frac{TP(CP,\ GP)}{TP(CP,\ GP)\ +\ FN(CP,\ GP)} \quad (14)$$

$$SPE(CP,\ GP) = \frac{TN(CP,\ GP)}{TN(CP,\ GP)\ +\ FP(CP,\ GP)} \quad (15)$$

where:

$TP(CP,\ GP)$—*True Positives:* intersection between segmentation and ground truth
$TN(CP,\ GP)$—*True Negatives*: part of the image beyond the intersection mentioned above
$FP(CP,\ GP)$—*False Positives:* segmented parts not overlapping the ground truth
$FN(CP,\ GP)$—*False negatives*: missed parts of the ground truth

As given by the Equations (14) and (15), the quantitative evaluation based on Sensitivity (SEN) and Specificity (SPE) were performed between the (GT) and the clustering result. (SEN) was the percentage of Region of Interest (ROI) recognized by the segmentation method. (SPE) was the percentage of non-ROI recognized by the segmentation method.

Measures based on (SEN) and (SPE) are commonly used for the semantic segmentation. In our work, the mentioned metrics are applied to evaluate the clustering performance for supervised topics where the number of classes and region contents are known.

For real images, cluster frequencies are unbalanced. Mentioned metrics are not appropriate for evaluating because they are biased by the dominant classes. To avoid this, we have conducted the evaluation per-class. The obtained results are averaged over the total number of classes.

For multiclass, Sensitivity and Specificity are called Average True Positive Rate (Av_TPR) and Average True Negative Rate (Av_TNR) and given by:

$$Av\_TPR = \frac{\sum_{k=1}^{nc} TP_k}{\sum_{k=1}^{nc} (TP_k + FN_k)} \tag{16}$$

$$Av\_TNR = \frac{\sum_{k=1}^{nc} TN_k}{\sum_{k=1}^{nc} (TN_k + FP_k)} \tag{17}$$

4. Experimental Results

Experimental Setting

For all the experiments discussed below, the particular parameters for the compared methods are summarized in Table 1. Only Gabor filtering parametrization is detailed in [29].

Table 1. Parameters of different applied methods.

Method	Pre-segmentation	Classification
SFFCM [9]	Min, Max radius: $(r_1, r_2) = (1, 10)$	$m = 2, \varepsilon = 10^{-3}$
Gabor-SLIC	Number of desired pixels: $s_k = 500$ Weighting factor: $s_m = 50$ Threshold for region merging: $s_s = 1$	$m = 2, \varepsilon = 10^{-3}$
Gabor-WT	Structured Element SE: a disk Radius: $r = 5$	$m = 2, \varepsilon = 10^{-3}$

In the first experiments, the tested images are synthetic with natural textures of Smoke, Fire, Sky, Sand, and Grass. For the second, we have tested on six real images from real scene of fire forest images. For the limited length of paper, in the last experiments, we only demonstrate the robustness and segmentation performance of our proposed method on a subset of twenty images from BSDS500 and MRSC datasets.

5. Application on Fire Forest Images

5.1. Results on Synthetic Images

At the first part of evaluation, we test the proposed method with the WT and the SFFCM algorithm on a set of six synthetic images shown by Figure 9a. For each class, (Fire, Smoke, Grass, Sand, Sky), the selected region is chosen from a random location in the original corresponding texture. All of the used synthetic images are with regions of regular boundaries. This is more suitable to manually generate the desired segmentation (see Figure 9b).

In this experiment, two types of noise are considered: Gaussian and Salt and Pepper. The robustness of each method is tested with four different densities of each kind of mentioned noises (10%, 20%, 30%, 40%). The quantitative segmentation results on the different blurred images achieved by using our developed method with WT and the SFFCM proposed by Lei 2019 [9]. Each experiment is repeated 10 times. All the obtained results are depicted in the boxplots in Figures 10 and 11, reporting both ED and SA metrics values given by Equations (12) and (13). The graphs of boxplots arranged similarly as the map of images (SI1–SI6) given by Figure 9 (e.g., the top left boxplot corresponds to the results on image (SI1)).

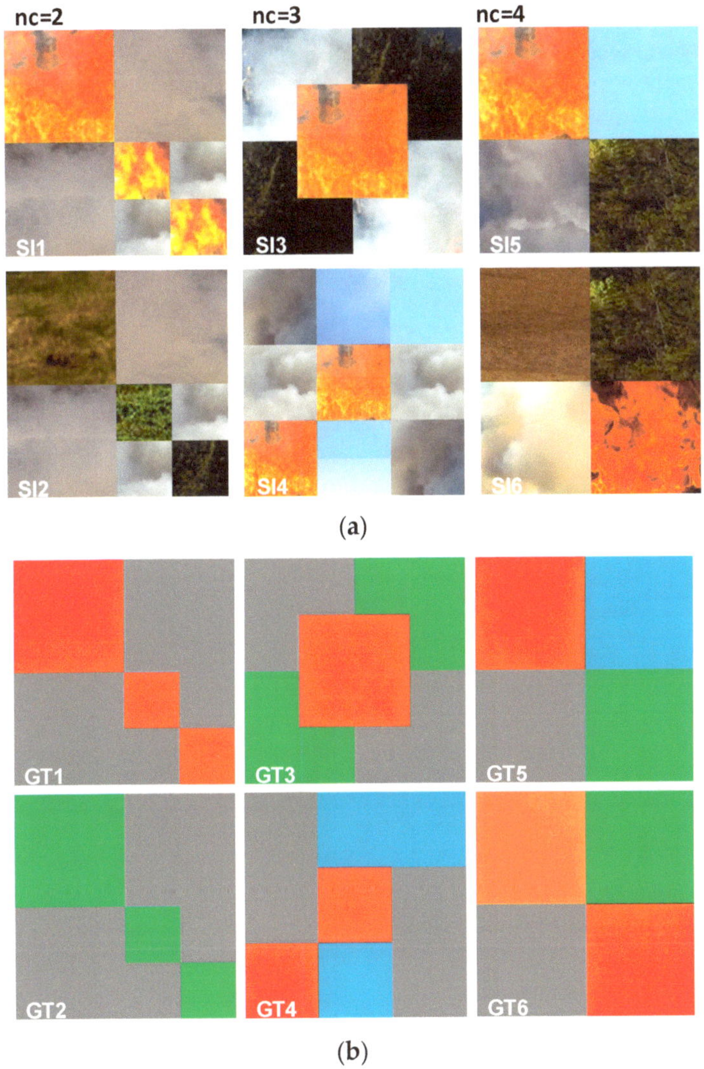

Figure 9. Synthetic test images of (256×256) pixels with manually created (GT). (**a**) Images with different regions of real contents. (**b**) The corresponding desired segmentation (GT).

In Figures 10 and 11, the lower and the upper bounds of each boxplot represent the first and third quartiles of the distribution, respectively. The mean values of used metrics (ED, SA) are represented by a black solid line and outliers are displayed as black diamonds. We observe that there is a greater variability of the SFFCM results compared to our proposed method. Moreover, the boxplots pertaining to the proposed method results present the lowest statistical dispersion in terms of box height and number of outliers, thus implying a lower standard deviation compared to the SFFCM method. Therefore, the use of the novel method allows for considerably robust and accurate segmentation results.

5.2. Results on Real Images

In addition to synthetic images, we shall evaluate the performance of our method on natural images. We apply the proposed method to the images from real fire forest sequences to examine the segmentation performance of our approach. The test images, given by Figure 12, are with different regions (fire, forest, smoke, cloud, grass, etc.). We shall assess the segmentation accuracy according to the visual inspection because no ground truths are available.

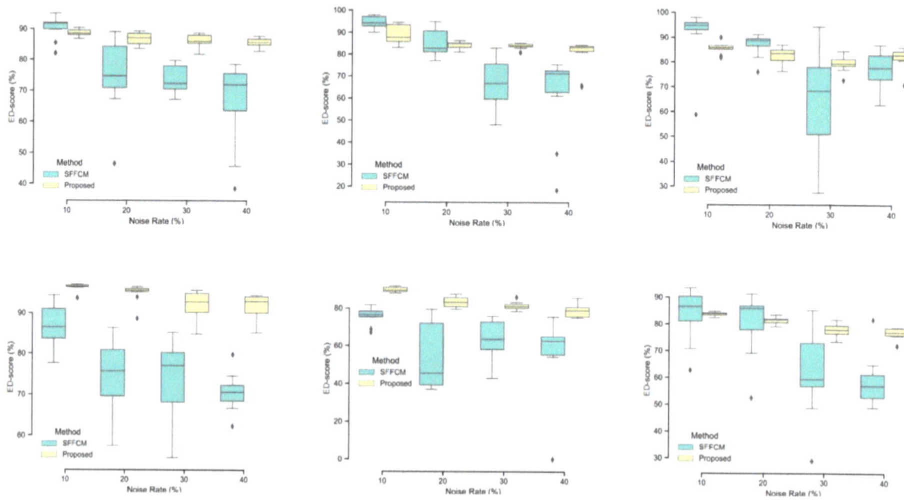

(**a**) Boxplots of Equality Degree (ED) metric obtained by 10 experiments

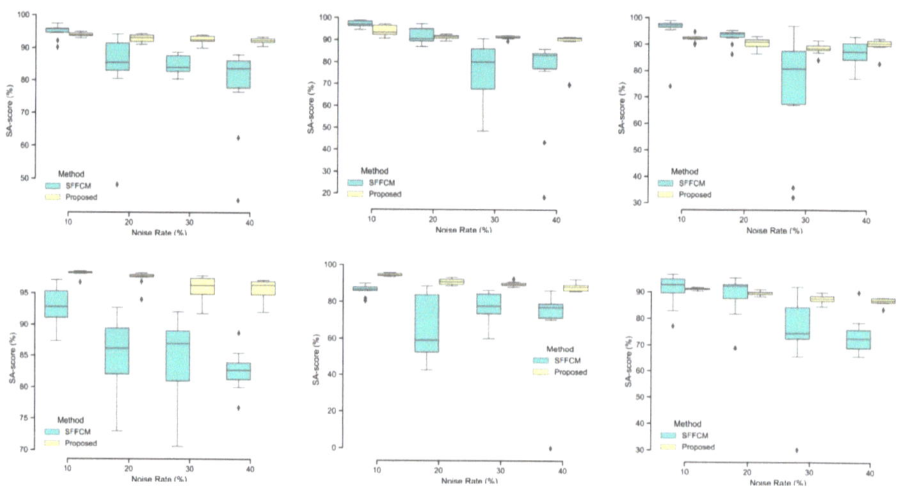

(**b**) Boxplots of Segmentation Accuracy (SA) metric obtained by 10 experiments

Figure 10. Comparison of SFFCM and our proposed method (*G-WT*) robustness. Application on the set of synthetic images SI1-SI6 (Figure 9) corrupted by the (10%, 20%, 30%, 40%) Gaussian noise.

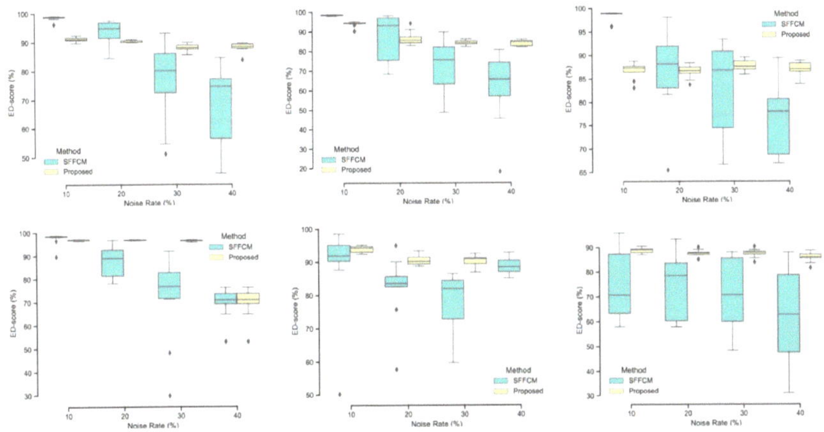

(**a**) Boxplots of ED metric obtained by 10 experiments

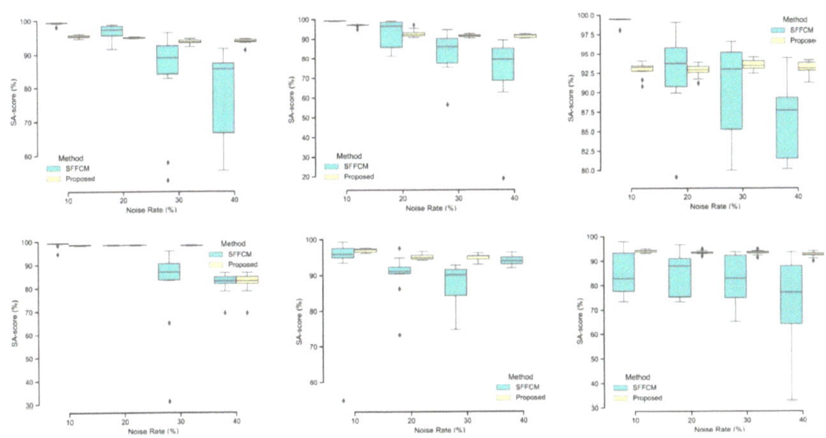

(**b**) Boxplots of SA metric obtained by 10 experiments

Figure 11. Comparison of SFFCM and our proposed method (*G-WT*) robustness. Application on the set of synthetic images SI1–SI6 (Figure 9 corrupted by the (10%, 20%, 30%, 40%) Salt and Pepper noise.

Figure 12. A set of real test fire forest images [21].

The difficulty of real image segmentation can be attributed to two reasons. The first is that image segmentation is a multiple solution problem. The number of clusters differs from a person to another. The second is that an image is always complex because of added noise and regions nonuniformity.

In this study, in order to address the first mentioned difficulty, we have shared the real test images with a group of 30 of our students in order to obtain their observations about the number of different observed clusters. The obtained statistics are summarized in Figure 13.

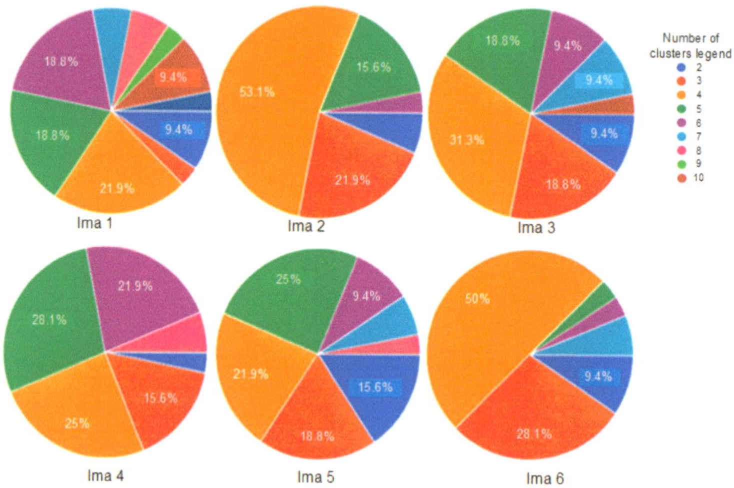

Figure 13. 30 humans' observations about the number of clusters of the real images given by Figure 12.

For each image, only the first three decisions with higher percentages were considered. i.e., as given by Figure 13, 53.1% of persons have considered that the image "Ima 2" is of 4 clusters, 21.9% have considered that the mentioned image is only of 3 clusters, and 15.6% observed that "Ima 2" is with 5 clusters. In our experiments, for mentioned image, we conduct the segmentation with 4, 3, and 5 clusters. All the obtained results are illustrated by Figures 14–19.

Figures 14–19 show the segmentation results of the real images depicted by Figure 12 and corrupted by the salt and pepper. In this experiment, we compare the SFFCM and our proposed method for two versions. The first by using the WT and for the second, we have introduced the SLIC pre-segmentation technique. By a visual inspection, for three compared methods, we notice that the region partition is satisfying. When the noise density is added, a lower performance is achieved. This is mainly due to the fact that high density of noise affects the texture structures, leading to the input image color degradation. Added noise affects the pre-segmentation performance and yields a lower classification performance. This clearly noticed with the SFFCM algorithm compared to the proposed method with WT and SLIC.

As given by Figures 14b, 15b, 16b, 17b, 18b and 19b, for different corrupted images, the obtained results with the proposed method using WT depicts that the different regions separation is more accurate than using the SLIC. For instance, we can see that the "fire" in Figure 14b and the "smoke" in Figures 15b, 16b and 17b are accurately segmented.

In summary, the segmentation results obtained by the proposed method, using WT or SLIC, are still more satisfying. This is due to the higher robustness of the multiresolution transform based on Gabor filters and the integration of the PCA in the pre-segmentation stage.

Figure 14. Comparison of segmentation results of original image Ima 1 (**a**) and corrupted by a 10% salt and pepper noise (**b**) obtained by: SFFCM (the first row), the proposed method based on Simple Linear Iterative Clustering (SLIC) (the second row), and the proposed method based on Watersheds Transform (WT) (the third row).

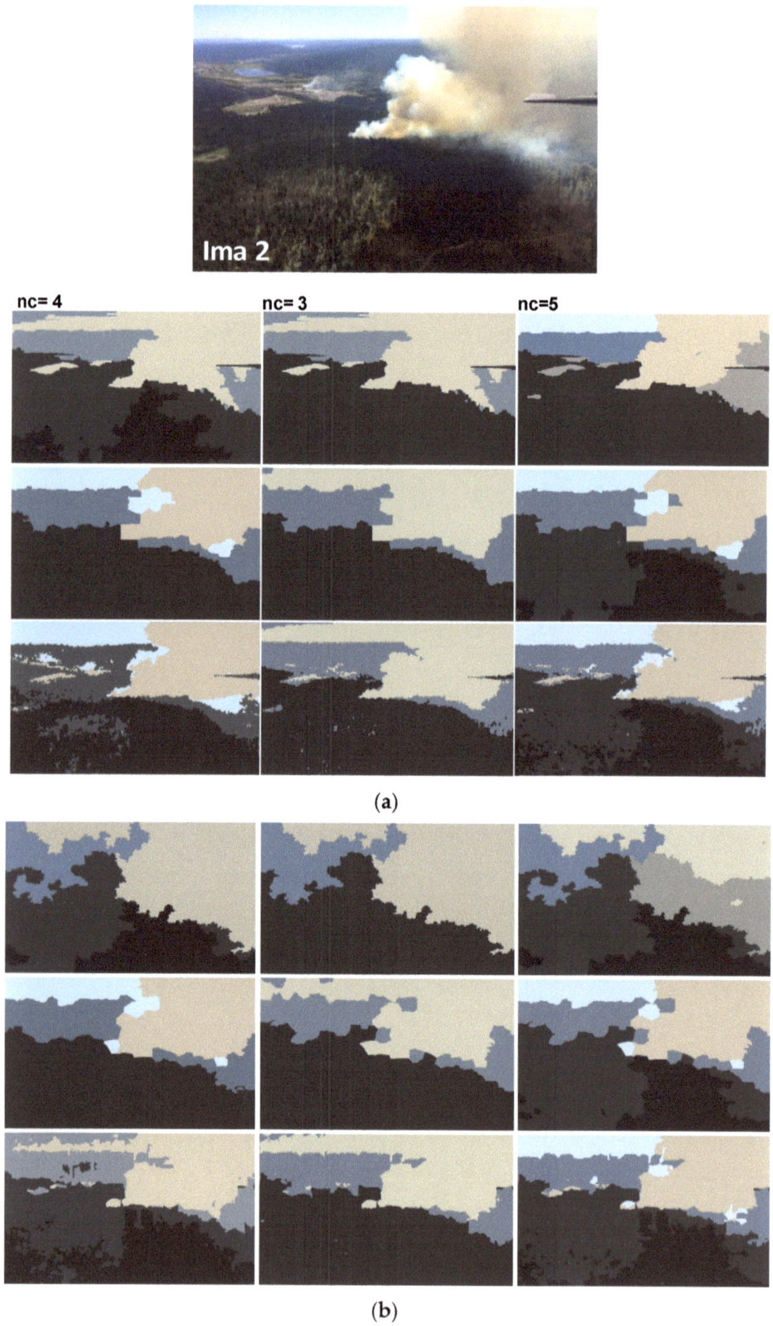

Figure 15. Comparison of segmentation results of original image Ima 2 (**a**) and corrupted by a 10% salt and pepper noise, (**b**) obtained by: SFFCM (the first row), the proposed method based on SLIC (the second row), and the proposed method based on WT (the third row).

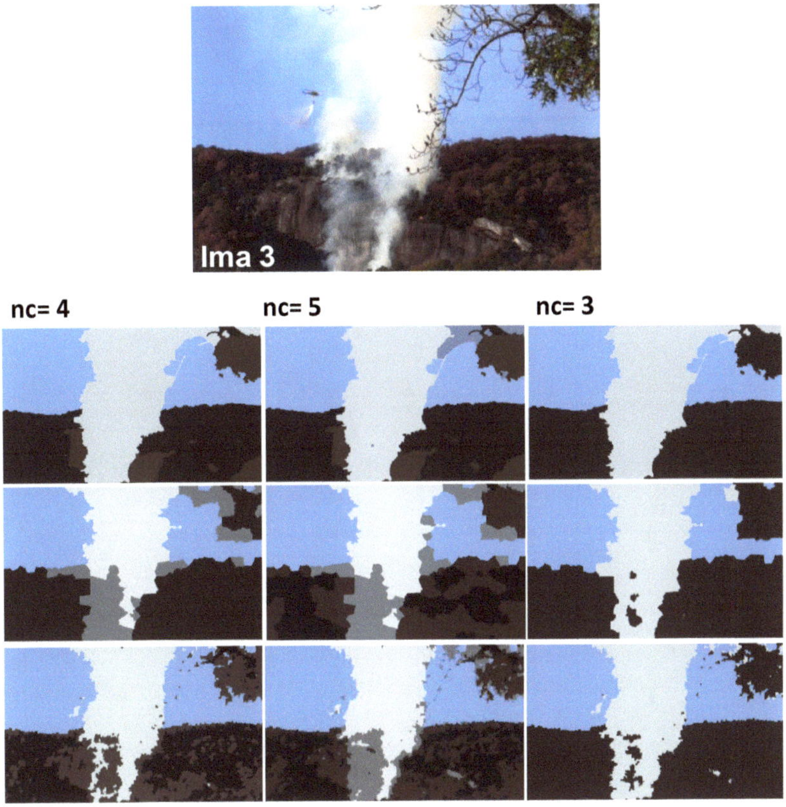

(a)

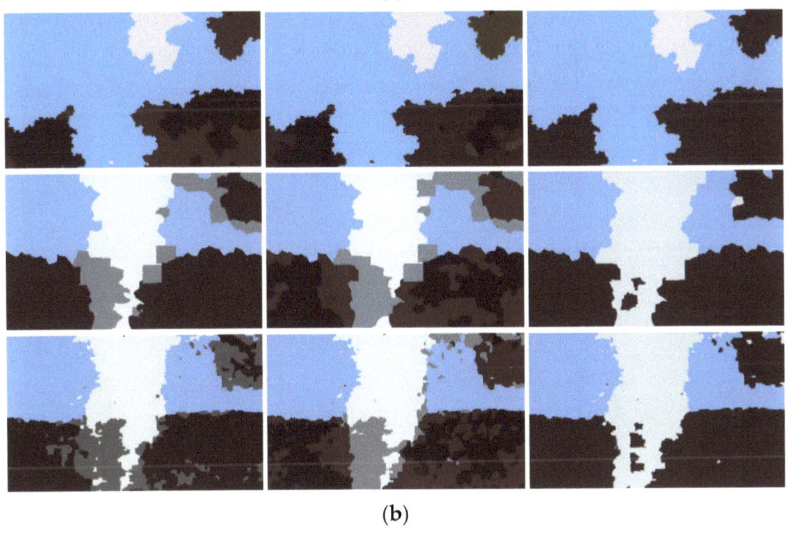

(b)

Figure 16. Comparison of segmentation results of original image Ima 3 (**a**) and corrupted by a 10% salt and pepper noise (**b**) obtained by: SFFCM (the first row), the proposed method based on SLIC (the second row), and the proposed method based on WT (the third row).

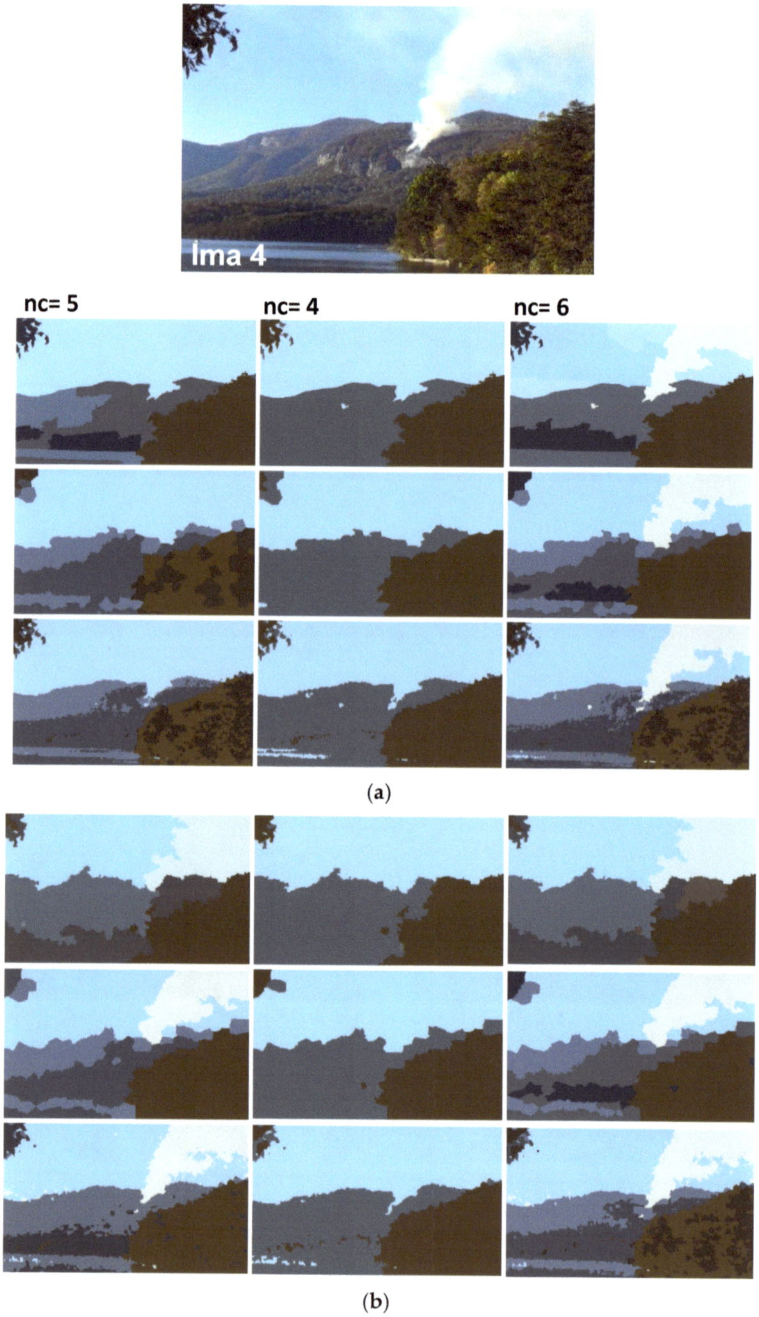

Figure 17. Comparison of segmentation results of original image Ima 4 (**a**) and corrupted by a 10% salt and pepper noise. (**b**) obtained by: SFFCM (the first row), the proposed method based on SLIC (the second row), and the proposed method based on WT (the third row).

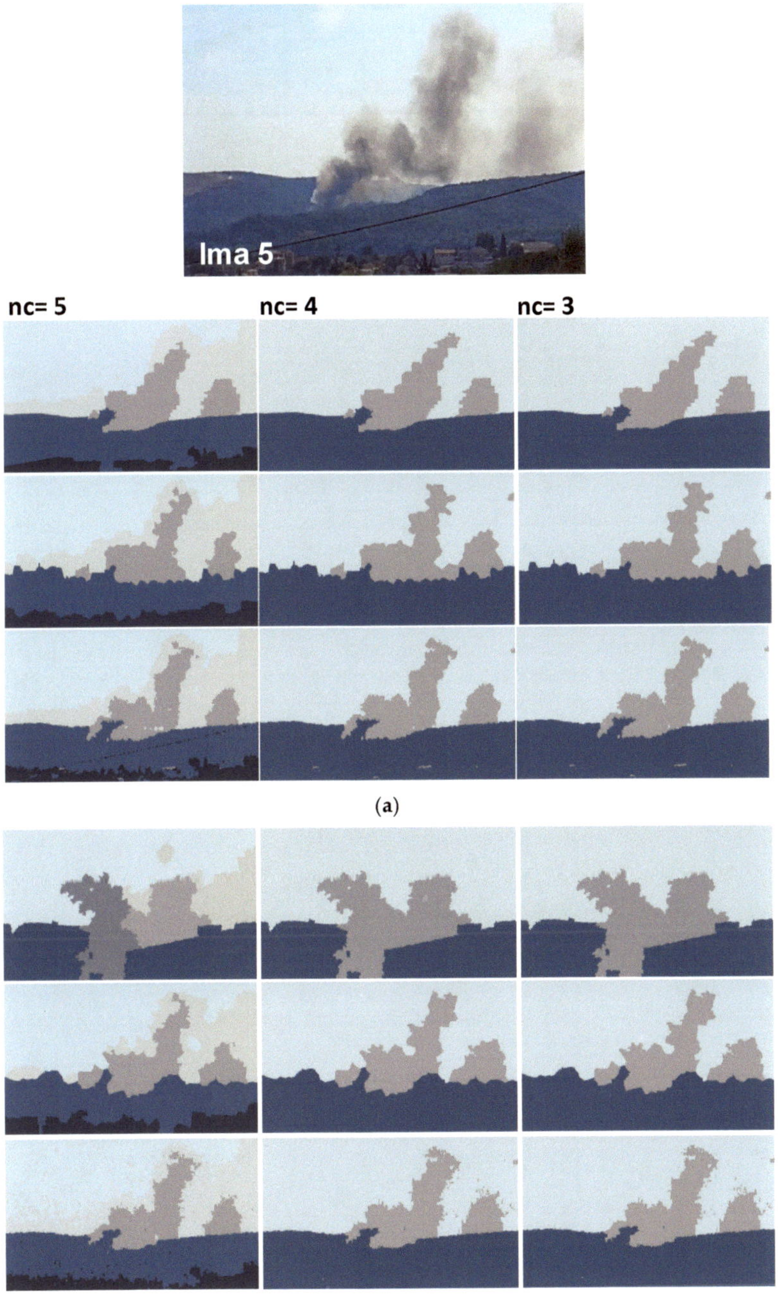

Figure 18. Comparison of segmentation results of original image Ima 5 (**a**) and corrupted by a 10% salt and pepper noise (**b**) obtained by: SFFCM (the first row), the proposed method based on SLIC (the second row), and the proposed method based on WT (the third row).

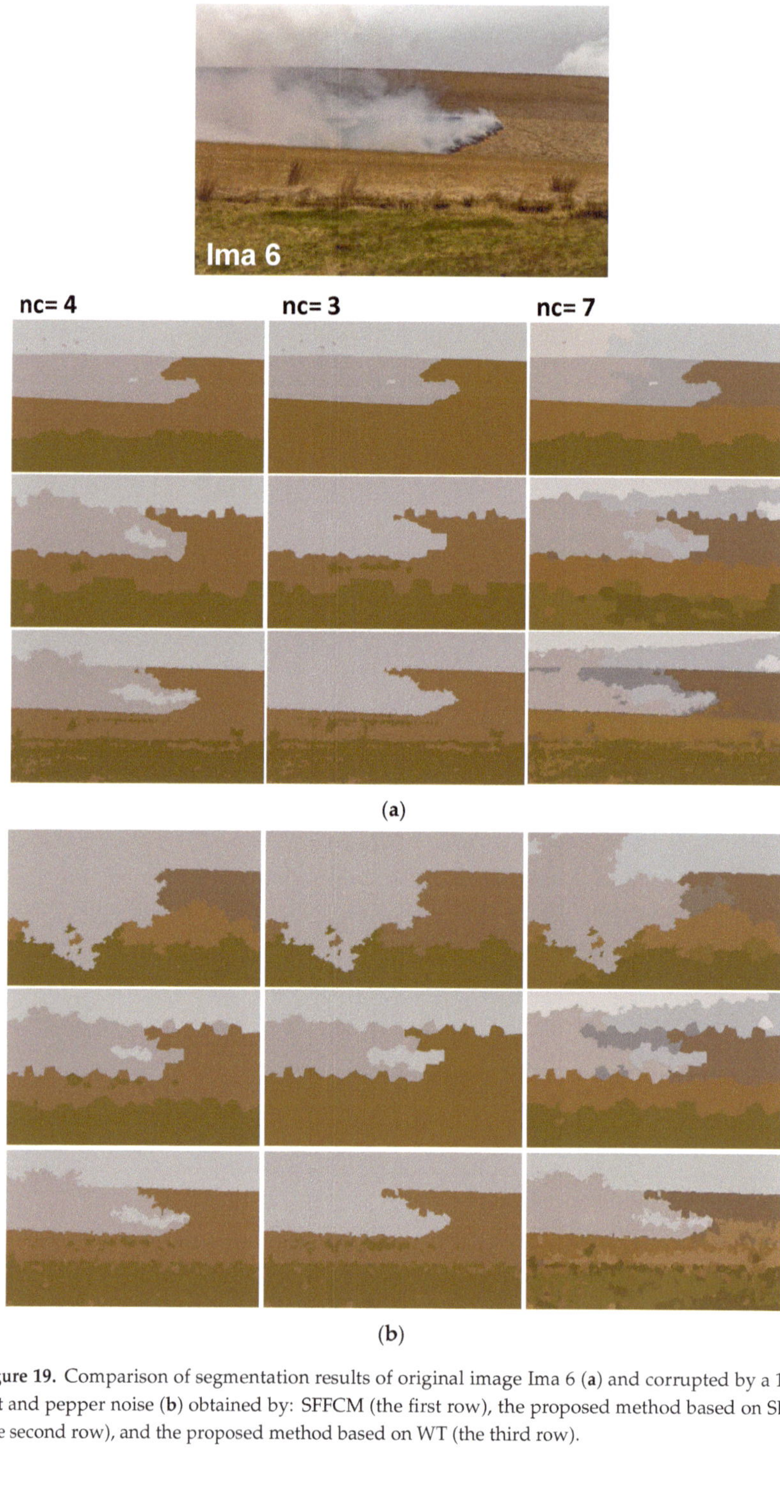

Figure 19. Comparison of segmentation results of original image Ima 6 (**a**) and corrupted by a 10% salt and pepper noise (**b**) obtained by: SFFCM (the first row), the proposed method based on SLIC (the second row), and the proposed method based on WT (the third row).

5.3. Application on Other Natural Images

To assess the performance of the proposed method, we further tested it on natural images from the BSDS and MSRC datasets (see Table 2). The both mentioned datasets are the most popular benchmarks and they are widely used by researchers for color image segmentation [27,39,40]. The results reported are averaged after 10 experiments and illustrated by Figure 20.

Table 2. Natural Images from BSDS 500 and MSRC datasets.

Image	Name	Dataset	Images
I1	"55067"	BSDS500	
I2	"41004"		
I3	"311068"		
I4	"3096"		
I5	"66075"		
I6	"5_26_s"	MSRC	
I7	"9_10_s"		
I8	"10_1_s"		
I9	"2_21_s"		
I10	"2_22_s"		
I11	"2_27_s"		
I12	"4_13_s"		
I13	"2_8_s"		
I14	"4_26_s"		
I15	"3_20_s"		
I16	"2_17_s"		
I17	"2_3_s"		
I18	"3_24_s"		
I19	"2_20_s"		
I20	"5_22_s"		

Referring to the barplots shown in Figure 20, a higher segmentation performance with the SFFCM is recorded. This superiority, of Sensitivity and Specificity, is clearly shown with original images (I1, I2, I3, I4, I6, I8, I12). For this image's subset, it can be seen that different classes are with homogenous microtexture regions. By adding the salt and pepper noise, we can notice the degradation of the SFFCM segmentation accuracy compared to the proposed method. This robustness limitations of SFFCM was previously illustrated by the boxplots (see Figures 10 and 11). Furthermore, it is clearly shown by Figure 20.

The selected images contain nonhomogeneous regions within the same class, and thus, grouping the superpixel regions in these cases would be a difficult task because these image blocks, which belong to the same group, are easily identified into two different groups. For instance, we can see the nonuniform texture patterns of "Trees" in images (I9, I10, I11, I13, I16, I17, I19). Nevertheless, the proposed method with WT (G-WT) reaches the higher degrees of true positive and true negative rates. This superiority is noted with original images and becomes greater for the case of blurred ones (see Figure 20). This is because our superpixel approach is based on Gabor filtering which is effective for macro texture characterization. This is was proved in our previous work [29].

The obtained results show that the segmentation with MMGR-WT proposed by Lei et al. gives the best Sensitivity and Specificity only for images with homogeneous regions. It is still with lower performance for images with textured regions of higher intensities variations (e.g., cloud, trees, grass). As a summary of all the obtained results, it is clearly noticed that the proposed is more performant for our application on fire forest images. Where, for the major cases, the different regions are with a large texture variety and higher nonhomogeneous regions (e.g., smoke, fire, trees, grass, etc.).

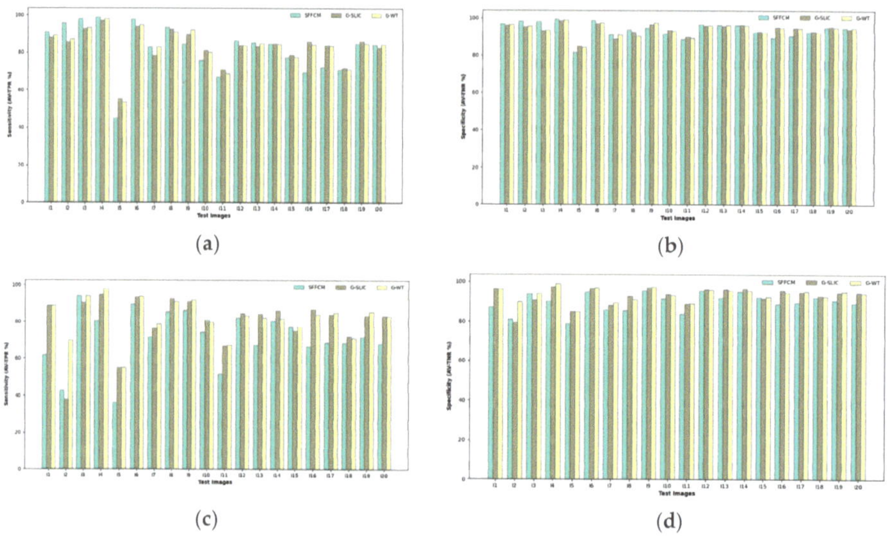

Figure 20. Comparison of SFFCM and our proposed method with WT and SLIC pre-segmentation techniques (G-WT, G-SLIC) based on averaged Sensitivity (**a**,**c**) and Specificity results (**b**,**d**) of 10 experiments. First row: test on natural images illustrated by Table 2. Second row: test on corrupted images with 10% Salt and Pepper noise.

6. Conclusions and Future Works

Segmentation is an important topic in the image processing community. In this study, we presented an end to end framework for application in fire forest image segmentation. The proposed approach is divided into two principle stages: the pre-segmentation and the fuzzy clustering. Our main contributions are in the pre-segmentation stage. First, we have applied a multiscale transformation based on Gabor filtering to improve the superpixel extractions. Second, for the variety of outputs generated by the different pairs of frequencies and orientations (24 filters), we have introduced the PCA to fulfill the dimensionality reduction. The goal is to keep only the most relevant output to improve the regional consistency at the end of presegmentation stage. The clustering is processed by the fuzzy method recently proposed by Lei et al.

The comparison results discussed above show the efficiency of the novel approach. This is clearly shown with images of nonhomogeneous regions. The robustness of the proposed method is experimentally justified by all the above segmentation results on a set of blurred images with different kinds and intensities of noise.

It is worth noting that, generally, our proposed method gives promising image segmentation performance, but it suffers from some shortcomings. First, a few parameters in the algorithm need to be selected appropriately so as to achieve satisfactory results (e.g., Gabor filter frequencies and orientation). Second, the first stage of pre-segmentation (i.e., Gabor filtering and PCA features reduction) is computationally expensive compared to the SFFCM method. Thus, it would be a future work on a fast and effective method can be used with fire forest images. Moreover, the fire and smoke are identified based on the range of color intensities. To improve the automatic fire and smoke detection, a semantic segmentation will be performed by introducing the Deep Learning techniques.

Author Contributions: Conceptualization, L.T.; Methodology, L.T., M.B. and M.T.; Software, L.T.; Validation, M.B. and M.T.; Formal Analysis, L.T.; Investigation, L.T.; Resources, L.T., M.B. and M.T.; Data Curation, L.T., M.B. and M.T.; Writing—Original Draft Preparation, L.T.; Writing—Review & Editing, L.T.; Visualization, L.T.; Supervision, M.S. and E.M.; Project Administration, M.S. and E.M.; Funding Acquisition, M.S. and E.M. All authors have read and agreed to the published version of the manuscript.

Funding: This research was funded by the "PHC Utique" program number 41,755XB of the French Ministry of Foreign Affairs and Ministry of higher education, research and innovation and the Tunisian Ministry of higher education and scientific research in the CMCU project number CMCU 19G1126 and the CARTT-IUT, University of Toulon, France.

Conflicts of Interest: The authors declare no conflict of interest.

References

1. Nemalidinne, S.M.; Gupta, D. Nonsubsampled contourlet domain visible and infrared image fusion framework for fire detection using pulse coupled neural network and spatial fuzzy clustering. *Fire Saf. J.* **2018**, *101*, 84–101. [CrossRef]
2. L-Dhief, F.T.A.; Sabri, N.; Fouad, S.; Latiff, N.M.A.; Albader, M.A.A. A review of forest fire surveillance technologies: Mobile ad-hoc network routing protocols perspective. *J. King Saud Univ. Comput. Inf. Sci.* **2019**, *31*, 135–146.
3. Ajith, M.; Martinez-Ramon, M. Unsupervised Segmentation of Fire and Smoke from Infra-Red Videos. *IEEE Access* **2019**, *7*, 182381–182394. [CrossRef]
4. Yuan, F.; Shi, J.; Xia, X.; Zhang, L.; Li, S. Encoding pairwise Hamming distances of Local Binary Patterns for visual smoke recognition. *Comput. Vis. Image Underst.* **2019**, *178*, 43–53. [CrossRef]
5. Gonçalves, W.N.; Machado, B.B.; Bruno, O.M. Spatiotemporal Gabor filters: A new method for dynamic texture recognition. *arXiv* **2012**, arXiv:1201.3612.
6. Dileep, R.; Appana, K.; Kim, J. Smoke Detection Approach Using Wavelet Energy And Gabor Directional Orientations. In Proceedings of the 12th IRF International Conference, Hyderabad, India, 26 June 2016.
7. Yuan, F. Video-based smoke detection with histogram sequence of LBP and LBPV pyramids. *Fire Saf. J.* **2011**, *46*, 132–139. [CrossRef]
8. Xu, G.; Zhang, Y.; Zhang, Q.; Lin, G.; Wang, Z.; Jia, Y.; Wang, J. Video smoke detection based on deep saliency network. *Fire Saf. J.* **2019**, *105*, 277–285. [CrossRef]
9. Lei, T.; Jia, X.; Zhang, Y.; Liu, S.; Meng, H.; Nandi, A.K. Superpixel-Based Fast Fuzzy C-Means Clustering for Color Image Segmentation. *IEEE Trans. Fuzzy Syst.* **2019**, *27*, 1753–1766. [CrossRef]
10. Guo, L.; Chen, L.; Chen, C.L.P.; Zhou, J. Integrating guided filter into fuzzy clustering for noisy image segmentation. *Digit. Signal Process. A Rev. J.* **2018**, *83*, 235–248. [CrossRef]
11. Miao, J.; Zhou, X.; Huang, Z.T. Local segmentation of images using an improved fuzzy C-means clustering algorithm based on self-adaptive dictionary learning. *Appl. Soft Comput. J.* **2020**, *91*, 106200. [CrossRef]
12. Li, C.; Huang, Y.; Zhu, L. Color texture image retrieval based on Gaussian copula models of Gabor wavelets. *Pattern Recognit* **2017**, *64*, 118–129. [CrossRef]
13. Dios, J.R.M.; Arrue, B.C.; Ollero, A.; Merino, L.; Gómez-Rodríguez, F. Computer vision techniques for forest fire perception. *Image Vis. Comput.* **2008**, *26*, 550–562. [CrossRef]
14. Wang, Y.; Chua, C.S. Face recognition from 2D and 3D images using 3D Gabor filters. *Image Vis. Comput.* **2005**, *231*, 1018–1028. [CrossRef]
15. Kaljahi, M.A.; Shivakumara, P.; Idris MY, I.; Anisi, M.H.; Lu, T.; Blumenstein, M.; Noor, N.M. An automatic zone detection system for safe landing of UAVs. *Expert Syst. Appl.* **2019**, *122*, 319–333. [CrossRef]
16. Parida, P.; Bhoi, N. 2-D Gabor filter based transition region extraction and morphological operation for image segmentation. *Comput. Electr. Eng.* **2017**, *62*, 119–134. [CrossRef]
17. Riabchenko, E.; Kämäräinen, J.K. Generative part-based Gabor object detector. *Pattern Recognit. Lett.* **2015**, *68*, 1–8. [CrossRef]

18. Wong, A.K.K.; Fong, N.K. Experimental study of video fire detection and its applications. *Procedia Eng.* **2014**, *71*, 316–327. [CrossRef]
19. Çetin, A.E.; Dimitropoulos, K.; Gouverneur, B.; Grammalidis, N.; Günay, O.; Habiboğlu, Y.H.; Verstockt, S. Video fire detection—Review. *Digit. Signal Process. Rev. J.* **2013**, *23*, 1827–1843. [CrossRef]
20. Sudhakar, S.; Vijayakumar, V.; Kumar, C.S.; Priya, V.; Ravi, L.; Subramaniyaswamy, V. Unmanned Aerial Vehicle (UAV) based Forest Fire Detection and monitoring for reducing false alarms in forest-fires. *Comput. Commun.* **2020**, *149*, 1–16. [CrossRef]
21. Toulouse, T.; Rossi, L.; Akhloufi, M.; Celik, T.; Maldague, X. Benchmarking of wildland fire colour segmentation algorithms. *IET Image Process.* **2015**, *92*, 1064–1072. [CrossRef]
22. Ganesan, P.; Sathish, B.S.; Sajiv, G. A comparative approach of identification and segmentation of forest fire region in high resolution satellite images. In Proceedings of the 2016 World Conference on Futuristic Trends in Research and Innovation for Social Welfare (Startup Conclave), Coimbatore, India, 29 February–1 March 2016; pp. 3–8.
23. Ko, B.; Jung, J.H.; Nam, J.Y. Fire detection and 3D surface reconstruction based on stereoscopic pictures and probabilistic fuzzy logic. *Fire Saf. J.* **2014**, *68*, 61–70. [CrossRef]
24. Wang, M.; Liu, X.; Gao, Y.; Ma, X.; Soomro, N.Q. Superpixel segmentation: A benchmark. *Signal Process. Image Commun.* **2017**, *56*, 28–39. [CrossRef]
25. Ma, J.; Li, S.; Qin, H.; Hao, A. Unsupervised Multi-Class Co-Segmentation via Joint-Cut over L1-Manifold Hyper-Graph of Discriminative Image Regions. *IEEE Trans. Image Process.* **2017**, *26*, 1216–1230. [CrossRef]
26. Shang, R.; Tian, P.; Jiao, L.; Stolkin, R.; Feng, J.; Hou, B.; Zhang, X. Metric Based on Immune Clone for SAR Image Segmentation. *IEEE J. Sel. Top. Appl. Earth Obs. Remote Sens.* **2016**, *9*, 1–13. [CrossRef]
27. Neubert, P.; Protzel, P. Compact Watershed and Preemptive SLIC: On Improving Trade-Offs of Superpixel Segmentation Algorithms. In Proceedings of the 2014 22nd International Conference on Pattern Recognition, Stockholm, Sweden, 24–28 August 2014; pp. 996–1001.
28. Shang, R.; Chen, C.; Wang, G.; Jiao, L.; Okoth, M.A.; Stolkin, R. A thumbnail-based hierarchical fuzzy clustering algorithm for SAR image segmentation. *Signal Process.* **2020**, *171*, 107518. [CrossRef]
29. Tlig, L.; Sayadi, M.; Fnaiech, F. A new fuzzy segmentation approach based on S-FCM type 2 using LBP-GCO features. *Signal Process. Image Commun.* **2012**, *27*, 694–708. [CrossRef]
30. Zhu, Z.; Jia, S.; He, S.; Sun, Y.; Ji, Z.; Shen, L. Three-dimensional Gabor feature extraction for hyperspectral imagery classification using a memetic framework. *Inf. Sci.* **2015**, *298*, 274–287. [CrossRef]
31. Tadic, V.; Popovic, M.; Odry, P. Fuzzified Gabor filter for license plate detection. *Eng. Appl. Artif. Intell.* **2016**, *48*, 40–58. [CrossRef]
32. Tan, X.; Triggs, B. Fusing gabor and LBP feature sets for kernel-based face recognition. *Lect. Notes Comput. Sci.* **2007**, *4778*, 235–249.
33. Kim, S.; Yoo, C.D.; Member, S.; Nowozin, S.; Kohli, P. Higher-Order Correlation Clustering. *IEEE Trans. Pattern Anal. Mach. Intell.* **2014**, *36*, 1761–1774. [CrossRef]
34. Zanaty, E.A. Determining the number of clusters for kernelized fuzzy C-means algorithms for automatic medical image segmentation. *Egypt. Inform. J.* **2012**, *13*, 39–58. [CrossRef]
35. Qu, F.; Hu, Y.; Xue, Y.; Yang, Y. A modified possibilistic fuzzy c-means clustering algorithm. *Proc. Int. Conf. Nat. Comput.* **2013**, *13*, 858–862.
36. Zheng, Y.; Jeon, B.; Xu, D.; Wu, Q.M.J.; Zhang, H. Image segmentation by generalized hierarchical fuzzy C-means algorithm. *J. Intell. Fuzzy Syst.* **2015**, *28*, 961–973. [CrossRef]
37. Gu, J.; Jiao, L.; Yang, S.; Zhao, J. Sparse learning based fuzzy c-means clustering. *Knowl.-Based Syst.* **2017**, *119*, 113–125. [CrossRef]
38. Stutz, D.; Hermans, A.; Leibe, B. Superpixels: An evaluation of the state-of-the-art. *Comput. Vis. Image Underst.* **2018**, *166*, 1–27. [CrossRef]

39. Gamino-Sánchez, F.; Hernández-Gutiérrez, I.V.; Rosales-Silva, A.J.; Gallegos-Funes, F.J.; Mújica-Vargas, D.; Ramos-Díaz, E.; Kinani, J.M.V. Block-Matching Fuzzy C-Means clustering algorithm for segmentation of color images degraded with Gaussian noise. *Eng. Appl. Artif. Intell.* **2018**, *73*, 31–49. [CrossRef]
40. Xu, G.; Li, X.; Lei, B.; Lv, K. Unsupervised color image segmentation with color-alone feature using region growing pulse coupled neural network. *Neurocomputing* **2018**, *306*, 1–16. [CrossRef]

Publisher's Note: MDPI stays neutral with regard to jurisdictional claims in published maps and institutional affiliations.

© 2020 by the authors. Licensee MDPI, Basel, Switzerland. This article is an open access article distributed under the terms and conditions of the Creative Commons Attribution (CC BY) license (http://creativecommons.org/licenses/by/4.0/).

sensors

MDPI

Article

Obscuration Threshold Database Construction of Smoke Detectors for Various Combustibles

Hyo-Yeon Jang and Cheol-Hong Hwang *

Department of Fire and Disaster Prevention, Daejeon University, 62 Daehak-ro, Dong-Gu, Daejeon 34520, Korea; fayahyo@gmail.com

* Correspondence: chehwang@dju.ac.kr; Tel.: +82-42-280-2592

Received: 8 September 2020; Accepted: 2 November 2020; Published: 4 November 2020

Abstract: The obscuration thresholds for various smoke detectors and combustibles, required as an input parameter in fire simulation, were measured to predict the accurate activation time of detectors. One ionization detector and nine photoelectric detectors were selected. A fire detector evaluator, which can uniformly control the velocity and smoke concentration, was utilized. Filter paper, liquid fuels, and polymer pellets were employed as smoke-generation combustibles. The nominal obscuration thresholds of the considered detectors were 15 %/m, but the ionization detectors activated at approximately 40 %/m and 16 %/m, respectively, on applying filter paper and kerosene. In contrast, the reverse obscuration thresholds were found quantitatively according to the combustibles in the photoelectric detector. This phenomenon was caused by differences in the color of the smoke particles according to the combustibles, which is explained by single-scattering albedo (ratio of light scattering to light extinction). The obscuration thresholds for liquid fuels (kerosene, heptane and toluene) as well as fire types of polymer plastic pellets were also measured for several photoelectric detectors. A database of obscuration thresholds was thereby established according to the detector and combustible types, and it is expected to provide useful information for predicting more accurate detector activation time and required safe egress time (REST).

Keywords: smoke detector; obscuration per meter (OPM); obscuration threshold; smoke color; performance-based fire safety design (PBD)

1. Introduction

To reduce the risk of fire due to the manhattanization, increased sizes and complexity of buildings, the number of countries introducing performance-based fire safety design (PBD) methods has been increasing [1]. The PBD approach generally assesses the fire risk based on a comparison between the available safe egress time (ASET) and the required safe egress time (RSET). This is based on a timeline analysis owing to the limitations of a complex review of the various factors that may affect the assessment. Specifically, the allowance of design uncertainties based on the ASET and RSET concepts can be expressed by a safety margin (ASET-RSET > safety margin) or a safety factor (ASET/RSET > safety factor). The RSET is evaluated as a relatively safe building from the impact of fire when it is larger than the ASET [2]. Therefore, to improve the reliability of fire risk assessment through PBD, the ASET and RSET must first be accurately calculated.

The RSET, defined as the time it takes for the occupants in a building to evacuate to a safe place after the occurrence of a fire, is the sum of the fire detection time of the detector, the response time of the occupant (including the alarm and cue, recognition, and time delay to begin action), and the movement time [3]. The response time and movement time of occupants are evaluated by various evacuation simulations or theories [4,5]. Figure 1 shows a schematic diagram of the role of fire and evacuation simulations for ASET and RSET calculation. The detection time, which is the activation time of the smoke or heat detector after the occurrence of a fire, is included in the RSET calculation; however, it is predicted through the fire simulation [6]. As a result, the detection time prediction of the detector is directly related to the calculation of the RSET. It is affected by the prediction accuracy of the fire simulation based on various fire scenarios or fire conditions. In particular, the input parameters of the numerical model for smoke and heat detectors included in the fire models are important for accurately predicting the detection time of the detector, calculating the RSET, and finally securing the reliability of the fire risk assessment for the PBD. For reference, a fire dynamics simulator (FDS) [7], which can analyze fire behavior over time in a three-dimensional space, is widely used as a representative fire model for fire risk assessment.

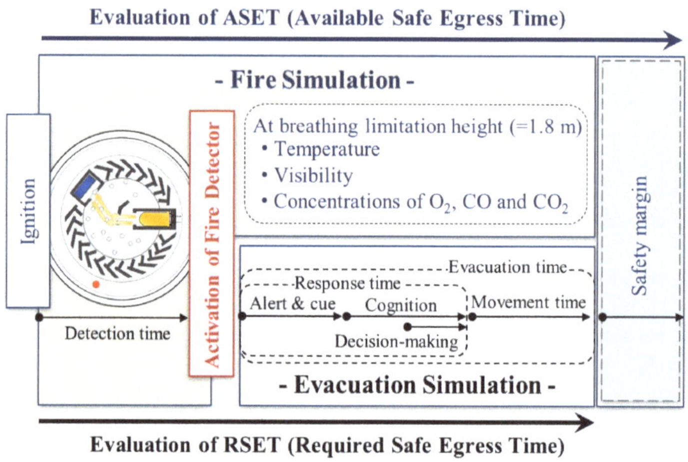

Figure 1. Classification of roles of fire and evacuation simulation for available safe egress time (ASET) and required safe egress time (RSET) evaluations.

The Heskestad and Cleary models have been applied to the FDS to predict the detection time of smoke detectors. These two numerical models can be classified as time lag methods in which a characteristic time (with a difference in smoke concentration existing between the housing and the sensing chamber), takes into account the entry resistance of smoke due to the shape of the inlet of the smoke detector [8]. In the Heskestad model [9,10], the delay time (or characteristic time) is the time required for the smoke to reach from outside of the detector to the inside. It is expressed as the ratio of the characteristic length (L, m) and the free stream velocity (U, m/s) [11,12], where L is an input parameter that must be determined experimentally. In the Cleary model [13], the mixing time inside the detector is also considered. It is the total delay time expressed by the sum of the dwell time (δt) required to flow into the sensing chamber from the outside and the mixing time (τ) inside. The smoke flow is divided into the plug flow and perfectly stirred flow. δt and τ are expressed as functions of U, and additional input parameters (α_e, β_e, α_c, β_c), which must be determined experimentally, are required. In other words, the input parameters for expressing the time lag of the detector activation differ according to the applied numerical model.

On the other hand, as a common input parameter to both the Heskestad and Cleary models, the obscuration threshold value—which is the obscuration per meter (OPM) at the moment the detector is activated—is required. The OPM is defined by Equation (1) and is signified by the ratio (%) of the intensity of light extinction by smoke particles per unit length (m).

$$\text{OPM} = \left(1 - (I/I_o)^{1/L_p}\right) \times 100\ (\%/\text{m}) \tag{1}$$

where I/I_0 is the ratio of the intensity of light extinction by smoke particles, and L_p is the light path length (m). The detailed concept and measurement methods of the individual and common input parameters required in the Heskestad and Cleary models can be found in previous studies [14,15].

Recently, our research team developed a test device—a fire detector evaluator (FDE)—to measure the input parameters required in the numerical model of smoke detectors. The FDS-based smoke detector input parameter values that are widely applied to PBD were presented [16]. Specifically, the input parameters of the Heskestad [9,10] and Cleary [13] numerical models that could predict the activation of specific photoelectric and ionization smoke detectors were measured. The fire simulation results that applied the input parameters obtained through the experiment correctly predicted the activation time of the detector measured under the same fire conditions, whereas large errors in the detector activation times occurred when the input parameters suggested in the FDS User's Guide were applied [14,15]. In addition, through sensitivity analysis of the input parameters required in the smoke detector numerical model, it was found that the obscuration threshold values, which are common parameters, had the greatest influence on detector activation compared to the development factors required in each numerical model [17]. Considering smoke detectors with a wide variety of structural characteristics and various types of combustibles, the results of the present study not only highlight the importance of measuring the input parameters of smoke detector numerical models but also provide information on important input parameters that must be initially considered.

In general, quantitatively accurate values were not provided for the obscuration thresholds of detectors. Only the application range or nominal values according to detector sensitivity tests were provided by the manufacturers. Therefore, in previous studies that predicted detector activation times using fire simulation, precise obscuration threshold measurements or additional verification processes were not specifically considered [18–20]. Furthermore, although the changes in smoke color due to the combustibles can produce a significant change in the light scattering signals applied to smoke detectors, the quantitative difference of the obscuration thresholds according to various combustibles has not yet been investigated.

For the ultimate purpose of improving the prediction accuracy of the detector activation time by constructing a database (DB) of input parameters required in the fire simulation for various smoke detectors and combustibles, measurement of the detector obscuration thresholds and DB construction was performed in this study. To this end, a total of ten detectors, mainly installed in South Korea, were reviewed. Firstly, the quantitative differences in the obscuration thresholds of the ionization and photoelectric detectors according to the colors of the smoke particles were examined. In addition, filter paper, various liquid fuels, and polymer pellets were considered as the combustibles. The differences in obscuration threshold values, which were based on the combustibles in photoelectric detectors to which the light scattering method was applied, were assessed. Performing the PBD through the DB provided in this study is expected to enable prediction of the activation time of detectors far more accurately by applying the appropriate input parameters according to the combustible material.

2. Experimental Conditions

2.1. Experimental Method and Procedures

To measure the obscuration threshold—a common input parameter of smoke detector numerical models—an FDE was used, as shown in Figure 2 [16]. Figure 2a shows the overall shape of the FDE, which was made with a square carbon steel duct with a cross-sectional area of 0.18 m^2 (0.6 m × 0.3 m). It was fastened with rubber packing. The inside of the duct of the FDE was designed to realize a uniform smoke flow over time and space. To reduce the intensity of turbulence, a smoke generator was placed in the front and a sirocco fan was placed in the rear of the FDE based on the detector and the measurement locations [21]. The flow rate was controlled using the sirocco fan, and the damper was controlled by means of an inverter. In addition, a honeycomb and mesh were installed to form a uniform smoke flow. As a smoke-generating device installed at the bottom of the FDE, a burner (optimized according to the type of combustibles) was installed to enable continuous smoke generation.

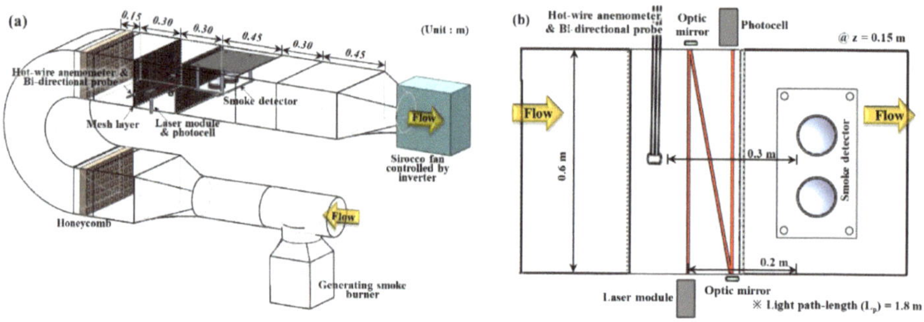

Figure 2. Schematics of (**a**) Fire Detector Evaluator (FDE) and (**b**) test section for measuring input parameters required in the numerical models of a smoke detector [16].

Figure 2b shows the measurement location of the velocity and smoke concentration located at the center of the duct at a vertical height of z = 0.15 m. The velocity was measured using a bi-directional probe calibrated by a hot wire anemometer (Testo 480) to accurately measure the low-speed flow and reproducibility of the experiment. The obscuration threshold for the detector was measured using a light-extinction method [22], detailed device descriptions can be confirmed in previous studies [23]. To minimize measurement errors of the OPM due to forward scattering, approximately 1.8 m of L_p between the laser module (650 nm) and the photocell was applied. Details of L_p selection and the FDE can be found in previous studies [16]. The power of the smoke detector was supplied from a p-type fire control panel, and the voltage signal to check the smoke concentration and the activation of the detector were recorded in real time at 1-s intervals using a DAQ (Graphtec, GL 820).

The measurement process of the obscuration threshold value, an input parameter that has the greatest influence on the detector activation time, is explained in Figure 3. The results in the figure show the voltage signals associated with the activation of the OPM and the detector over time after kerosene ignition using a lamp wick. In the detector sensitivity test, considering that the smoke flow velocity was 0.2–0.4 m/s [24], the average velocity (U) inside the FDE was fixed at 0.3 m/s. The delay time, which is the time at which the smoke OPM was measured to the moment the smoke flowed into the detector (based on the moment when the voltage increased in the detector's internal circuits), was 4.0 s. The OPM of the smoke reaching the test section inside the FDE increased relatively linearly with time and reached a quasi-steady state at approximately 55 %/m. The voltage signal associated with the detector's activation showed that the concentration of the smoke gradually increased, and a voltage drop occurred when it exceeded the obscuration threshold. In other words, the OPM was synchronized over time and the voltage signal related to the detector activation had a delay time of 4.0 s. Consequently, the obscuration threshold value of 36.3 %/m, which was the OPM at the moment the detector was activated, could be obtained. Obscuration threshold measurements were repeated at least five times under the same experimental conditions, and the average value and standard deviation expressed in the form of a vertical error bar are shown in the graph.

2.2. Selection of Smoke Detectors and Smoke Generators for Various Combustibles

Smoke detectors have a wide variety of structures, determined by the shape of the housing according to the manufacturer, or the principle of fire detection, the structure of the optical maze to form the dark room inside the sensing chamber, the porosity of the mesh to prevent the inflow of foreign substances, and the angles of the light-emitting and the light-receiving parts. Therefore, considering the efficiency of the DB construction for the obscuration thresholds of the detectors, one ionization and nine photoelectric smoke detectors were selected. All smoke detectors reviewed in this study were of the two-class kind, and the nominal obscuration threshold was 15 %/m based on the information provided by the manufacturers [24]. The photoelectric detectors were named A through I for convenience. They were selected through the investigation of a number of installations in South Korea over the past three years and by leveraging the expertise of approximately 30 related specialists. In practice, even if the same detector is applied, a difference in the obscuration threshold may occur owing to the differences in light scattering and light-extinction characteristics of the smoke particles according to the combustibles. Therefore, various combustibles, such as filter paper, liquid fuels, and polymer pellets, were considered.

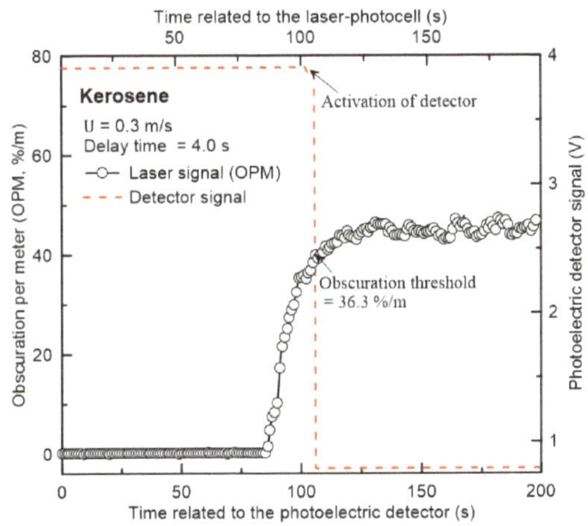

Figure 3. Procedure for measuring the obscuration threshold of smoke detector using FDE.

Figure 4 shows a photograph and schematic diagram of a smoke-generating device based on the type of combustible material. Different shapes of burners were applied according to the combustibles to enable control of the amount of smoke generated. Filter paper, produced by ADVANTEC, is a standard combustible applied to the sensitivity test of detectors in South Korea [24]. In general, to generate white smoke from filter paper, smoldering combustion is induced by placing it on a 400 °C hot plate or by cutting a long piece of paper and burning it in a receptacle (or burner) [25–27]. However, it is difficult to control a uniform amount of smoke over time, and it was found that the high-temperature smoke flow due to intermittent flaming combustion caused thermal stratification inside the FDE. Since the occurrence of thermal stratification causes fluctuations in smoke concentration and flow velocity inside the FDE, it can lead to different detection times and low measurement reproducibility, depending on the vertical height of the detector [28]. Therefore, in this study, 20 g of shredded filter paper (approximately 10 × 40 mm) was irregularly stacked into a receptacle. Then, a stainless-steel plate with a 65-mm diameter was installed on the top, as shown in Figure 4a. When the paper was ignited through a hole located at the bottom of the receptacle, the steel plate moved to the bottom very slowly. As a result, the occurrence of flaming combustion due to sufficient air contact of the paper was suppressed, and a sufficient amount of smoke at a very low temperature was generated for the duration of the experiment.

In general, a pool fire—which can easily control the scale of a fire owing to changes in the surface area of the fire source—is applied to liquid fuels that generate black smoke; however, high-temperature smoke flow due to the flame can cause significant thermal stratification. Therefore, in this study, a cuboid burner with an inserted lamp wick, as shown in Figure 4b, was applied to minimize the heat generated and to control the amount of smoke generated [16]. Smoke generation using a wick has the advantage that it is easy to control the amount of smoke produced through changes in the diameter, exposed length, and the number of wicks. Kerosene ($C_{12}H_{24}$), heptane (C_7H_{16}), and toluene (C_7H_8)—with a relatively high soot tendency—were considered the liquid fuels, and the number of wicks was changed according to each liquid fuel to generate smoke that could reach the obscuration threshold of the detector.

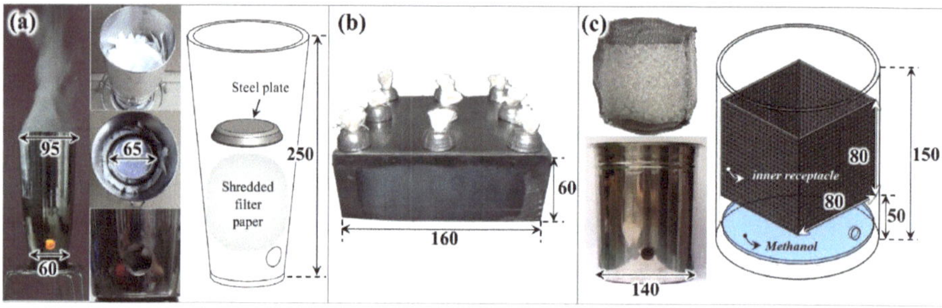

Figure 4. Photographs and schematics of smoke generators using (**a**) filter paper, (**b**) liquid fuel and (**c**) polymer pellets (unit:mm).

Figure 4c shows a photo and schematic diagram of a smoke generator of polymer plastics that can be easily identified as combustibles in modern buildings. For ease of ignition and continuous burning of solid combustibles, pellets of approximately 5 mm in diameter were applied. For initial ignition, 10 mL of methanol was supplied to the bottom of the receptacle, and 2 g of pellets were supplied to the inner receptacle of an 80 mm long cube made of steel mesh. Methanol flame can supply sufficient heat for the pyrolysis of the pellets; however, it may not have a significant effect on the obscuration threshold (on account of the burning of polymers) because of the small quantity of soot generated. Thus, since the supply of methanol and pellets was very small in this study, thermal stratification did not occur in the FDE. This was clearly observed through the uniformity of the smoke concentration along the vertical height in the test section.

Figure 5 shows photographs of the polymer pellets—polyethylene (PE), polypropylene (PP), polyurethane (PU), polyvinyl chloride (PVC), and polystyrene (PS). These polymers are representative components of combustibles, such as mattresses, sofas, and home appliances, which are common in building fire scenarios.

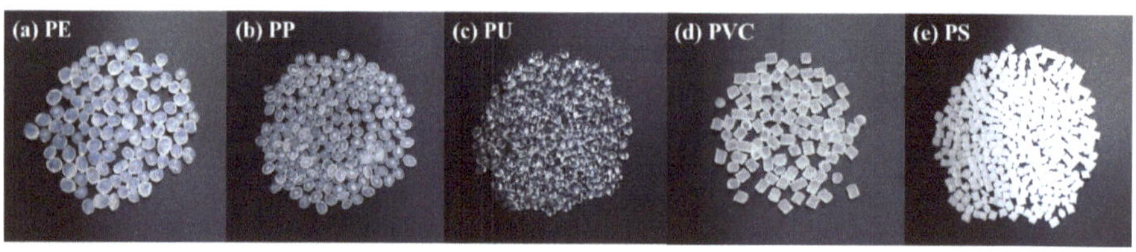

Figure 5. Photographs of polymer pellets (**a**) polyethylene, (**b**) polypropylene, (**c**) polyurethane, (**d**) polyvinyl chloride and (**e**) polystyrene.

3. Results and Discussion

3.1. Obscuration Threshold of Ionization and Photoelectric Detectors according to Smoke Particle Colors

Smoke detectors can generally be classified into ionization-type and photoelectric-type detectors based on the detection method. Figures 6 and 7 show the obscuration thresholds of ionization and photoelectric detectors for filter paper and kerosene, in which the smoke particles are white and black, respectively. The measurement of the obscuration threshold for each combustible material with the same detector was repeated ten times, as shown in Figure 3. A new detector was then installed that had not been used in any preceding experiment.

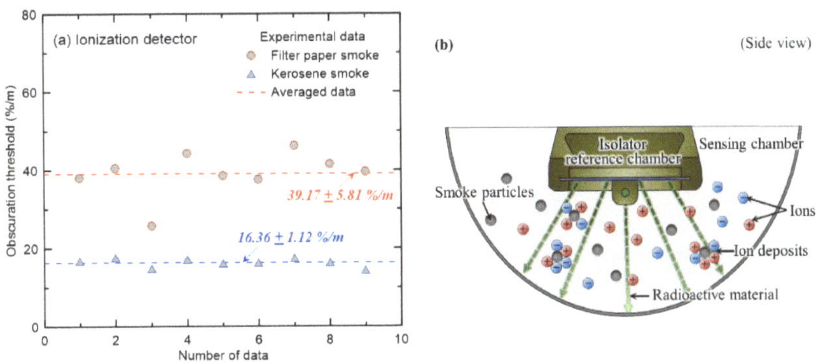

Figure 6. For the ionization smoke detector, (**a**) obscuration thresholds with filter paper and kerosene and (**b**) the conceptual diagram of the detector.

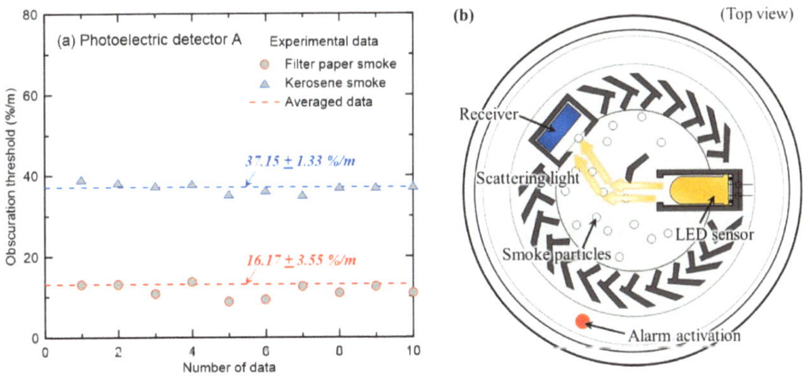

Figure 7. For the photoelectric smoke detector, (**a**) obscuration thresholds with filter paper and kerosene and (**b**) the conceptual diagram of the detector.

Firstly, the results of the ionization-type detector in Figure 6a show that the average obscuration threshold obtained from repeated experiments is 39.17 %/m, and the standard deviation is 5.81 %/m when filter paper is applied. The average obscuration threshold of kerosene is 16.36 ± 1.12 %/m, showing a significant difference compared to that of the filter paper. This can be explained by the structure and principle of the ionization detector shown in Figure 6b. The structure of the ionization detector is divided into a sensing chamber (external ion chamber) and an isolator reference chamber (internal ion chamber). The air in the sensing chamber is ionized by a small amount of radioactive material emitted from the isolator reference chamber. When smoke-containing soot generated from a fire enters the sensing chamber, it deposits ions and increases the electrical resistance inside the sensing chamber. In other words, when the flow of electric current due to the inflow of smoke particles decreases below a set value, the detector is activated. Water vapor accounts for the largest proportion of white smoke generated from filter paper, and black smoke generated from kerosene accounts for a substantial proportion of soot generated by the bonding and growth of carbon particles. Consequently, it can be expected that the detector activates at a relatively low OPM as the soot in the black smoke deposits ions more easily compared to the water vapor in white smoke [16].

Figure 7a shows the results of photoelectric detector A under the same combustibles and experimental conditions as shown in Figure 6. On average, the obscuration threshold of smoke generated by the smoldering combustion of filter paper is 16.17 ± 3.55 %/m, and the obscuration threshold was measured to be 37.15 ± 1.33 %/m under the flaming combustion conditions of kerosene. In other words, contrary to the ionization detector results, the detector was activated at a relatively high OPM under the flaming combustion conditions of kerosene, which generated black smoke. Although the principles of smoke detection by ionization and photoelectric detectors are different, it was interesting to discover that the obscuration threshold had the opposite quantitative result based on the colors of the smoke particles. To analyze the cause of the above phenomenon, the structure and principle of the photoelectric detector were assessed. As shown in Figure 7b, the detector was activated when a certain amount of scattered light (incident light from the LED sensor, which was scattered by smoke particles) reached the receiver (photocell). Based on the incident light extinction through the scattering and absorption by smoke particles, black smoke is expected to have less light extinction through scattering than white smoke. This phenomenon can be clearly explained through a review of single-scattering albedo (SSA) [29], which is expressed as the ratio of light scattering to light extinction (sum of scattering and absorption). The SSA factor was studied to examine the scattering characteristics of soot according to the intrinsic optical properties of smoke generated from various fuels. If the SSA = 1, it means that the light extinction of smoke particles is only due to scattering, whereas if the SSA = 0, it means that light extinction is caused primarily by absorption of smoke particles. Specifically, black smoke particles (generated from kerosene) have been reported as having an SSA = 0.3 [29] at a wavelength of 532 nm, and white smoke particles (mainly generated from paper or wood fires) have been reported as having an SSA = 0.6~0.8 [30] at a wavelength of 500 nm. This is why the detector activated at a higher OPM when kerosene was applied than when filter paper was used.

Summarizing the results of Figures 6 and 7, quantitatively similar obscuration thresholds were found under the conditions of white smoke in the ionization detector and black smoke in the photoelectric detector owing to the differences in the activating principles according to the detector type. The value was very high at approximately 40 %/m. The smoke detectors reviewed in this study were of the two-class types [24] with a nominal obscuration threshold of 15 %/m. Thus, to accurately predict the detector activation time, an additional review of various combustibles with different smoke particle colors had to be considered carefully in addition to the detector type.

3.2. Database Construction of the Obscuration Thresholds with Liquid Fuels and Polymer Pellets for Various Models of the Photoelectric Smoke Detector

As mentioned above, the nominal obscuration threshold for several types of photoelectric smoke detectors applied in the activation and non-activation tests for the performance test of smoke detectors was equally implemented at 15 %/m based on the two-class type [24]. However, activation of the photoelectric detector was determined by the correlation among the smoke characteristics, smoke transport, and detector characteristics. Therefore, to consider the optical properties of the smoke particles, a review of various fuels was required. In addition, because there was a difference in various structures and sensitivity inside the sensing chambers, depending on the detector manufacturer, it was necessary to calculate an accurate obscuration threshold for several detectors with a high frequency of use. Since ionization detectors containing radioactive substances have high costs of disposal after their useful lifetimes, their production and usage are very low compared to those of photoelectric detectors. Therefore, only photoelectric detectors were considered for the examination of the obscuration threshold based on different combustibles.

Figure 8 shows the obscuration threshold when the liquid fuels kerosene, heptane, and toluene were applied as the combustibles targeting photoelectric detectors (A–I) from the different manufacturing companies. Five replicates were performed on the same detector models, and the symbol and vertical error bars in Figure 8 represent the mean and the standard deviation of each, respectively. Figure 8a shows the result of kerosene, and it was found that even if the same detector model was used, the standard deviation of the obscuration threshold varied considerably depending on the manufacturer. The average obscuration threshold of each detector model was also significantly different. The average value for all detectors was 38.41 %/m, showing a standard deviation of 3.85 %/m. In other words, even if the same experimental conditions and combustibles were considered, the obscuration threshold had a quantitative difference that could not be ignored, not only between the detector models of different manufacturers but also between detectors from the same manufacturer. Nevertheless, the provision of an average obscuration threshold for a large number of detectors with a high frequency of use can have a significant meaning for PBDs that evaluate fire risk using fire simulation. The detector model of a specific manufacturer is not determined specifically in the PBD process, and obscuration thresholds, which are common input parameters of the detector model for fire simulation, are arbitrarily set by the user. Therefore, the application of the average obscuration threshold measured for multiple detectors is an appropriate approach for a more accurate detector activation time and RSET calculation from a statistical point of view.

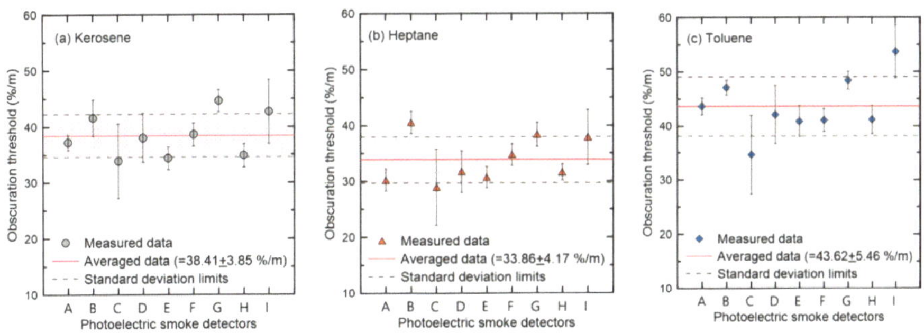

Figure 8. Comparison of obscuration thresholds with liquid fuels ((**a**) kerosene, (**b**) heptane and (**c**) toluene) for various models of the photoelectric detector (A~I).

As shown in Figure 8b,c with heptane and toluene applied, the average obscuration thresholds of the considered detectors were 33.86 and 43.62 %/m, respectively. As shown in Figure 7, it was found that the detector activated at a smoke concentration much higher than the nominal obscuration threshold of 15 %/m for these black-smoke-generating liquid fuels. In terms of the quantitative differences in the obscuration threshold of the considered liquid fuels, the highest values are observed in the order of toluene–kerosene–heptane. The cause can be analyzed by comparing the SSA and mass specific extinction coefficient (K_m) [31,32]. To this end, a detailed comparative review of the optical properties of smoke particles from each fuel should be performed. A direct comparison of the values provided in the literature has a clear limit owing to the quantitative difference between the SSA and K_m (the wavelength of incident light). Therefore, an investigation of the differences in the obscuration threshold for liquid fuels that generate visually similar black smoke particles was not considered in the scope of this study.

Figure 9 shows the obscuration thresholds of photoelectric detectors A–C when five types of polymer plastic (PP, PVC, PU, PE, and PS) pellets, which can be easily identified as combustibles in modern buildings, were applied. The average values of the three manufacturers' detectors based on the composition of the polymer pellets are simultaneously shown as shades and numerical values in the figure. The obscuration threshold of the polymer pellets differed significantly, from a minimum of approximately 19 %/m to a maximum of approximately 33 %/m, depending on the pellets. The obscuration thresholds of PP and PE were 19.72 and 18.97 %/m, respectively, showing quick response characteristics regardless of the detector model. On the other hand, the obscuration thresholds of PVC and PS were 30.38 and 32.67 %/m, respectively. It was expected that the detector activation could be slower under the same assumed fire conditions. In Figure 8, the differences in the obscuration thresholds of the polymer plastics according to their respective compositions are shown. Detailed analysis is difficult owing to limited data on the SSA and K_m.

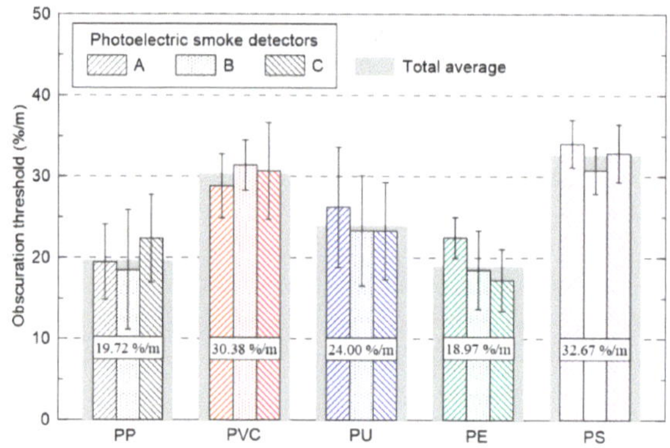

Figure 9. Comparison of obscuration thresholds with various polymer pellets (polypropylene (PP), polyvinyl chloride (PVC), polyurethane (PU), polyethylene (PE) and polystyrene (PS)) for the photoelectric detector models (A–C).

However, a schematic analysis of the differences among them, depending on the colors of the smoke particles, may be possible through the analytical method shown in Figure 7.

Figure 10 shows an instantaneous picture of the flame and smoke generated when the PP and PS pellets were burned under the same ventilation conditions. When the PP pellets were burned, white smoke particles were generated, and the flames were visually gray and black. On the other hand, PS pellets generated a large quantity of soot, and black smoke was clearly identified. Although the color of smoke cannot be clearly defined visually, gray and black smoke was visible in the PP and PE samples, and black smoke was visible in the PVC and PS samples. As a result, it was expected that the PP and PE containing white smoke particles would have a lower obscuration threshold owing to the higher ratio of scattered light in the extinction light. However, it was presumed that the PVC and PS samples, in which light extinction was mainly caused by absorption, would have a relatively high obscuration threshold.

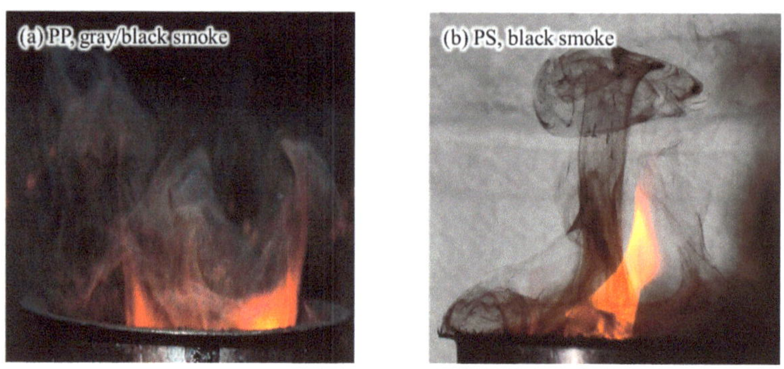

Figure 10. Photographs of smoke flow generated from burning of (**a**) PP and (**b**) PS pellets.

Table 1 summarizes the obscuration thresholds of filter paper, liquid fuels, and polymer pellets for ionization and photoelectric detectors, respectively, as shown in Figures 6–9. The obscuration threshold of the proposed ionization detector is an average value that was obtained by repeating the tests five times for the same detector model. In addition to the repeated experiments, the average value of the nine photoelectric detectors was simultaneously measured. As a result, the measured obscuration thresholds presented for various combustibles could be assessed to have had sufficient reliability (from a statistical point of view) to be used as an important input DB for predicting the activating time of a detector using fire simulations. In addition, more specific device properties of detectors required as input parameters for fire simulations can be found on the Fire Technology Solution DB website, which is openly activated for engineers performing PBD [33].

Table 1. Summary of obscuration thresholds (%/m) of filter paper, liquid fuels and polymer pellets (for the photoelectric smoke detector, the average value of all considered detectors is presented).

	Filter Paper	Kerosene	Heptane	Toluene	PP	PVC	PU	PE	PS
Ionization detector	39.2	16.4	11.1	-	-	-	-	-	-
Photoelectric detector	16.2	38.4	33.9	43.6	19.7	30.4	24.0	19.0	32.7

4. Conclusions

The objective of this study was to improve the prediction accuracy of the detector activation time by establishing a DB of input parameters required for the fire simulation of various smoke detectors and combustibles. The obscuration threshold, which is the OPM (%/m) at the moment the detector is activated, was measured. To this end, one ionization detector and nine photoelectric detectors, which are frequently used in South Korea, were selected. In addition, a fire detector evaluator (FDE) that could uniformly control the velocity and smoke concentration was used. Filter paper, liquid fuels, and polymer pellets were employed as the combustibles for smoke generation. The main results are described below.

The nominal obscuration thresholds of the considered smoke detectors were all 15 %/m; however, the ionization detectors operated at approximately 40 and 16 %/m, respectively, when the filter paper and kerosene were applied. The reverse obscuration thresholds were quantitatively determined according to the combustibles in the photoelectric detector. This phenomenon was caused by the differences in smoke particles, which were white and black, based on the combustibles. It was specifically explained through the SSA, which is defined as the ratio of light scattering to light extinction.

The average obscuration thresholds of the liquid fuels, such as kerosene, heptane, and toluene, were measured for nine photoelectric detectors. Even under the same experimental conditions and combustibles, significant differences in obscuration thresholds were found, not only between the detectors of different manufacturers but also between the detectors of the same manufacturer. As a result, the application of the average obscuration threshold measured for multiple detectors can be deemed an approach to predicting a more efficient detector activation time at the PBD stage where the detector model has not yet been determined.

The obscuration threshold of the photoelectric detector was measured for five types of polymer plastic (PP, PVC, PU, PE, and PS) pellets that can be easily identified as combustibles in modern buildings. The lower obscuration thresholds of PP and PE compared to those of PVC and PS were analyzed through the visualization of gray and black smoke particles, including white smoke.

In conclusion, a DB was constructed (based on the detector type and combustible type) of the obscuration threshold required as the detector input information for a fire simulation. The results of this study are expected to provide useful information for achieving more accurate detector activation times and RSET predictions. More specific effects of smoke particle characteristics according to ventilation and burning type such as flaming and flameless (smoldering) combustion on the obscuration threshold of smoke detector will be carried out in future studies.

Author Contributions: Conceptualization, C.-H.H.; methodology, H.-Y.J. and C.-H.H.; software, C.-H.H.; validation, H.-Y.J. and C.-H.H.; formal analysis, C.-H.H.; resources, C.-H.H.; data curation, H.-Y.J.; writing—original draft preparation, H.-Y.J.; writing—review and editing, C.-H.H.; visualization, H.-Y.J.; supervision, C.-H.H.; project administration, C.-H.H.; funding acquisition, C.-H.H. All authors have read and agreed to the published version of the manuscript.

Funding: This work was partly supported by the National Fire Agency of Korea through the field-based firefighting activity supporting technology development project (MPSS-Fire Safety-2015-66); and Institute of Information and Communications Technology Planning Evaluation (IITP) grant funded by the Korea government (MIST) (No. 2020-0-00012, Development of intelligent fire detection equipment based on smoke particle spectrum analysis).

Conflicts of Interest: The authors declare no conflict of interest.

Abbreviations

The following abbreviations are used in the manuscript:

PBD	Performance-Based Fire Safety Design	L_p	Light Path Length [m]
ASET	Available safe egress time	FDE	Fire detector evaluator
RSET	Required safe egress time	DB	database
FDS	Fire dynamics simulator	PE	Polyethylene
L	Characteristic length [m]	PP	Polypropylene
U	Free stream velocity [m/s]	PU	Polyurethane
δt	Dwell time	PVC	Polyvinyl chloride
τ	Mixing time	PS	Polystyrene
α_e, β_e, α_c, β_c	Additional input parameters	SSA	Single scattering albedo
OPM	Obscuration per meter [%/m]	K_m	Mass specific extinction coefficient [m^2/g]

References

1. Beck, V.R. Performance-based fire engineering design and its application in Australia. *Fire Saf. Sci.* **1997**, *5*, 23–40. [CrossRef]
2. Purser, D.A. ASET and RSET: Addressing some issues in relation to occupant behavior and tenability. *Fire Saf. Sci.* **2003**, *7*, 91–102. [CrossRef]
3. Proulx, G. Movement of people: The evacuation timing. In *SFPE Handbook of Fire Protection Engineering*, 3rd ed.; Section 3; DiNenno, P.J., Drysdale, D., Beyler, C.L., Walton, W.D., Custer, R.-P., Hall, J.R., Watts, J.M., Eds.; Society of Fire Protection Engineers: Gaithersburg, MD, USA, 2002; pp. 342–366.
4. Poon, S.L. A dynamic approach to ASET/RSET assessment in performance based design. *Proc. Eng.* **2014**, *71*, 173–181. [CrossRef]
5. Yan, Z.; Han, X.; Li, M. Accurate assessment of RSET for building fire based on engineering calculation and numerical simulation. In Proceedings of the MATEC Web of Conference, APOP2016, Shanghai, China, 28–29 May 2016; Volume 61, p. 04024. [CrossRef]
6. Xing, Z.-X.; Zhao, X.-F.; Song, H.; Gao, W.-L. Applied research of performance-based fire protection design in a large building. *Proc. Eng.* **2011**, *11*, 566–574. [CrossRef]
7. McGrattan, K.; Hostikka, S.; McDermott, R.; Floyd, J.; Vanella, M.; Weinschenk, C.; Overholt, K. *Fire Dynamics Simulator User's Guide*, 6th ed.; NIST Special Publication 1019; National Institute of Standard and Technology: Gaithersburg, MD, USA, 2017.
8. Zhang, W.; Klassen, M.; Roby, R. Numerical prediction of smoke detector activation accounting for aerosol characteristics. *Fire Saf. Sci.* **2005**, *8*, 1543–1554. [CrossRef]
9. Heskestad, G. Generalized characteristics of smoke entry and response for products-of-combustion detectors. In Proceedings of the 7th International Conference on Problems of Automatic Fire Detection, Rheinish-Westfalischen Technischen Hochschule Aachen, Aachen, Germany, 31 March–1 April 1975; pp. 267–310.
10. Björkman, J.; Kokkalal, M.A.; Ahola, H. Measurement of the characteristic lengths of smoke detectors. *Fire Technol.* **1992**, *28*, 99–106. [CrossRef]
11. Mowrer, F.W. Lag times associated with fire detection and suppression. *Fire Technol.* **1990**, *26*, 244–265. [CrossRef]
12. Marrion, C.E. Lag Time Modeling and Effects of Ceiling Jet Velocity on the Placement of Optical Smoke Detectors. Master's Thesis, Worcester Polytechnic Institute, Center for Fire Safety Studies, Worcester, UK, 1989.
13. Cleary, T.; Chernovsky, A.; Grosshandler, W.; Anderson, M. Particulate entry lag in spot-type smoke detectors. *Fire Saf. Sci.* **2000**, *6*, 779–790. [CrossRef]
14. Kim, K.H.; Hwang, C.H. Measurement of the device properties of a ionization smoke detector to improve predictive performance of the fire modeling. *J. Kor. Inst. Fire Sci. Eng.* **2013**, *27*, 27–34. [CrossRef]

15. Cho, J.H.; Mun, S.Y.; Hwang, C.H.; Nam, D.G. Measurement of the device properties of photoelectric smoke detector for the fire modeling. *Fire Sci. Eng.* **2014**, *28*, 62–68. [CrossRef]
16. Jang, H.Y.; Hwang, C.H. Revision of the input parameters for the prediction models of smoke detectors based on the FDS. *Fire Sci. Eng.* **2017**, *31*, 44–51. [CrossRef]
17. Jang, H.Y.; Hwang, C.H. Methodology for DB construction of input parameters in FDS-based prediction models of smoke detector. *J. Mech. Sci. Technol.* **2020**. in review.
18. D'souza, V.T.; Sutula, J.; Olenick, S.; Zhang, W.; Roby, R. Predicting smoke detector activation using the fire dynamics simulator. *Fire Saf. Sci.* **2003**, *7*, 187–195. [CrossRef]
19. Kulkami, V.; Lilley, D.G. FDS: Application of the fire dynamics simulator code to a three-room structure with experimental fires and 60 smoke detectors. In Proceedings of the 47th AIAA Aerospace sciences meeting including the new horizons forum and aerospace exposition, Orlando, Florida, 4–7 January 2011; pp. 1–21. [CrossRef]
20. Miller, J.-T. Analyzing Photo-Electric Smoke Detector Response Based on Aspirated Smoke Detector Obscuration. M.S. Thesis, University of Maryland, College Park, 2010. Available online: http://hdl.handle.net/1903/10455 (accessed on 22 March 2018).
21. Kametani, Y.; Fulkagata, K.; Orlu, R.; Scahlatter, P. Effect of uniform blowing/suction in a turbulent boundary layer at moderate reynolds number. *Int. J. Heat Fluid Flow* **2015**, *55*, 132–142. [CrossRef]
22. Mulholland, G.W.; Johnsson, E.L.; Fernandez, M.G.; Shear, D.A. Design and testing of new smoke concentration meter. *Fire Mater.* **2000**, *24*, 231–243. [CrossRef]
23. Kim, B.J.; Cho, J.H.; Hwang, C.H.; Park, S.H. A study on the development of a low-cost device for measuring the optical smoke density. *Fire Sci. Eng.* **2015**, *29*, 81–88. [CrossRef]
24. Korea Fire Institute (KFI); National Fire Agency. *Detailed Inspection Regulations for Model Approval & Inspection Technical Standards for Fire Detectors*; National Standards of the Republic of Korea, Korea Fire Institute (KFI); National Fire Agency: Yongin-si, Korea, 2019; pp. 31–34.
25. George, W.M. Smoke production and Properties. In *SFPE Handbook of Fire Protection Engineering*, 3rd ed.; Philip, J.D., Craig, L.B., Eds.; Society of Fire Protection Engineers: Gaithersburg, MD, USA, 2002; Sec. 2–13; pp. 473–484.
26. Underwriters Laboratories Inc. *UL 217, Single and Multiple Station Smoke Detectors*, 4th ed.; Underwriters Laboratories Inc., UL: Northbrook, IL, USA, 1993; pp. 50–69.
27. Underwriters Laboratories Inc. *UL 268, Standard for Smoke Detectors for Fire Protective Signaling Systems*, 4th ed.; Underwriters Laboratories Inc., UL: Northbrook, IL, USA, 2003.
28. Matthew, B.; Frederick, W.M.; James, A.M.; Pravinray, G. *Validation of a Smoke Detection Performance Prediction Methodology, Characterization of Incipient Fire Sources*; Fire Protection Research Foundation: Quincy, MA, USA, 2008; Volume 1, pp. 6–7.
29. Patrick Arnott, W.; Hans, M.; John, W.W. Nitrogen dioxide and kerosene-flame soot calibration of photoacoustic instruments for measurement of light absorption by aerosols. *Rev. Sci. Instrum.* **2020**, *71*, 4545–4552. [CrossRef]
30. Sujeeta, S.; Marc, N.F.; Solomon, B. Measurement of size-dependent single scattering albedo of fresh biomass burning aerosols using the extinction-minus-scattering technique with combination of cavity ring-down spectroscopy and nephelometry. *Atmos. Chem. Phys.* **2016**, *16*, 13491–13507. [CrossRef]
31. Mulholand, G.W.; Croakin, C. Specific extinction coefficient flame generated smoke. *Fire Mater.* **2000**, *24*, 227–230. [CrossRef]
32. Mulholland, G.W.; Choi, M.Y. Measurement of the mass specific extinction coefficient for acetylene and ethane smoke using the large agglomerate optics facility. *Symp. Combust. Proc.* **1998**, *27*, 1515–1522. [CrossRef]
33. Fire Technology Solution DB. Available online: https://www.kfiredb.com (accessed on 22 March 2018).

Publisher's Note: MDPI stays neutral with regard to jurisdictional claims in published maps and institutional affiliations.

© 2020 by the authors. Licensee MDPI, Basel, Switzerland. This article is an open access article distributed under the terms and conditions of the Creative Commons Attribution (CC BY) license (http://creativecommons.org/licenses/by/4.0/).

sensors

Article

Development of the User Requirements for the Canadian WildFireSat Satellite Mission

Joshua M. Johnston [1,*], Natasha Jackson [2], Colin McFayden [3], Linh Ngo Phong [2], Brian Lawrence [2], Didier Davignon [4], Martin J. Wooster [5], Helena van Mierlo [2], Dan K. Thompson [1], Alan S. Cantin [1], Daniel Johnston [1], Lynn M. Johnston [1], Meghan Sloane [1], Rebecca Ramos [1] and Tim J. Lynham [1]

1 Canadian Forest Service, Great Lakes Forestry Centre, 1219 Queen St. E., Sault Ste. Marie, ON P6A 2E5, Canada; daniel.thompson@canada.ca (D.K.T.); alan.cantin@canada.ca (A.S.C.); daniel.johnston@canada.ca (D.J.); lynn.johnston@canada.ca (L.M.J.); meghan.sloane@canada.ca (M.S.); rebecca.catherine.ramos@gmail.com (R.R.); tim.lynham@gmail.com (T.J.L.)
2 Canadian Space Agency, 6767 Route de L'aéroport, Longueuil, QC J3Y 8Y9, Canada; natasha.jackson@canada.ca (N.J.); linh.ngophong@canada.ca (L.N.P.); brian.lawrence@canada.ca (B.L.); helena.vanmierlo@canada.ca (H.v.M.)
3 Ontario Ministry of Natural Resources and Forestry, Aviation, Forest Fire and Emergency Services, 95 Ghost Lake Rd, Dryden, ON P8N 2Z5, Canada; Colin.McFayden@ontario.ca
4 Environment and Climate Change Canada, Canadian Meteorological Centre Operations, 2121 route Transcanadienne, Dorval, QC H9P 1J3, Canada; didier.davignon@canada.ca
5 Department of Geography, Leverhulme Center for Wildfires, Environment and Society, NERC National Centre for Earth Observation, King's College London, London WC2B 4BG, UK; martin.wooster@kcl.ac.uk
* Correspondence: Joshua.Johnston@canada.ca

Received: 10 August 2020; Accepted: 31 August 2020; Published: 7 September 2020

Abstract: In 2019 the Canadian Space Agency initiated development of a dedicated wildfire monitoring satellite (WildFireSat) mission. The intent of this mission is to support operational wildfire management, smoke and air quality forecasting, and wildfire carbon emissions reporting. In order to deliver the mission objectives, it was necessary to identify the technical and operational challenges which have prevented broad exploitation of Earth Observation (EO) in Canadian wildfire management and to address these challenges in the mission design. In this study we emphasize the first objective by documenting the results of wildfire management end-user engagement activities which were used to identify the key Fire Management Functionalities (FMFs) required for an Earth Observation wildfire monitoring system. These FMFs are then used to define the User Requirements for the Canadian Wildland Fire Monitoring System (CWFMS) which are refined here for the WildFireSat mission. The User Requirements are divided into Observational, Measurement, and Precision requirements and form the foundation for the design of the WildFireSat mission (currently in Phase-A, summer 2020).

Keywords: wildfire; wildfire management; satellite design; wildfire monitoring; wildfire detection; air quality; carbon emissions; user requirements; wildland fire; forest fire; earth observation; remote sensing

1. Introduction

Global monitoring of wildfire emissions is supported by the network of geo-stationary weather satellites [1]. Finer resolution polar orbiting systems provide further support by correcting for observational biases [2–4], and are more commonplace in direct wildfire management applications. Even still, the use of satellite data in real-time emergency management decision-making remains rare, partially due to the latency of satellite wildfire data [5–7]. Furthermore, at high latitudes the geostationary network is challenged by rapidly degrading spatial resolution and atmospheric transmission [1,4], leaving wildfire-prone northern boreal regions with limited wildfire monitoring capacity from polar-orbiting systems and a heavy reliance on monitoring from aircraft.

Efforts to enhance the uptake of satellite data in wildfire management have been pursued through satellite design (e.g., [8–10]) and purpose-built information systems (e.g., Global Wildfire Information System [GWIS] [11]); additionally, various commercial proposals have been proposed. However, to date no system has successfully delivered end-to-end operational support to address the specific needs of fire managers. In part this is a result of the broad range of wildfire management practices globally and the resulting variation in specific regional requirements for wildfire monitoring. Inability to accurately define the requirements of end-users presents a broad reaching barrier to operational implementation.

Responses to wildfires vary across Canada [12] and range from a "Full Response" (immediate, aggressive initial/sustained attack), to "Monitored Response" (observation and periodic reassessment; [13]), guided either by zonation, or wildfire specific conditions termed "appropriate response" [14]. In situations with increased wildfire activity, the suppression capacity can be rapidly overwhelmed, resulting in escaped wildfires that may burn very large areas [15,16]. These larger wildfires represent only 3% of the number of wildfires in Canada, yet they account for 97% of the area burned [12,17], and require substantially more resources to manage [18,19].

In many higher risk locations, Canadian wildfire management agencies generally rely on the Initial Attack model where wildfires are detected early and suppressed small. The early detection and suppression of wildfires is critical to successful wildfire control resulting in fewer escaped wildfires, therefore reducing impacts and response costs [20]. Wildfire remote sensing has been recognized for its capacity to detect wildfires (e.g., [21]). However, there is a significant gap between what is required for "early" detection (e.g., identifying small sub-canopy wildfires) for wildfires that require suppression and what can be accomplished reliably with satellite remote sensing [22]. Beyond detection, during periods of escalated wildfire activity it is useful to have current intelligence about all ongoing wildfires. Reliable wildfire intelligence is critical for situational awareness and informed decision-making including prioritization and strategic and tactical wildfire response.

Wildfires that are threatening communities and critical infrastructure are prioritized for suppression action over remote wildfires where there is more opportunity for the natural ecological role of wildfire on the landscape [23]. Therefore, these wildfires more frequently grow larger and are generally managed through modified tactics (e.g., continuous mapping and monitoring; [14]). Although satellites are not typically helpful for early detection [22], there is an emerging requirement for large wildfire and regional intelligence gathering to maintain situational awareness during periods of escalated wildfire activity. This intelligence requirement can be met through the proper application of satellite technology, and is broadly described as wildfire "monitoring" here.

Wildfire monitoring is also an essential component of Canada's efforts to track and predict smoke dispersion from active wildfires. In recent years, wildfire smoke has been the dominant cause of poor air quality for large portions of Canada [24]. The impacts of smoke on communities can necessitate an evacuation, even without a direct threat from wildfire [25–27]. This led to the development of methods to derive emissions from satellite-detected wildfires, and to use the emission estimates in smoke dispersion [28,29] and air quality [30] forecast systems. These automated systems require the provision of timely and reliable wildfire activity data, with an emphasis on wildfire events that produce large long-range transportation of smoke, or wildfire events near communities. With additional development, these systems can evolve to incorporate Fire Radiative Power (FRP; MW) measurements as an additional source for the estimation of wildfire emissions (e.g., [31,32]), as is becoming increasingly common throughout the world (e.g., Global Fire Assimilation System (GFAS); [33].

Under climate change a substantial increase in the frequency and intensity of wildfires is expected [34,35]. In particular, northern regions such as Canada are expected to see an increase of wildfire activity related to increases in conditions conducive to extreme wildfire weather [19,36–41]. Consequently, frequency of extreme burning days where wildfires are able to escape is also projected to increase [41,42].

In 2019 the Canadian Space Agency (CSA) initiated the development of a dedicated wildfire monitoring satellite "WildFireSat" mission [43]. WildFireSat (WFS) intends to leverage uncooled microbolometer technology developed by the CSA and Institut National d'Optique (INO). An earlier version of this technology called the New InfraRed Sensor Technology (NIRST) was the first mid-wave infrared (IR) microbolometer used in space-based wildfire remote sensing on the 2011 Aquarius SAC-D mission [44]. Following the NIRST experiment, the same technology was used in a feasibility study (referred to as "Phase 0") to demonstrate the technical feasibility of a cost-effective, dedicated Canadian Wildland Fire Monitoring System (CWFMS) [45]. Since then, the detector technology has continued to evolve (e.g., [46]), while new Low Earth Orbit (LEO) wildfire products (e.g., from the Visible Infrared Imaging Radiometer Suite (VIIRS) and Sea and Land Surface Temperature Radiometer (SLSTR); [47,48]) have filled some temporal coverage gaps (Figure 1), which will improve the feasibility of a targeted wildfire monitoring mission.

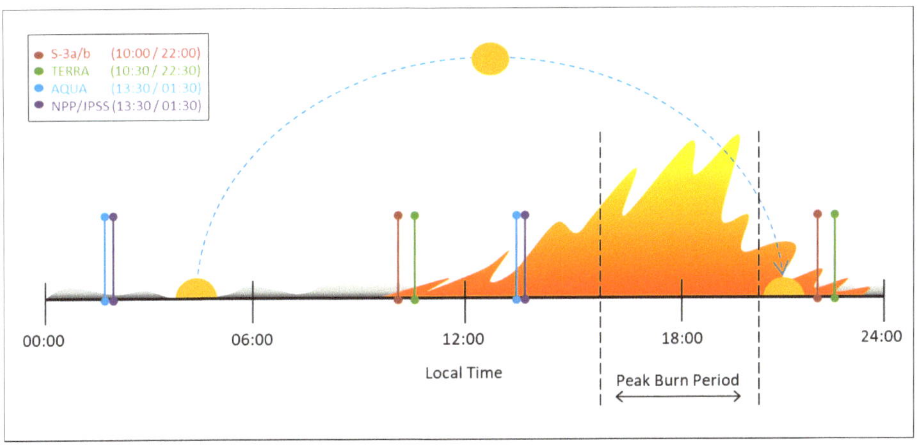

Figure 1. Active-fire satellite overpass time in relation to the diurnal wildfire activity cycle. Wildfire activity varies throughout the day in relation to the changing solar elevation and resulting changes in atmospheric moisture, ambient temperatures, and wind speeds. In general, wildfire activity is dominated by smoldering combustion overnight and in the early morning, and peaks in the late afternoon period centered around 18:00 local time known as the "peak burn period". Notably, currently available active-fire low-earth-orbiting satellite observations from instruments such as SLSTR (Sentinel-3a/b), MODIS (TERRA/AQUA), and VIIRS (NPP/JPSS) fail to observe wildfires during the most active portions of the day.

Phase-A of the WFS mission is driven by the Mission Requirements [49] which extend from the User Requirements. However, the User Requirements defined in CWFMS [50] required substantial revisions to accommodate the new context of this mission. The aim of this study is to trace the process used to update and re-scope the CWFMS User Requirements for WFS through consideration of emerging science and ongoing end-user consultation (e.g., [51]). This study presents the WFS User Requirements and provides cross reference to their heritage in CWFMS where applicable. We trace the two primary phases of this process: (1) Canadian wildfire management needs are assessed through direct engagement of wildfire management end-users, leading to a set of key Fire Management Functionalities (FMFs); (2) User Requirements for the WFS mission are refined through the integration of the wildfire management needs with the best available scientific techniques. The result of this study is the definition of User Requirements for the first dedicated operational wildfire monitoring satellite, forming the foundation for later stages of mission development.

2. Wildfire Management Needs Assessment

Wildfire management agencies generally employ a risk-based approach where the potential impact(s), likelihood, and resulting expected loss or benefit are assessed at the appropriate scale according to the complexity of the wildfire situation [26,52,53]. Decisions often involve multiple decision-makers and stakeholders with varying perspectives concerning risk [23,26]. Decisions are not static and are frequently updated through an iterative process of determining and taking actions, monitoring outcomes, and revising actions until the situation is resolved [53]. When assessing progress in the decision-making cycle, decision-makers require different types of intelligence.

We define wildfire intelligence as information which is collected to support wildfire management activities. This may include current or forecasted information such as: Wildfire behavior, location, size, shape, spatial context (e.g., fuels, topography, proximity to areas of concern, etc.), firefighting resource allocation and use, and wildfire effects and impacts (e.g., social, economic). The type, precision, accuracy, and timeliness of intelligence required varies depending on whether tactical or strategic planning is being conducted.

2.1. Wildfire Management Engagement

In the first step of defining the Mission Requirements, Canadian wildfire managers were surveyed to better understand the relative importance of the various wildfire monitoring products and the constraints for their relevance as a source of intelligence in both tactical (e.g., same-day/near-term operations) and strategic (e.g., longer-term preparedness, large wildfire planning) decision-making. Respondents were posed a series of questions regarding potential Earth Observation (EO) data products and asked to consider each in the context of both tactical and strategic decision-making. They were asked to provide the optimal and maximal data latencies (i.e., the time lag from collection to receipt of data when it has most value and the point at which it no longer has value), as well a rating of the importance of each product using a Likert scale (1–5; low–high); respondents were also invited to provide comments to aid in the interpretation of their responses.

Survey Results

In total, 55 senior staff from 7 of 13 wildfire management agencies in Canada responded to the survey (Table 1). Most responses were completed individually while a few were coordinated efforts by groups. This sample size is consistent with comparable engagement efforts in similar communities (e.g., [54]). The format of some responses required the data to be reformatted prior to analysis. Group responses were weighted according to the number of people contributing to ensure proportional representation.

Products identified as the most important in both tactical and strategic decision-making were active-fire products; these products are typically derived from thermal observations of active wildfire events, and contribute to assessments of location, spread rate, and intensity (Table 2). Whereas post-fire mapping products (e.g., burned area and severity) were primarily valuable at the strategic level (as well as for non-response management activities, e.g., forest inventory). Tactical intelligence was generally required "as quickly as possible" for all products, with the median of responses indicating 30 min or less (Table 2). For strategic uses the same general preference for active-fire intelligence is found (Table 2); however, slightly longer latencies are acceptable. Data latencies of up to 2 h were indicated as the thresholds for tactical and strategic decision-making. In many cases the information continued to have some value to managers for several hours up to 24 h, but not necessarily for tactical or strategic decision-making. Notably, this consultation process also revealed that the timing of data delivery during daily operations was also a key factor in data utility, due to the cyclical timing of daily decision-making.

Table 1. Summary of roles in wildfire management and experience among survey respondents.

Role in Wildfire Management	Description	Percent of Respondents
Aerial Operations	Roles range from Air Operations Branch Director, Air Attack Officers, and Aerial Detection Leaders. In some agencies, these staff coordinate high-level infrared services and other mapping/scanning roles.	17%
Incident Commander (IC) Types 1 and 2	The IC has overall authority and responsibility for conducting incident operations and is responsible for the management of all operations. Levels 1 and 2 are those that lead the most complex wildfire situations.	22%
Incident Commander (IC) Types 3 to 5	Same as above, however the wildfires are less complex.	9%
Plans Section	Roles range from Planning Section Chiefs, Fire Behaviour Analysts, Situation Unit Leaders, GIS, prediction and forecasting services. In some agencies, these staff coordinate airborne infrared services and other mapping operations.	13%
Group Responses	Consisting of agency-selected individuals including skills from roles above.	39%

Table 1: Wildfire management experience levels ranged from 10–39 years with a median of 25 years. Of the 13 Canadian fire management agencies, 7 agencies responded to the survey (specifically: Agencies from Northwest Territories, British Columbia, Alberta, Saskatchewan, Ontario, Québec, and Parks Canada).

2.2. *Summary of Wildfire Manager Needs*

In order to maximize the value of a satellite system for wildfire management, certain features were highlighted through additional comments provided during the end-user consultation. These features included: Fast and consistent data delivery, mapping of active and inactive wildfire areas, smoke and air quality information, wildfire behavior, and threat estimates, as well as detection in remote regions.

2.2.1. Fast and Consistent Data Delivery

Daily wildfire management activities follow planning cycles which depend on the scale of management occurring. For example, an incident command team responsible for planning and carrying out wildfire operations on a large wildfire may have different needs for the frequency and timeliness of information than those planning strategic response at a regional, provincial, or national scale. Generally speaking, in order for intelligence to be incorporated into daily planning activities data must reflect the current situation (i.e., low/short latency), but it is also important to receive the information at a consistent time of day to facilitate routine integration.

2.2.2. Mapping of Active and Inactive Wildfire Areas

Although there is a definite interest in the actively spreading portion, intelligence is required for the entire wildfire. The full perimeter of the burned area as well as the active and previously burned area are valuable in wildfire operations. Previously burned areas may still be smoldering and require prolonged suppression, while unburned "islands" in these areas pose a threat for re-burning. Managers also indicated that they were satisfied with the 375 m spatial resolution of the VIIRS I-band wildfire products [47,55], for general applications (though for high complexity incidents fine resolution airborne mapping may also be necessary).

Table 2. Summary of median latency requirements (minutes) for tactical and strategic decision making. FGM refers Fire Growth Modeling. Ideal and threshold times reflect the median desired data latency and the time at the data was considered no longer valuable for the intended purpose. Likert scoring (1–5; unimportant to very important) is presented as the mode value of the respondent rankings.

Intelligence Type	Product	Tactical Decision Making			Strategic Decision Making		
		Likert Ranking	Ideal Time	Threshold Time	Likert Ranking	Ideal Time	Threshold Time
Active-Fire Observation	Wildfire detection and clustering	5	1	120	5	15	1440
	Rate and direction of spread	4	15	120	4	17.5	300
	Fire intensity	5	15	720	5	12.5	240
Active-Fire Modeling	Smoke mapping	2	30	720	3	45	560
	Suppression effectiveness	3	10	660	3	17.5	660
	Proximity to interface zones	5	5	120	5	15	240
	Risk	3	60	190	3	60	360
	Assimilation into FGM	5	5	720	5	5	720
	FGM prediction	3	20	720	3	17.5	720
	Fuel type verification	3	22.5	720	3	20	720
Pre/Post-Fire Observation	Burned area	3	60	1440	5	60	1440
	Burn severity	2	120	1440	3	90	1440
	Arrival time	4	60	840	3	60	1440

2.2.3. Wildfire Behavior and Threat Estimates

The proximity and threat to interface zones was identified as critical intelligence in the survey (Table 2). Proximity to these zones is achievable through accurate detection and mapping in conjunction with national interface maps [54].

Wildfire behavior observations were considered to have both tactical and strategic value, particularly in terms of estimating the potential threat of a wildfire. Information relating to the rate and direction of wildfire spread (ROS (m s^{-1}) and DIR (deg); [56,57]) as well as the Fire Intensity (FI, (kW m^{-1})), are essential to characterizing the behavior of actively spreading wildfires. Johnston et al. (2017) [58] demonstrated that FI can be estimated directly from IR measurements of FRP. Wildfire behavior is of particular interest during the late afternoon "peak burn" period (Figure 1). This information should ideally be paired with the spatial context (e.g., adjacent fuels, topography, and proximity to areas of concern; [59]) to provide estimates of proximity and threat to these areas.

2.2.4. Detection in Remote Regions

Wildfire management practices in Canada vary dramatically across the landscape, and generally in relation to population distributions (e.g., [9,10]). In vast remote areas, wildfire managers do not typically conduct dedicated detection activities due to the decreased likelihood for negative impacts from wildfire, and the higher cost and operational complexity of these patrols (e.g., [60]). Space-based EO is particularly suitable for gathering intelligence in these situations [22]. The value of EO-derived detection of wildfires identified in the survey (Table 2) was highlighted as particularly valuable in these regions in the comments.

2.2.5. Smoke and Air Quality Information

Although smoke management tools were not identified as critical to tactical decision-making in the survey responses, this information is critical for other emergency management operations. Non-fire management users require smoke-related intelligence for critical operations such as evacuation planning [61–63], public health forecasting [64], and aviation visibility [65]. Smoke forecasting using tools such as FireWork [30,66] and BlueSky [29,67] are dependent on wildfire size and location information. Other smoke monitoring applications (e.g., Global Fire Assimilation System (GFAS); [33]) require FRP [31,32] as a primary input.

The delivery of operational smoke and air quality forecast is a highly automated process. As of 2020, Environment and Climate Change Canada will launch a new air quality forecast twice a day (initiated at 00 and 12 UTC), and smoke forecast every 6 h (00, 06, 12, 18 UTC). Each execution is updated with the latest wildfire information available. The scheduling of forecast executions is tied to the availability of new weather and wildfire emission data, and computing resources. Due to this scheduling, the data latency requirements are less stringent than for wildfire management applications, but the requirement to focus on the peak burn overpass period still persists. Additionally, smoke and air quality applications emphasize a strong interest in smaller wildfire detection, an interest in thermodynamic parameters controlling plume rise and height (e.g., [68,69]), and a larger coverage (e.g., North America).

2.3. Fire Management Functionalities:

The needs identified above were translated into a set of key Fire Management Functionalities (FMF) necessary to define the User Requirements:

(1) The Area of Interest (AoI) is defined as the whole vegetated Canadian landmass (Figure 2);
(2) Daily (or better) coverage of the AoI at a specific and consistent time of day including peak burn (1600–2000 local time), with data delivery (to end-users) before the start of the operational response period (~0700 local) for overnight observations and before the end of day planning period (~1900 local) for peak burn observations;

(3) Detection and mapping of wildfires and their plumes, specifically:

   i. The ability to detect wildfires with comparable or improved sensitivity to existing satellite systems, and to serve as an early-detection system for remote access wildfires;
   ii. There must be sufficient spatial resolution and geolocation accuracy for locating and mapping wildfires in relation to their previous position and other landscape features;

(4) Estimation of wildfire behavior, specifically:

   i. The ability to collect FRP measurements;
   ii. The ability to characterize sub-pixel wildfire features (e.g., temperature and area);

(5) Compatibility with other available EO data sources and formats;
(6) Near-real-time data, with tactical products to be delivered within 30 min, and a 2-h latency for all end-user products as threshold for utility.

The interconnectivity of the FMFs with the requirements laid out in the User Requirements Document [49] and the Mission Requirements Document [50] is summarized in Table 3. The development and rationale for these requirements is described in the following sections.

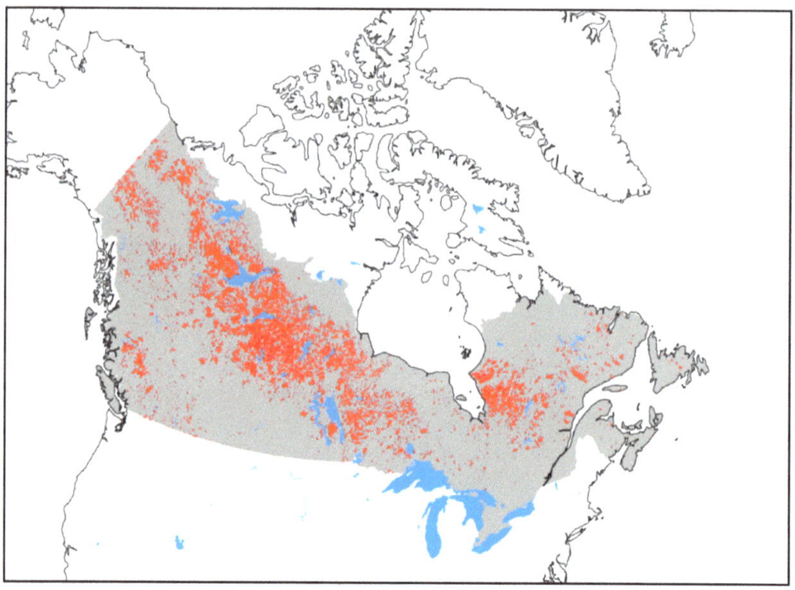

Figure 2. Spatial extent of the WildFireSat Area of Interest (AoI), which includes the entire continuously vegetated extent of Canada (gray). For context, the historic burned area on public, primarily forested lands (1980–2019) is overlaid in red [17]. Although the full AoI is not prone to frequent wildfire activity, the vegetated areas shown here are burnable. The spatial distribution of wildfire within the AoI is expected to change under climate change. The distribution of agricultural burning (primarily confined to the southern half of the AoI) is not shown.

Table 3. Primary relationships (not exhaustive) between Fire Management Functionalities (FMFs) and User Requirements from the User Requirements Document [49] (URD). FRP refers to Fire Radiative Power (e.g., [32]).

FMFs	URD Parent References	Rationale
Daily (or better) coverage of the Canadian landmass, with consistent timing and peak burn observation (FMF-1,2)	Temporal resolution (CWFM-URD-0020)	To ensure a minimum of daily coverage is provided
	Peak burn observation (CWFM-URD-0030/40)	To ensure that the daytime overpass falls in the peak burn period
Detection and mapping of wildfires and smoke plumes (FMF-3)	Spatial resolution (CWFM-URD-0080-90)	Constraints are applied to the spatial resolution to ensure adequate ability to resolve active-fire area and to detect change in its position
	Swath (CWFM-URD-0110/120)	Minimum swath width to increase the probability of containing the full wildfire within a single observation
	Spectral bands, dynamic ranges, and sensitivities (CWFM-URD-0150-251)	Minimum spectral bands and performance requirements to conduct cloud masking, wildfire detection and characterization, and burned area mapping
	Band co-registration (CWFM-URD-0260-280)	Wildfire detection, characterization, and mapping requires multispectral measurements. Band co-registration requirements are defined to ensure cross-band analysis is possible across the swath
	Geo-coding (CWFM-URD-0300)	Geographical positioning requirements to enable change detection, comparison to landscape features, and response operations
Measurement of wildfire behaviour (FMF-4)	Spatial resolution (CWFM-URD-0080-90)	Spatial resolution is constrained for optimal ROS measurement [70], and FRP accuracy [71]
	Swath (CWFM-URD-0110/120)	Minimum swath width to increase the probability of containing the full wildfire within a single observation
	Spectral bands, dynamic ranges, and sensitivities (CWFM-URD-0150-251)	Minimum spectral bands and performance requirements to conduct wildfire detection, sub-pixel characterization, and collect FRP measurements
	Band co-registration (CWFM-URD-0260-280)	Sub-pixel wildfire characterization requires multispectral measurements. Band co-registration requirements are defined to ensure cross-band analysis is possible across the swath
Compatibility with other EO systems (FMF-5)	Spatial resolution (CWFM-URD-0080-90)	The spatial resolution range required to meet FMF-3 and FMF-4 is broad, the specific requirements are chosen to closely match VIIRS I-Band spatial resolution
	Peak burn observation (CWFM-URD-0030/40)	A peak burn overpass is required to ensure sufficient temporal offset from the VIIRS overpass time (~ 13:00 local) for optimal Rate of Spread measurement at the specified spatial resolution
Near-real-time data (FMF-6)	Data latency (CWFM-URD-0050)	A data latency of no more than 30 min from overpass to end user delivery is required
	Downlink priority (CWFM-URD-0070)	In the event that not all data can be delivered in near-real-time, priority is given to spectral bands required for active-fire detection and characterization

Table 3: Where URD refers to the User Requirements Document [49], and FRP refers to Fire Radiative Power (e.g., [32]).

3. Definition of the User Requirements

The FMFs were considered in order to define EO User Requirements during the CWFMS feasibility study [49] taking into account space system capabilities and limitations, including payload technology capabilities, but without targeting a specific mission scope. With the initiation of the WFS mission, the CWFMS User Requirements were further refined to correspond to a specific mission scope and matching level of funding to produce the Mission Requirements specific for WFS [50]. A simple approach to defining these requirements would be to guarantee success by over prescribing the needs of the mission. However, this is not a programmatically feasible approach as it inflates the mission cost and complexity. In this study, we acknowledge that programmatic constraints will necessitate trade-off analysis during the implementation of the mission. As such, many of the User Requirements are stated as both "goal" (*SHOULD*) and "threshold" (*SHALL*) requirements to reflect optimal and minimum required performances. In this Section the critical WFS User Requirements required to fulfill each of the six FMFs are described in terms of Observation, Measurement, and Precision Requirements for the WFS mission.

3.1. Observation Requirements

In this section we describe the critical requirements necessary to observe the target area and report data necessary to achieve the FMFs described in the previous section. Observation requirements described here include those which describe the required coverage and data latency.

3.1.1. Coverage Requirements

Given the success of geostationary wildfire monitoring it could be argued that continuous observations of actively burning wildfires are required in order to rapidly detect new starts. However, persistent observation is typically associated with coarse spatial resolution, which negatively impacts small wildfire detection sensitivity and geographic mapping precision (e.g., [4]). Furthermore, given that FMF-3 does not address detection in high risk areas, this functionality can be met without persistent observation (Table 3). Specifically, one or more satellites with moderate spatial resolution in LEO could accommodate FMF-3.

The ability to observe wildfire behavior (i.e., FI and ROS; FMF-4) is also linked to observational frequency. ROS measurements are a function of spatial and temporal precision and thus the minimum required revisit time can be estimated based on the spatial resolution and the speed of the wildfire being observed [56,71,72]. A ROS-driven coverage requirement was considered in the feasibility study for the CWFMS which proposed a constellation of 9 satellites providing ~500 m spatial resolution with a 2-h revisit period [45]. However, given the overpass times of currently available LEO active-fire observations (Figure 1), a stand-alone constellation is not essential to comply with the ROS driven coverage requirement. A strategically positioned satellite with peak burn overpass (FMF-2), that is compatible with the existing systems (FMF-5) would also meet FMF-4 (Table 3).

Daily coverage during the peak burning period for the entire Canadian landmass (FMF-1) with consistent mapping and wildfire behavior data (FMF-3, 4) delivered prior to the end of the day planning period (~19:00 local time; FMF-2) is also necessary to address the full spectrum of these functionalities. These FMFs also suggest the requirement for consistent overpass times to ensure predictable data delivery times. This implies a preference towards the use of a sun-synchronous polar orbit with a Local Time of Ascending Node or Local Time of Descending Node, i.e., local overpass time of ~18:00 (FMF-2, 6). Notably, the requirement to consistently observe the full AoI (Figure 2) further suggests that the required revisit period cannot be met through instrument pointing or satellite maneuvering as this will cause coverage gaps elsewhere in the AoI.

The User Requirements for coverage are summarized in Table 4. Temporal resolution requirements are specified with the goal of covering the full AoI (Figure 2) on a daily basis. Allowances are made for incidental coverage gaps provided they do not persist for multiple days (Table 4). The temporal resolution is further constrained to ensure that the daily coverage is provided strategically during the peak burn period (Table 4; Figure 1). Additionally, a minimum swath width is defined to ensure that each overpass is capable of observing a complete wildfire area (centrally positioned in the frame), to further support the mapping requirements of FMF-3 (Table 4).

Table 4. Summary of key Observational (coverage and latency) Requirements as outlined in the User Requirements Document (URD), and refined in the Mission Requirements for WildFireSat. AoI refers to the Area of Interest (Figure 2). See Table 5 for spectral band definitions.

Type	URD Parent Reference	Requirement
Coverage & Temporal Resolution	CWFM-URD-0020	The mission SHALL provide the capability to observe at minimum 97% of all points within the AoI at least once per any 48-h period, and all points within the AoI at least once per 72-h period The mission SHALL provide the capability to observe on average 85% of all points within the AoI at least once per any 24-h period As a goal, the mission SHOULD provide the capability to observe all points within the AoI at least once per any 24-h period
Peak Burn Observation	CWFM-URD-0030/40	The design SHALL provide observation during each peak burning period
Data Latency	CWFM-URD-0050	Time lag between data acquisition and delivery to user SHALL not exceed 30 min for the MWIR, LWIR, NIR, and VIS (red) for 90% of observations and 24 h for all data Time lag between data acquisition and delivery to user SHOULD not exceed 30 min for all data
Downlink Priority	CWFM-URD-0070	Priority downlink SHALL be given to MWIR, LWIR, NIR, and VIS (red) bands
Swath Width	CWFM-URD-0110/0120	Swath width for all spectral bands SHALL be no less than 200 km

3.1.2. Latency Requirements

The time sensitivity of active-fire data has been consistently highlighted by the end-users and is reflected in FMF-6 (Table 3). The requirement for near-real-time (NRT) data latency is also necessary to ensure that peak burn observations (e.g., ~18:00 local) are delivered to end-users prior to the end of the daily planning period. A threshold requirement for data latency of 30 min is applied to meet these requirements. This latency was selected as a threshold as it was deemed technically achievable; however, the goal latency is to approach real-time if possible. A downlink priority band list (based on minimum requirements for active-fire detection calculations) is provided in the event that not all data can be downlinked in NRT (Table 4).

3.2. Measurement Requirements

In this section we define the measurements required from the payload to provide sufficient data to meet the FMFs. Specifically, this section explores instrument band requirements, as well as their associated spatial resolutions and dynamic ranges (Table 5). Information pertaining to band sensitivity and performance is presented in Section 3.3 (Precision Requirements).

In order to meet FMF-3 and -4 the instrument payload must meet all the requirements for active-fire monitoring. This necessitates the ability to detect new wildfires, as well as map their active wildfire areas and behavior, thereby supporting smoke and air quality modeling. This requires the instrument to collect thermal observations to conduct wildfire detection and characterization analysis (e.g., [32,73,74]). It also requires multi-spectral data from a suite of Mid-, and Long-Wave infrared (MWIR and LWIR) combined with Near-infrared (NIR) and Visible (VIS) bands at a minimum (e.g., [47,74,75]). The IR and VIS-NIR bands which observe primarily emitted and reflected energy, respectively, are treated as two complimentary data sets, and are described separately here.

Table 5. Summary of key Measurement Requirements as outlined in the User Requirements Document (URD), and refined in the Mission Requirements for WildFireSat. TOA refers to Top of Atmosphere based on MODTRAN atmospheric simulations.

Type	URD Parent Reference	Requirement
Spatial Resolution	CWFM-URD-0080	When the spacecraft is nadir-pointing, the design SHALL provide imagery with spatial resolution no larger than: • 400 (MWIR, LWIR) and 200 m (VIS, NIR), at the geodetic sub-satellite point • 800 (MWIR, LWIR) and 400 m (VIS, NIR), for all pixels
	CWFM-URD-0090	When the spacecraft is nadir-pointing, the design SHOULD provide imagery with spatial resolution no larger than: • 300 (MWIR, LWIR) and 150 m (VIS, NIR), at the geodetic sub-satellite point
Spectral Bands	CWFM-URD-0150	The payload SHALL provide at least one band in each of the following channels: • MWIR: (3.4–4.2 μm), including include 3.9 μm • LWIR: (8.0–9.5 μm) OR (10.4–12.3 μm; preferred) • NIR: (0.8–0.9 μm) • VIS: (0.6–0.7 μm; red)
	CWFM-URD-0151	The payload SHOULD provide at least one band in each of the following channels: • MWIR: (3.5–4.2 μm), including 3.9 μm • LWIR: (10.4–12.3 μm) • NIR: (0.8–0.9 μm) • VIS: (0.6–0.7 μm; red), (0.5–0.6 μm; green), AND (0.4–0.5 μm; blue)
IR Dynamic Ranges	CWFM-URD-0180	Brightness temperatures retrieved in the MWIR band(s) SHALL be in the range of at least 300 to 720 K at 200 m spatial resolution, or 300 to 635 K at 400 m, at surface level
	CWFM-URD-0190	Brightness temperatures retrieved in the LWIR band(s) SHALL be in the range of at least 300 to 580 K at 200 m spatial resolution, or 300 to 470 K at 400 m, at surface level
VIS-NIR Dynamic Ranges	CWFM-URD-0200	The NIR and VIS TOA radiance range SHALL be a minimum of: • NIR: 4.0–46.4 (W m^{-2} sr^{-1} μm^{-1}) • VIS (red): 4.1–25.2 (W m^{-2} sr^{-1} μm^{-1}) • VIS (green): 2.5–25.8 (W m^{-2} sr^{-1} μm^{-1}) • VIS (blue): 0.7–16.9 (W m^{-2} sr^{-1} μm^{-1})
		The NIR and VIS TOA radiance range SHALL be a minimum of: • NIR: 4.0–139.0 (W m^{-2} sr^{-1} μm^{-1}) • VIS (red): 4.1–75.5 (W m^{-2} sr^{-1} μm^{-1}) • VIS (green): 2.5–77.3 (W m^{-2} sr^{-1} μm^{-1}) • VIS (blue): 0.7–50.5 (W m^{-2} sr^{-1} μm^{-1})

3.2.1. MWIR and LWIR Band Requirements

The MWIR and LWIR bands are central to the detection of wildfires using their thermal radiance. These bands are also essential in the measurement of FRP (MW), which is a key parameter in wildfire behavior estimation (e.g., [58,76]), and smoke plume emissions (e.g., [31]). FRP calculation can be achieved with single band measurements in the MWIR [32]. However, the LWIR remains essential in the detection of wildfire pixels for FRP analysis as well as in interrogating sub-pixel characteristics of wildfire pixels (e.g., effective wildfire temperature and area; [73]).

Optimal MWIR band placement for wildfire detection and FRP measurement is a narrow window of observation centred at 3.9 μm (e.g., MODIS, Band 21). Wider spectral bands have also been demonstrated to be effective (e.g., SLSTR, Figure 1), provided that they enter into the CO_2 band at 4.2 μm, and avoid encroaching significantly below 3.5 μm where solar reflection becomes a stronger contributor. These parameters are reflected in the MWIR spectral band requirements in Table 5. The LWIR band's function of supporting contextual detection analysis and sub-pixel analysis affords more freedom in precise band placement. The LWIR band is specified to ensure that it falls within an atmospheric window in the 8–14 μm range (Table 5), providing sufficient spectral separation for MWIR-LWIR differential analysis.

FRP measurement and sub-pixel characterization also requires measurement over the full dynamic range of the scene. Dynamic ranges for the MWIR and LWIR bands (Table 5) were defined in Brightness Temperatures (BT; K) at surface level (i.e., without atmospheric attenuation) as the precise spectral bands are not yet specified. Both bands are required to make ambient (~300 K) surface temperature measurements. Given that this satellite will be designed to observe Canadian wildfires during the peak burn period, the saturation point was defined in anticipation of observing the extreme intensities of Canadian boreal crown wildfires (e.g., [77]) during their peak period. This is a condition not yet achieved through LEO satellites. Therefore, a sub-pixel scene was modeled in which a 50-m deep flame front with net blackbody emission equivalent to ~900 K crosses the pixel area diagonally, while the background area remains ~ 300 K. Under such conditions the saturation point for the MWIR and LWIR spectral bands vary in accordance with their spectral band and with respect to the sub-pixel area of the wildfire. As such, the required dynamic ranges are bound by spectral band selection and spatial resolution (Table 5; Figure 3). Ideally the dynamic ranges for these bands would be stated in spectral radiance ranges at Top of Atmosphere (TOA); however, until the Spectral Response Functions (SRF) and spatial resolutions are known, the modeling exercise in Figure 3 cannot be completed. Atmospheric transmittance modeling as performed for the VIS and NIR bands should be replicated for MWIR and LWIR when appropriate.

Given the interdependency of spatial resolution and dynamic range it is not necessarily advantageous to target unnecessarily fine spatial resolution imagery in the MWIR and LWIR bands. Recent experiences in analysis of the VIIRS active-fire products have demonstrated increasing errors of commission (i.e., false positives) associated with finer scale active-fire detections [55,78]. Johnston (2016) [70] found that ROS measurements could be reasonably estimated under crown wildfire conditions at spatial resolutions as coarse as 500 m. Given FMF-5, MWIR and LWIR spatial resolutions were defined with the intent to be comparable to the VIIRS I-bands (~375 vs. 400 m for WFS) in order to serve as a complimentary data source to the VIIRS data which is captured a few hours earlier in the day (Figure 1).

3.2.2. VIS and NIR Requirements

Although the VIS and NIR bands support contextual wildfire detection (e.g., [74]), their primary contribution is to cloud masking and false positive identification. These applications are less constraining on band selection and dynamic range than the secondary applications of these bands. Although post-fire mapping products are only of strategic value, the ability to map the burned area of a wildfire adjacent to its active portions is an integral part of FMF-3, which requires measurements in the VIS and NIR bands at a minimum (and ideally would include the short-wave IR, see Discussion). VIS and NIR bands are typically used for measures of vegetation "greenness" (e.g., Normalized Difference Vegetation Index; NDVI) to infer vegetation health metrics. Detection of sudden changes in these metrics can be attributed to vegetation disturbance such as wildfires (e.g., [79]). Although numerous systems are capable of mapping global burned area at a relatively fine scale (e.g., [80]), this capability is not typically available as an NRT product in conjunction with active-fire observations as required by FMF-3. Requirements for spectral bands were specified as VIS-red and NIR bands at a minimum. However, the goal band list includes all VIS spectral bands required for forestry observation (Table 5).

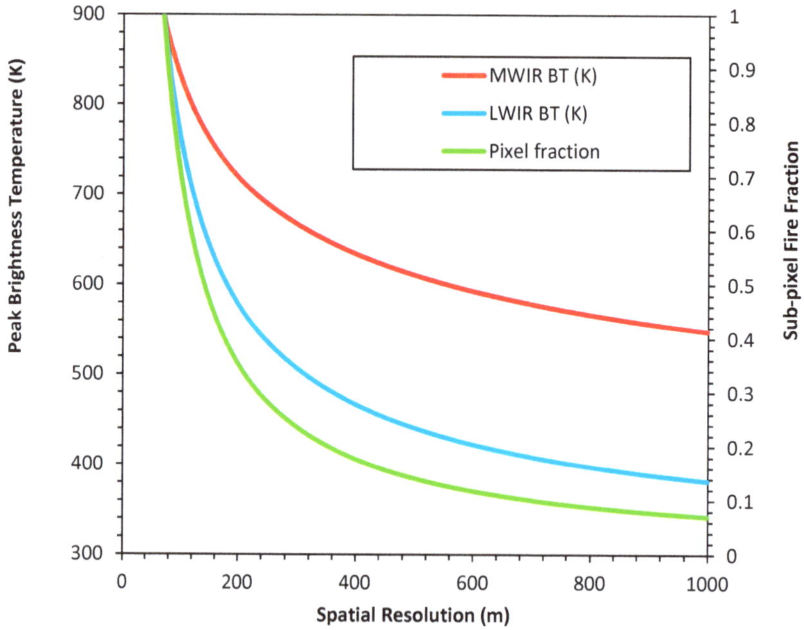

Figure 3. Simulated Infrared (IR) band saturation temperatures as a function of pixel spatial resolution and sub-pixel wildfire area. This simulation assumes a flame front depth of 50 m with mean temperature of 900 K extending diagonally across the pixel. Sub-pixel wildfire area was computed and the saturation temperatures were defined based on the peak pixel brightness temperatures (BT) at a given spatial resolution (Table 5). The saturation temperatures change with spatial resolution in both the LWIR and MWIR, e.g., at 250 m the MWIR and LWIR saturation temperatures are ~690 and ~540 K, respectively, while at 500 m the MWIR and LWIR saturation temperatures are ~610 and ~440 K, respectively. No atmospheric effects on the signal are included in this simulation.

Due to the unique overpass time (and consequently unique solar angles) it was not possible to infer dynamic ranges for these bands based off of other burned area mapping systems (which typically have late morning overpasses). To ensure the bands would cover the full dynamic range, a series of MODTRAN simulations were conducted. In these simulations it was assumed that the VIS and NIR bands are expected to be used for imaging surface reflectances during the late afternoon period over the boreal forest. Solar radiation was modeled using a mid-latitude summer atmosphere, with 407 ppmv CO_2, for a central location in the Canadian landmass (~ latitude 58.0°, longitude 101.0°), at 18:00 local time. Typical boreal forest albedos range from 0.1 to 0.25, while burn scar albedos (~0.05) represent the minimum reflectance of interest, and maximum scene reflectances of ~0.9 are expected from clouds. The threshold VIS and NIR dynamic ranges were defined as those required to observe burned areas up to the peak reflectance of boreal forest, while the goal requirements are defined to include cloud measurements within this range (Table 5).

The objective of burned area mapping with this system is to provide a product suitable for complimenting active-fire detection in FMF-3, not to replace robust techniques used in regional and global carbon accounting (e.g., [81–84]). With this in mind, fine spatial resolution is not necessary and requirements are relaxed to reduce the volume of data to facilitate NRT calculation. However, as is commonplace in existing systems (e.g., [74,75]), it was deemed desirable to maintain finer scale measurements in the VIS and NIR as compared to the MWIR and LWIR to improve sub-pixel context for cloud masking and resolving cross band misalignments.

3.3. Precision Requirements

Precision requirements are defined here to ensure that the measurements provided are of sufficient quality for the analysis required to meet FMF-3 and -4. Table 6 describes instrument band performance requirements, as well as band co-alignment requirements for multi-spectral analysis and geo-location requirements to enable users to locate ground targets.

Although it is clear that the goal is to achieve maximum precision, threshold values for these requirements were defined to support flexibility in design at the mission level as well as overall mission feasibility and affordability. Throughout this analysis the threshold was not intended to represent the optimal design but rather the minimum performance at which the intended analyses and production of the intended data products would be possible (although potentially limited in scope compared with the optimal design).

3.3.1. Sensitivity and Noise Requirements

The goal for all spectral bands is to minimize noise (i.e., maximize signal to noise ratio; SNR), and achieve the highest degree of sensitivity possible (Table 6). However, such requirements have limited value in conducting trade-off analysis and identifying candidate systems and approaches. Given that the primary aim of the mission is to support monitoring activities and only provide detection capacity in remote regions, the sensitivity requirements were not driven by detection requirements. Wildfire managers described satisfaction with the detection capacity of MODIS active-fire products [74] for use in remote regions. Initial screening of candidate pixels in wildfire detection products with comparable sensitivity to MODIS was used to define the thresholds for the MWIR and LWIR bands. The noise and sensitivity requirements in the MWIR also govern FRP accuracy, which was similarly derived with a threshold of achieving the minimum detector performance required to replicate MODIS quality detections. Keeping in mind that the measurement requirements (Table 5) include significantly finer spatial resolution than MODIS, the sensitivity and noise requirements could be relaxed further and still meet the user requirements (Table 6). Additionally, FRP integrity is known to be linked to viewing geometry (e.g., [85]), warranting specification of constraints for off-nadir FRP accuracy (Table 6).

VIS and NIR bands were also defined in terms of their minimum functionality. In this case the threshold criteria were the performance required to adequately generate cloud masks and support contextual wildfire detection calculations. For reference, the threshold SNR values described for the VIS and NIR are comparable to the lower end of what Landsat ETM+ produced [86]. At this performance level the requirement to support active-fire detections will be achieved and would be able to provide data for entry level mapping products.

3.3.2. Co-Registration Requirements

Multi-spectral analysis is essential to the delivery of all of the active- and post-fire products required for the mission. Band-to-band co-registration requirements are defined to minimize band remapping requirements prior to data processing (Table 6), helping to facilitate NRT data delivery under FMF-6. Particularly strong co-registration requirements are imposed on the MWIR and LWIR bands due to their use in bi-spectral sub-pixel analysis [73], which is known to be highly sensitive to band mis-registration errors [87]. Although FRP calculations no longer require the use of the Dozier method [32], the sub-pixel descriptors it produces are required for complete wildfire behavior analysis.

Table 6. Summary of key Precision Requirements as outlined in the User Requirements Document (URD), and refined in the Mission Requirements for WildFireSat. SNR refers to the Signal to Noise Ratio, NESR refers to the Noise Equivalent Spectral Radiance, and TOA refers to Top of Atmosphere.

Type	URD Parent Reference	Requirement
Noise	CWFM-URD-0207	The VIS/NIR noise SHALL be: • SNR > 30 over the full dynamic range (Table 5; CFWM-URD-0200) for the VIS bands • SNR > 40 over the full dynamic range (Table 5; CFWM-URD-0200) for the NIR band
	CWFM-URD-0208	The LWIR band SHALL have NESR < 0.12 (W m^{-2} sr^{-1} μm^{-1}) when computed for a TOA scene at 300 K
	CWFM-URD-0209	The MWIR band SHALL have NESR < 0.22 (W m^{-2} sr^{-1} μm^{-1}) when computed for a TOA scene at 300 K
	CWFM-URD-0210	The band noise SHOULD be: • SNR > 90 over the full dynamic range (Table 5; CFWM-URD-0200) for the VIS bands • SNR > 110 over the full dynamic range (Table 5; CFWM-URD-0200) for the NIR bands • LWIR and MWIR NESR should be minimized
Sensitivity	CWFM-URD-0220/30	The design SHALL be capable of measuring the background temperature in the MWIR and LWIR channels with a 2σ uncertainty of no more than 5% at TOA for the sub-satellite pixel The design SHOULD be capable of measuring the background temperature in the MWIR and LWIR channels with a 2σ uncertainty of no more than 3% at TOA for the sub-satellite pixel
	CWFM-URD-0240	Minimum FRP detectable SHOULD be 5 MW at nadir and 40 MW at edge of swath (up to 45 degree off-nadir)
	CWFM-URD-0250	The uncertainty of the FRP of the center pixel when nadir-pointing SHALL be less than 15% of the FRP, or less than 5 MW (whichever value is larger).
	CWFM-URD-0251	The uncertainty of the FRP of pixels at edge of swath (45 degree off-nadir) when nadir-pointing SHOULD be less than 15% of the FRP, or less than 5 MW (whichever value is larger).
Co-registration	CWFM-URD-0260	Co-registration between MWIR and LWIR bands SHALL be provided within 1/3 MWIR pixel accuracy for all pixels in each of the bands
	CWFM-URD-0270/71	Co-registration between LWIR band and VIS/NIR bands SHALL be provided within 1 LWIR pixel accuracy for all pixels in each of the bands, and SHOULD be within 0.5 LWIR pixel accuracy
	CWFM-URD-0272/73	Co-registration between MWIR band and VIS/NIR bands SHALL be provided within 1 MWIR pixel accuracy for all pixels in each of the bands, and SHOULD be within 0.5 MWIR pixel accuracy
	CWFM-URD-0280	Co-registration between NIR and VIS bands SHOULD be provided within 1-pixel accuracy of either the VIS or NIR band (whichever has the smallest spatial resolution at nadir), for all pixels in each of the bands
Image quality	CWFM-URD-0290	The Modulation Transfer Function for all bands SHOULD be >0.3 for all frequencies below the Nyquist frequency
Geo-location	CWFM-URD-300	Data provided to the users SHALL be tagged with geo-referencing information accurate to within 0.5-pixel resolution

3.3.3. Image Quality and Geo-Location Requirements

A threshold Modulation Transfer Function requirement for the IR optics is defined in Table 6. In order to collect accurate FRP measurements, the MWIR and LWIR bands must collect spatially explicit radiometric measurements. The intent of this requirement is to ensure that the majority of the observed energy originates from within the pixel foot print area.

Wildfire management staff relying on the products from a satellite in an operational setting require that the data is accurately geo-referenced to ensure the location of target wildfires is known with a high degree of certainty (Table 6). Furthermore, FMF-4 and -5 require the data to be compared to observations made from other sources; this cross-platform data fusion necessitates precise geo-location to enable accurate comparisons [88].

4. Discussion

The aim of this study was to define requirements for a wildfire monitoring satellite system explicitly to support wildfire management. To do so we surveyed experienced wildfire managers with various specialties in order to better capture their needs in the form of the six FMFs. These functionalities were then used to guide the definition of the WFS User Requirements based on current scientific and technical capabilities.

This study does not describe all factors or analyses considered throughout the process of defining the User Requirements for WFS. Some decisions were made in defining the scope and thus impacted the entire process. For example, the assumption of a LEO satellite mission was taken as the baseline. However, it could be argued that Highly Elliptical Orbiting (HEO) satellite constellations are better suited for the Canadian AoI (e.g., [89,90]), although they are far more costly.

Further, many active- and post-fire products exploit the availability of Short-Wave infrared (SWIR) spectral bands for overnight active wildfire detection and burn severity mapping. Although this spectral band has been considered it was not deemed essential to deliver the FMFs so it was not included as an essential requirement for the mission, though it does remain a goal.

The greatest challenge throughout this process has been defining threshold requirements. It is a simple task to identify the ideal system, payload, and detector. However, determining the limit beyond which a system will no longer be functional is challenging. There is little precedence for marginal systems which can be drawn from. Definition of end-user "usefulness" happens gradually and there is rarely a single threshold of usefulness, particularly when introducing new capabilities. Ultimately, numerous trade-off analyses must be carried out by the Space Team and the User and Science Team where prioritization of the various goal and threshold requirements is necessary for mission development.

The inclusion of trade-off criteria that underpin these requirements is essential as budgetary, technical, and practical limitations will inevitably limit the ability to achieve all of the goal requirements. For example, finer spatial resolution will either reduce the swath and therefore coverage or increase the number of pixels and data volume, challenging the latency requirements. Balancing the consequences of competing requirements is difficult to prescribe a priori. Generally, the trade-off criteria hold coverage of the AoI and IR payload performance as the highest priorities, although overall ability to meet the FMFs is the guiding need. As such, the requirements presented here are a documentation of process at this point in time, but are expected to evolve throughout the mission. The intent is for these requirements to be interpreted in close coordination with the User and Science Team throughout the full mission development.

5. Conclusions

In this study we provided an overview of the approach taken to understand Canadian wildfire management EO needs and transcribe User Requirements which can be used to develop a purpose-built wildfire monitoring satellite to meet their needs. The User Requirements presented here originate in the requirements for CWFMS [49], which were refined through consultation of scientific (e.g., [51]) and wildfire management users to produce the User Requirements for the WFS mission.

The translation of the Wildfire Management Needs into User Requirements for WFS is a foundational step in Phase-A of the mission (Figure 4). Through this process we developed qualitative FMFs based on operational wildfire management needs. Considering technical capabilities and limitations allows User Requirements to be defined for a non-specific space system to address the FMFs. When financial and scope considerations are applied to the User Requirements, the WFS Mission Requirements [50] can be specified. Ultimately, Phase-A of WFS culminates by extending the Mission Requirements into detailed technical specifications for the satellite in the System Requirements. As Figure 4 illustrates, although each stage of Phase-A becomes more technically specific, all of the System Requirements are traceable to their origins in wildfire management needs.

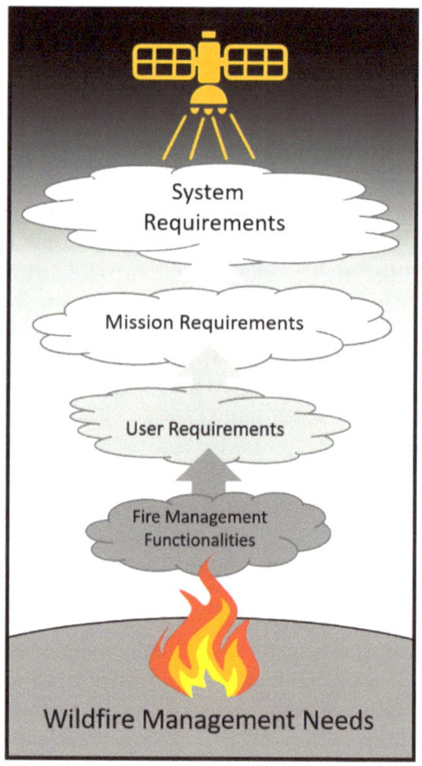

Figure 4. Illustration of the process of defining satellite system requirements based on wildfire management needs in Phase-A of the WildFireSat mission. Fire Management Functionalities (FMFs) are defined qualitatively to capture the Wildfire Management Needs without consideration of space system capabilities or limitations. The FMFs are then translated into User Requirements, which take into consideration space system capabilities required to deliver the necessary data, but do not target any specific mission configuration or scenario. The Mission Requirements define a specific mission scope to address the User Requirements under a specific funding envelope. The Mission Requirements are then used to produce detailed System Requirements describing the technical design of the satellite system. At each level of refinement, the original source of each requirement can be traced back to the end-user's needs thereby ensuring that the technical design of the satellite system is fundamentally driven by the Wildfire Management Needs.

WildFireSat aims to deliver a purpose-built operational wildfire monitoring satellite to support wildfire managers as the primary users. To that end, despite the technical and scientific challenges of the mission, the key to operational success remains in the hands of the wildfire management community. In order to achieve meaningful impact in wildfire management operations, the end-user engagement described in this study must continue for the duration of the mission to ensure that wildfire management needs continue to be heard and that wildfire managers develop a sense of ownership in the mission.

Author Contributions: Conceptualization, J.M.J., T.J.L., H.v.M., D.D., M.J.W.; analysis, J.M.J., N.J., L.N.P., B.L.; investigation, J.M.J., C.M., N.J., A.S.C., D.J., D.D., D.K.T.; data curation, C.M., R.R.; writing—original draft preparation, J.M.J., N.J., C.M., L.M.J., M.S.; writing—review and editing, all authors; supervision, J.M.J., H.v.M., T.J.L.; project administration, J.M.J., D.J. All authors have read and agreed to the published version of the manuscript.

Funding: The WildFireSat mission is a joint Canadian federal government mission led by the Canadian Space Agency. The space segment of the mission is funded by the Canadian Space Agency, while ground segment and user activities are supported by Natural Resources Canada and Environment and Climate Change Canada. Author contributions to this study have been provided through in-kind contribution on behalf of their affiliate agencies. Martin Wooster's contribution is supported by NERC NC funding to the National Centre for Earth Observation (NE/R016518/1).

Acknowledgments: The authors would like to thank the WildFireSat User and Science Team for their ongoing support and guidance for the mission, as well as Den Boychuk and Jordan Evens for their insightful comments. We would also like to thank Jinkai Zhang, Brady Holliday, Jonathan Boucher, Melanie Wheatley, and the Canadian Interagency Forest Fire Centre for their assistance in surveying wildfire managers.

Conflicts of Interest: The authors declare no conflict of interest.

References

1. Hall, J.V.; Zhang, R.; Schroeder, W.; Huang, C.; Giglio, L. Validation of GOES-16 ABI and MSG SEVIRI active fire products. *Int. J. Appl. Earth Obs. Geoinf.* **2019**, *83*, 101928. [CrossRef]
2. Freeborn, P.; Wooster, M.J.; Roberts, G. Addressing the spatiotemporal sampling design of MODIS to provide estimates of the fire radiative energy emitted from Africa. *Remote Sens. Environ.* **2011**, *115*, 475–489. [CrossRef]
3. Freeborn, P.H.; Wooster, M.J.; Roberts, G.; Malamud, B.D.; Xu, W. Development of a virtual active fire product for Africa through a synthesis of geostationary and polar orbiting satellite data. *Remote Sens. Environ.* **2009**, *113*, 1700–1711. [CrossRef]
4. Sirish, U.; Changyong, C.; Xi, S. Geo-Leo intercalibration to evaluate the radiometric performance of NOAA-20 VIIRS and GOES-16 ABI. *Proc. SPIE* **2019**, 11127. [CrossRef]
5. Lindley, T.; Anderson, A.; Mahale, V.; Curl, T.; Line, W.; Lindstrom, S.; Bachmeier, S. Wildfire detection notifications for impact-based decision support services in Oklahoma using geostationary super rapid scan satellite imagery. *J. Oper. Meteorol.* **2016**, *4*, 182–191. [CrossRef]
6. MNP. *A Review of the 2016 Horse River Wildfire: Alberta Agriculture and Forestry Preparedness and Response*; MNP LLP: Edmonton, AB, Canada, 2017; Available online: https://open.alberta.ca/publications/a-review-of-the-2016-horse-river-wildfire-alberta-agriculture-and-forestry-preparedness-and-response (accessed on 24 August 2020).
7. Schmidt, C. Chapter 13—Monitoring Fires with the GOES-R Series. In *The GOES-R Series*; Goodman, S.J., Schmit, T.J., Daniels, J., Redmon, R.J., Eds.; Elsevier: Amsterdam, The Netherlands, 2020; pp. 145–163. [CrossRef]
8. Fischer, C.; Klein, D.; Kerr, G.; Stein, E.; Lorenz, E.; Frauenberger, O.; Borg, E. Data Validation and Case Studies Using the TET-1 Thermal Infrared Satellite System. *Int. Arch. Photogramm. Remote Sens. Spat. Inf. Sci.* **2015**, *XL-7/W3*, 1177–1182. [CrossRef]
9. Zhukov, B.; Briess, K.; Lorenz, E.; Oertel, D.; Skrbek, W. Detection and analysis of high-temperature events in the BIRD mission. *Acta Astronaut.* **2005**, *56*, 65–71. [CrossRef]
10. Zhukov, B.; Lorenz, E.; Oertel, D.; Wooster, M.; Roberts, G. Spaceborne detection and characterization of fires during the bi-spectral infrared detection (BIRD) experimental small satellite mission (2001–2004). *Remote Sens. Environ.* **2006**, *100*, 29–51. [CrossRef]

11. GEO. Global Wildfire Information System (GWIS). 2020. Available online: https://gwis.jrc.ec.europa.eu/ (accessed on 24 August 2020).
12. Stocks, B.J.; Mason, J.A.; Todd, J.B.; Bosch, E.M.; Wotton, B.M.; Amiro, B.D.; Flannigan, M.D.; Hirsch, K.G.; Logan, K.A.; Martell, D.L.; et al. Large forest fires in Canada, 1959–1997. *J. Geophys. Res.* **2002**, *108*. [CrossRef]
13. CIFFC Glossary Task Team and Training Working Group. *Canadian Wildland Fire Management Glossary*; Canadian Interagency Forest Fire Centre: Winnipeg, MB, Canada, 2017; Available online: https://www.ciffc.ca/sites/default/files/2019-03/CIFFC_Canadian_Wildland_Fire_Mgmt_Glossary_2017_10_24.pdf (accessed on 24 August 2020).
14. Tymstra, C.; Stocks, B.J.; Cai, X.; Flannigan, M.D. Wildfire management in Canada: Review, challenges and opportunities. *Prog. Disaster Sci.* **2020**, *5*, 100045. [CrossRef]
15. McAlpine, R.S.; Hirsch, K.G. LEOPARDS—Level of Protection Analysis Software. *For. Chron.* **1998**, *75*, 615–621. [CrossRef]
16. Wotton, B.M.; Stocks, B.J. Fire Management in Canada: Vulnerability and Risk Trends. In *Canadian Wildland Fire Strategy: Background Synthesis, Analysis, and Perspectives*; Hirsch, K., Fuglem, P., Eds.; Canadian Council of Forest Ministers, Natural Resources Canada, Canadian Forest Service, Northern Forestry Centre: Edmonton, AB, Canada, 2006; pp. 49–55.
17. Hanes, C.C.; Wang, X.; Jain, P.; Parisien, M.-A.; Little, J.M.; Flannigan, M.D. Fire-regime changes in Canada over the last half century. *Can. J. For. Res.* **2018**, *49*, 256–269. [CrossRef]
18. Podur, J.; Wotton, M. Will climate change overwhelm fire management capacity? *Ecol. Mod.* **2010**, *221*, 1301–1309. [CrossRef]
19. Wotton, B.M.; Flannigan, M.D.; Marshall, G.A. Potential climate change impacts on fire intensity and key wildfire suppression thresholds in Canada. *Environ. Res. Lett.* **2017**, *12*, 095003. [CrossRef]
20. Kourtz, P. The Need for Improved Forest Fire Detection. *For. Chron.* **1987**, *63*, 272–277. [CrossRef]
21. Flannigan, M.D.; Vonder Haar, T.H. Forest fire monitoring using NOAA satellite AVHRR. *Can. J. For. Res.* **1986**, *16*, 975–982. [CrossRef]
22. Johnston, J.M.; Johnston, L.M.; Wooster, M.J.; Brookes, A.; McFayden, C.; Cantin, A.S. Satellite detection limitations of sub-canopy smouldering wildfires in the North American boreal forest. *Fire* **2018**, *1*, 28. [CrossRef]
23. McFayden, C.B.; Boychuk, D.; Woolford, D.G.; Wheatley, M.J.; Johnston, L. Impacts of wildland fire effects on resources and assets through expert elicitation to support fire response decisions. *Int. J. Wildland Fire* **2019**, *28*, 885–900. [CrossRef]
24. Munoz-Alpizar, R.; Pavlovic, R.; Moran, M.D.; Chen, J.; Gravel, S.; Henderson, S.B.; Ménard, S.; Racine, J.; Duhamel, A.; Gilbert, S.; et al. Multi-Year (2013–2016) PM2.5 Wildfire Pollution Exposure over North America as Determined from Operational Air Quality Forecasts. *Atmosphere* **2017**, *8*, 179. [CrossRef]
25. Beverly, J.L.; Bothwell, P. Wildfire evacuations in Canada 1980–2007. *Nat. Hazards* **2011**, *59*, 571–596. [CrossRef]
26. Johnston, L.M.; Wang, X.; Erni, S.; Taylor, S.W.; McFayden, C.B.; Oliver, J.A.; Stockdale, C.; Christianson, A.; Boulanger, Y.; Gauthier, S.; et al. Wildland fire risk research in Canada. *Environ. Rev.* **2020**, *28*, 164–186. [CrossRef]
27. McGee, T.; McFarlane, B.; Tymstra, C. Wildfire: A Canadian Perspective. In *Wildfire Hazards, Risks, and Disasters*; Paton, D., Ed.; Elsevier: Waltham, MA, USA, 2015. [CrossRef]
28. D'Amours, R.; Malo, A.; Flesch, T.; Wilson, J.; Gauthier, J.-P.; Servranckx, R. The Canadian Meteorological Centre's Atmospheric Transport and Dispersion Modelling Suite. *Atmos. Ocean* **2015**, *53*, 176–199. [CrossRef]
29. Schigas, R.; Stull, R. BlueSky Canada Part 3—BlueSky Canada Wildfire Smoke: Status at UBC. *Can. Smoke Newsl.* **2013**, *2013*, 29–32.
30. Pavlovic, R.; Chen, J.; Anderson, K.; Moran, M.D.; Beaulieu, P.-A.; Davignon, D.; Cousineau, S. The FireWork air quality forecast system with near-real-time biomass burning emissions: Recent developments and evaluation of performance for the 2015 North American wildfire season. *J. Air Waste Manag. Assoc.* **2016**, *66*, 819–841. [CrossRef] [PubMed]
31. Mota, B.; Wooster, M.J. A new top-down approach for directly estimating biomass burning emissions and fuel consumption rates and totals from geostationary satellite fire radiative power (FRP). *Remote Sens. Environ.* **2018**, *206*, 45–62. [CrossRef]

32. Wooster, M.J.; Roberts, G.; Perry, G.L.W.; Kaufman, Y.J. Retrieval of biomass combustion rates and totals from fire radiative power observations: FRP derivation and calibration relationships between biomass consumption and fire radiative energy release. *J. Geophys. Res.* **2005**, *110*, D24311. [CrossRef]
33. Kaiser, J.W.; Heil, A.; Andreae, M.O.; Benedetti, A.; Chubarova, N.; Jones, L.; Morcrette, J.-J.; Razinger, M.; Schultz, M.G.; Suttie, M.; et al. Biomass burning emissions estimated with a global fire assimilation system based on observed fire radiative power. *Biogeosciences* **2012**, *9*, 527–554. [CrossRef]
34. Coogan, S.C.P.; Robinne, F.-N.; Jain, P.; Flannigan, M.D. Scientists' warning on wildfire—A Canadian perspective. *Can. J. For. Res.* **2019**, *49*, 1015–1023. [CrossRef]
35. Flannigan, M.D.; Krawchuk, M.A.; de Groot, W.J.; Wotton, B.M.; Gowman, L.M. Impacts of changing climate for global wildland fire. *Int. J. Wildland Fire* **2009**, *18*, 483–507. [CrossRef]
36. Boulanger, Y.; Parisien, M.-A.; Wang, X. Model-specification uncertainty in future area burned by wildfires in Canada. *Int. J. Wildland Fire* **2018**, *27*, 164–175. [CrossRef]
37. Flannigan, M.D.; Wotton, B.M.; Marshall, G.A.; de Groot, W.J.; Johnston, J.; Jurko, N.; Cantin, A.S. Fuel moisture sensitivity to temperature and precipitation: Climate change implications. *Clim. Change* **2016**, *134*, 59–71. [CrossRef]
38. Krawchuk, M.A.; Moritz, M.A.; Parisien, M.-A.; Van Dorn, J.; Hayhoe, K. Global Pyrogeography: The Current and Future Distribution of Wildfire. *PLoS ONE* **2009**, *4*, e5102. [CrossRef] [PubMed]
39. Woolford, D.G.; Dean, C.; Martell, D.L.; Cao, J.; Wotton, B. Lightning-caused forest fire risk in Northwestern Ontario, Canada, is increasing and associated with anomalies in fire weather. *Environmetrics* **2014**, *25*, 406–416. [CrossRef]
40. Wotton, B.M.; Nock, C.A.; Flannigan, M.D. Forest fire occurrence and climate change in Canada. *Int. J. Wildland Fire* **2010**, *19*, 253–271. [CrossRef]
41. Xianli, W.; Kala, S.; Marc-Andre, P.; Steve, W.T.; Jean-Noël, C.; Yan, B.; Mike, D.F. Projected changes in fire size from daily spread potential in Canada over the 21st century. *Environ. Res. Lett.* **2020**. [CrossRef]
42. Wang, X.; Thompson, D.K.; Marshall, G.A.; Tymstra, C.; Carr, R.; Flannigan, M.D. Increasing frequency of extreme fire weather in Canada with climate change. *Clim. Chang.* **2015**, *130*, 573–586. [CrossRef]
43. Van Mierlo, H. *Business Case for WildFireSat*; Canadian Space Agency, Ed.; Space Utilization: St Hubert, QC, Canada, 2019.
44. Ngo Phong, L.; Pancrati, O.; Marchese, M.; Châteauneuf, F. Spaceborne linear arrays of 512 × 3 microbolometers. *Proc. SPIE* **2013**, *8614*. [CrossRef]
45. Hamel, J.-F.; St-Amour, A.; de Lafontaine, J.; Van Mierlo, M.; Lawrence, B.; Ngo Phong, L.; Lynham, T.J.; Johnston, J.M.; Cain, J.; Lihou, M.; et al. The innovative microsatellite-based Canadian wildland fire monitoring system. In Proceedings of the 66th International Astronautical Congress, Jerusalem, Israel, 12–16 October 2015; pp. 1–7.
46. Ngo Phong, L.; Dufour, D.; Johnston, J.; Chevalier, C.; Côté, P.; Fisette, B.; Wang, M.; Châteauneuf, F. A low resource imaging radiometer for nanosatellite based fire diagnosis. *Proc. SPIE* **2018**, *10765*. [CrossRef]
47. Schroeder, W.; Oliva, P.; Giglio, L.; Csiszar, I.A. The New VIIRS 375 m active fire detection data product: Algorithm description and initial assessment. *Remote Sens. Environ.* **2014**, *143*, 85–96. [CrossRef]
48. Xu, W.; Wooster, M.J.; He, J.; Zhang, T. First study of Sentinel-3 SLSTR active fire detection and FRP retrieval: Night-time algorithm enhancements and global intercomparison to MODIS and VIIRS AF products. *Remote Sens. Environ.* **2020**, *248*, 111947. [CrossRef]
49. Lynham, T.J.; Johnston, J.M.; Ball, D.; Phong, L.N.; Ng, A.; Lawrence, B. *Canadian Wildland Fire Monitoring System User Requirements Document*, version 4.0; Canadian Space Agency: St Hubert, QC, Canada, 2015; in press.
50. Jackson, N.; Johnston, J.M. *WildFireSat: Mission Requirements Document Rev-A.*; Canadian Space Agency: St Hubert, QC, Canada, 2020; p. 51.
51. Johnston, J.M.; van Mierlo, H. Wildfire Remote Sensing Workshop, Report. In Proceedings of the EO Summit 2017, Montreal, QC, Canada, 11 April 2018; p. 54. Available online: https://crss-sct.ca/wp-content/uploads/2018/04/EO-Summit-2017-Wildfire-Remote-Sensing-Workshop-Report-2018-04-11.pdf (accessed on 24 August 2020).
52. Scott, J.H.; Thompson, M.P.; Calkin, D.E. *A Wildland Fire Risk Assessment Framework for Land and Resource Management*; RMRS-GTR-315; U.S. Department of Agriculture, Forest Service, Rocky Mountain Research Station: Fort Collins, CO, USA, 2013. [CrossRef]

53. Taber, M.A.; Elenz, L.M.; Langowski, P.G. *Decision Making for Wildfires: A Guide for Applying a Risk Management Process at the Incident Level*; General Technical Report RMRS-GTR-298; U.S. Department of Agriculture, Forest Service, Rocky Mountain Research Station: Fort Collins, CO, USA, 2013. [CrossRef]
54. Rapp, C.; Rabung, E.; Wilson, R.; Toman, E. Wildfire decision support tools: An exploratory study of use in the United States. *Int. J. Wildland Fire* **2020**, *29*, 581–594. [CrossRef]
55. Oliva, P.; Schroeder, W. Assessment of VIIRS 375m active fire detection product for direct burned area mapping. *Remote Sens. Environ.* **2015**, *160*, 144–155. [CrossRef]
56. Johnston, J.M.; Wheatley, M.J.; Wooster, M.J.; Paugam, R.; Davies, G.M.; DeBoer, K.A. Flame-Front Rate of Spread Estimates for Moderate Scale Experimental Fires Are Strongly Influenced by Measurement Approach. *Fire* **2018**, *1*, 16. [CrossRef]
57. McRae, D.J.; Jin, J.-Z.; Conard, S.G.; Sukhinin, A.I.; Ivanova, G.A.; Blake, T.W. Infrared characterization of fine-scale variability in behaviour of boreal forest fires. *Can. J. For. Res.* **2005**, *35*, 2194–2206. [CrossRef]
58. Johnston, J.M.; Wooster, M.J.; Paugam, R.; Wang, X.; Lynham, T.J.; Johnston, L.M. Direct Estimation of Byram's fire intensity from infrared remote sensing imagery. *Int. J. Wildland Fire* **2017**, *26*, 668–684. [CrossRef]
59. Johnston, L.M.; Flannigan, M.D. Mapping Canadian wildland fire interface areas. *Int. J. Wildland Fire* **2018**, *27*, 1–14. [CrossRef]
60. McFayden, C.B.; Woolford, D.G.; Stacey, A.; Boychuk, D.; Johnston, J.M.; Wheatley, M.J.; Martell, D.L. Risk assessment for wildland fire aerial detection patrol route planning in Ontario, Canada. *Int. J. Wildland Fire* **2020**, *29*, 28–41. [CrossRef]
61. Kochi, I.; Donovan, G.H.; Champ, P.A.; Loomis, J.B. The economic cost of adverse health effects from wildfire-smoke exposure: A review. *Int. J. Wildland Fire* **2010**, *19*, 803–817. [CrossRef]
62. Krstic, N.; Henderson, S.B. Use of MODIS data to assess atmospheric aerosol before, during, and after community evacuations related to wildfire smoke. *Remote Sens. Environ.* **2015**, *166*, 1–7. [CrossRef]
63. Richardson, L.A.; Champ, P.A.; Loomis, J.B. The hidden cost of wildfires: Economic valuation of health effects of wildfire smoke exposure in Southern California. *J. For. Econ.* **2012**, *18*, 14–35. [CrossRef]
64. Yao, J.; Henderson, S.B. An empirical model to estimate daily forest fire smoke exposure over a large geographic area using air quality, meteorological, and remote sensing data. *J. Expo. Sci. Environ. Epidemiol.* **2014**, *24*, 328–335. [CrossRef] [PubMed]
65. Lee, S.-Y.; Gan, C.; Chew, B.N. Visibility deterioration and hygroscopic growth of biomass burning aerosols over a tropical coastal city: A case study over Singapore's airport. *Atmos. Sci. Lett.* **2016**, *17*, 624–629. [CrossRef]
66. Chen, J.; Anderson, K.; Pavlovic, R.; Moran, M.D.; Englefield, P.; Thompson, D.K.; Munoz-Alpizar, R.; Landry, H. The FireWork v2.0 air quality forecast system with biomass burning emissions from the Canadian Forest Fire Emissions Prediction System v2.03. *Geosci. Model Dev.* **2019**, *12*, 3283–3310. [CrossRef]
67. Larkin, N.K.; O'Neill, S.M.; Solomon, R.; Raffuse, S.; Strand, T.; Sullivan, D.C.; Krull, C.; Rorig, M.; Peterson, J.; Ferguson, S.A. The BlueSky smoke modeling framework. *Int. J. Wildland Fire* **2009**, *18*, 906–920. [CrossRef]
68. Gonzi, S.; Palmer, P.I.; Paugam, R.; Wooster, M.; Deeter, M.N. Quantifying pyroconvective injection heights using observations of fire energy: Sensitivity of spaceborne observations of carbon monoxide. *Atmos. Chem. Phys.* **2015**, *15*, 4339–4355. [CrossRef]
69. Paugam, R.; Wooster, M.J.; Atherton, J.; Freitas, S.R.; Schultz, M.G.; Kaiser, J.W. Development and optimization of a wildfire plume rise model based on remote sensing data inputs—Part 2. *Atmos. Chem. Phys. Discuss.* **2015**, *15*, 9815–9895. [CrossRef]
70. Johnston, J.M. Infrared Remote Sensing of Fire Behaviour in Canadian Wildland Forest Fuels. Ph.D. Thesis, King's College London, London, UK, 2016.
71. Sperling, S.; Wooster, M.J.; Malamud, B.D. Influence of Satellite Sensor Pixel Size and Overpass Time on Undercounting of Cerrado/Savannah Landscape-Scale Fire Radiative Power (FRP): An Assessment Using the MODIS Airborne Simulator. *Fire* **2020**, *3*, 11. [CrossRef]
72. Johnston, J.M.; Paugam, R.; Whitman, E.; Schiks, T.; Cantin, A.S. Remote Sensing of Fire Behavior. In *Encyclopedia of Wildfires and Wildland-Urban Interface (WUI) Fires*; Manzello, S.L., Ed.; Springer International Publishing: Cham, Switzerland, 2019; pp. 1–5. Available online: https://link.springer.com/content/pdf/10.1007%2F978-3-319-51727-8_174-1.pdf (accessed on 24 August 2020). [CrossRef]
73. Dozier, J. A Method for Satellite Identification of Surface Temperature Fields of Subpixel Resolution. *Remote Sens. Environ.* **1981**, *11*, 221–229. [CrossRef]

74. Giglio, L.; Schroeder, W.; Justice, C.O. The Collection 6 MODIS Active Fire Detection Algorithm and Fire Products. *Remote Sens. Environ.* **2016**, *178*, 31–41. [CrossRef]
75. Wooster, M.J.; Xu, W.; Nightingale, T. Sentinel-3 SLSTR active fire detection and FRP product: Pre-launch algorithm development and performance evaluation using MODIS and ASTER datasets. *Remote Sens. Environ.* **2012**, *120*, 236–254. [CrossRef]
76. Smith, A.M.S.; Wooster, M.J. Remote classification of head and backfire types from MODIS fire radiative power and smoke plume observations. *Int. J. Wildland Fire* **2005**, *14*, 249–254. [CrossRef]
77. Wooster, M.J.; Zhang, Y.H. Boreal forest fires burn less intensely in Russia than in North America. *Geophys. Res. Lett.* **2004**, *31*. [CrossRef]
78. Zhang, T.; Wooster, M.J.; Xu, W. Approaches for synergistically exploiting VIIRS I- and M-Band data in regional active fire detection and FRP assessment: A demonstration with respect to agricultural residue burning in Eastern China. *Remote Sens. Environ.* **2017**, *198*, 407–424. [CrossRef]
79. Whitman, E.; Johnston, J.M.; Schiks, T.; Paugam, R.; Cantin, A.S. Imaging Postfire Environments. In *Encyclopedia of Wildfires and Wildland-Urban Interface (WUI) Fires*; Manzello, S.L., Ed.; Springer International Publishing: Cham, Switzerland, 2019; pp. 1–6. Available online: https://link.springer.com/content/pdf/10.1007%2F978-3-319-51727-8_175-1.pdf (accessed on 24 August 2020). [CrossRef]
80. Giglio, L.; Randerson, J.T.; van der Werf, G.R. Analysis of daily, monthly, and annual burned area using the fourth-generation global fire emissions database (GFED4). *J. Geophys. Res. Biogeosci.* **2013**, *118*, 317–328. [CrossRef]
81. Chuvieco, E.; Mouillot, F.; van der Werf, G.R.; San Miguel, J.; Tanase, M.; Koutsias, N.; García, M.; Yebra, M.; Padilla, M.; Gitas, I.; et al. Historical background and current developments for mapping burned area from satellite Earth observation. *Remote Sens. Environ.* **2019**, *225*, 45–64. [CrossRef]
82. Hall, R.J.; Skakun, R.S.; Metsaranta, J.M.; Landry, R.; Fraser, R.H.; Raymond, D.; Gartrell, M.; Decker, V.; Little, J. Generating annual estimates of forest fire disturbance in Canada: The National Burned Area Composite. *Int. J. Wildland Fire* **2020**. [CrossRef]
83. Roy, D.P.; Huang, H.; Boschetti, L.; Giglio, L.; Yan, L.; Zhang, H.H.; Li, Z. Landsat-8 and Sentinel-2 burned area mapping—A combined sensor multi-temporal change detection approach. *Remote Sens. Environ.* **2019**, *231*, 111254. [CrossRef]
84. van der Werf, G.R.; Randerson, J.T.; Giglio, L.; van Leeuwen, T.T.; Chen, Y.; Rogers, B.M.; Mu, M.; van Marle, M.J.E.; Morton, D.C.; Collatz, G.J.; et al. Global fire emissions estimates during 1997–2016. *Earth Syst. Sci. Data* **2017**, *9*, 697–720. [CrossRef]
85. Freeborn, P.; Wooster, M.J.; Hao, W.M.; Ryan, C.A.; Nordgren, B.L.; Baker, S.P.; Ichoku, C. Relationships between energy release, fuel mass loss, and trace gas and aerosol emissions during laboratory biomass fires. *J. Geophys. Res.* **2008**, *113*. [CrossRef]
86. Irons, J.R.; Dwyer, J.L.; Barsi, J.A. The next Landsat satellite: The Landsat Data Continuity Mission. *Remote Sens. Environ.* **2012**, *122*, 11–21. [CrossRef]
87. Giglio, L.; Kendall, J.D. Application of the Dozier retrieval to wildfire characterization: A sensitivity analysis. *Remote Sens. Environ.* **2001**, *77*, 34–49. [CrossRef]
88. Crowley, M.A.; Cardille, J.A.; White, J.C.; Wulder, M.A. Multi-sensor, multi-scale, Bayesian data synthesis for mapping within-year wildfire progression. *Remote Sens. Lett.* **2019**, *10*, 302–311. [CrossRef]
89. Duncan, B.N.; Ott, L.E.; Abshire, J.B.; Brucker, L.; Carroll, M.L.; Carton, J.; Comiso, J.C.; Dinnat, E.P.; Forbes, B.C.; Gonsamo, A.; et al. Space-Based Observations for Understanding Changes in the Arctic-Boreal Zone. *Rev. Geophys.* **2020**, *58*, e2019RG000652. [CrossRef]
90. Kidder, S.Q.; Vonder Haar, T.H. On the Use of Satellites in Molniya Orbits for Meteorological Observation of Middle and High Latitudes. *J. Atmos. Ocean. Technol.* **1990**, *7*, 517–522. [CrossRef]

© 2020 by the authors. Licensee MDPI, Basel, Switzerland. This article is an open access article distributed under the terms and conditions of the Creative Commons Attribution (CC BY) license (http://creativecommons.org/licenses/by/4.0/).

sensors

MDPI

Article

Computationally Efficient Wildfire Detection Method Using a Deep Convolutional Network Pruned via Fourier Analysis

Hongyi Pan *, Diaa Badawi and Ahmet Enis Cetin *

Department of Electrical and Computer Engineering, University of Illinois at Chicago, Chicago, IL 60607, USA; dbadaw2@uic.edu

* Correspondence: hpan21@uic.edu (H.P.); aecyy@uic.edu (A.E.C.)

Received: 10 April 2020; Accepted: 15 May 2020; Published: 20 May 2020

check for updates

Abstract: In this paper, we propose a deep convolutional neural network for camera based wildfire detection. We train the neural network via transfer learning and use window based analysis strategy to increase the fire detection rate. To achieve computational efficiency, we calculate frequency response of the kernels in convolutional and dense layers and eliminate those filters with low energy impulse response. Moreover, to reduce the storage for edge devices, we compare the convolutional kernels in Fourier domain and discard similar filters using the cosine similarity measure in the frequency domain. We test the performance of the neural network with a variety of wildfire video clips and the pruned system performs as good as the regular network in daytime wild fire detection, and it also works well on some night wild fire video clips.

Keywords: wildfire detection; block-based analysis; transfer learning; Fourier analysis; pruning and slimming

1. Introduction

Early wildfire detection is of utmost importance to combat the unprecedented scale of wildfires happening all over the world. Recently, there has been a notable interest in developing real-time algorithms to detect wildfires using regular video-based surveillance systems [1–20]. Video-based forest fire detection can be used to replace traditional point-sensor type detectors because a single pan-tilt-zoom type camera can monitor a wide area, detect forest fire and smoke immediately after the start of the wildfire—as long as the smoke is within the viewing range of the camera. Nowadays, with the development of 5G communication [21,22], unmanned aerial vehicles (UAVs) have also become a good option for wildfire surveillance tasks because of their flexibility compared to fixed surveillance towers. However, all of the traditional video-based methods [1–20] rely on choosing features manually. In contrast, deep neural networks can extract relevant features by themselves given sufficient training data.

In recent years, deep learning has been widely used in a myriad of computer vision applications because of its high recognition capability. To the best of our knowledge, Gunay et al. [23] is the first paper to use deep learning in dynamic texture recognition including wildfires. In early deep learning based forest detection, researchers designed blank convolutional neural networks and trained them with collected or synthesized images [24–28]. For example, Zhao et al. designed a 15-layer convolutional neural network (CNN) to detect the forest fire [25]. What is more, to locate the fire and smoke in frames, Barmpoutis et al. [29], Huang et al. [30] and Chaoxia et al. [31] proposed R-CNN -based fire detection method. However, neural network training requires huge amounts of data, while such data can be very expensive and infeasible. If the training data is insufficient, the network may not be robust. To overcome this problem, some researchers adopted transfer learning [32] to design neural networks. Transfer learning is a very efficient method and it is widely used in recognition tasks because of its advantage that we only need to train only the final several layers instead of the whole network. Transfer learning requires a smaller dataset for training and can avoid overfitting. Typical examples are AlexNet-based CNN by Muhammad et al. [33] and YOLO -based CNN by Park et al. [34] and Jiao et al. [35]. Inspired by them, in this work, we propose to use transfer learning from MobileNet-V2 [36] for forest fire detection. Moreover, after updating the network with the available data, we notice that some kernels (filters) have very low energy. Computing these low-energy kernels is a waste of resources since their output is insignificant compared with the outputs of other kernels. We also observe that some kernels have very similar frequency responses, which means storing only one from each pair of similar kernels can significantly reduce storage memory for edge devices. We take advantage of these two facts and prune the kernels according to their frequency response via Fourier analysis. Fourier analysis not only trims the deep neural network but also removes the unnecessary convolutional units. The trimmed network provides as good results as the regular network. Fourier domain trimming is not specific to the wildfire detection task, it can be used in trimming other networks as well. We take advantage of these two facts of low-energy and similar kernels and prune the kernels according to their frequency response via Fourier analysis.

Compared with other deep learning methods [23–31,33–35], the major advantage is that, after transfer learning, we prune the network via Fourier analysis. More specifically, we have the following advantages:

- Our method takes advantage of MobileNet-V2 [36] in efficiency. Details of this aspect are provided in Section 3.
- We prune and slim the convolutional and dense layers according to frequency response of kernels using Fourier analysis in order to accelerate the inference of the neural network and save storage. Details of this aspect are provided in Sections 3.1 and 3.2.
- We detect wildfire in overlapping windows so we can easily detect smoke even if it exists near the edge of a frame. We achieve this by dividing the frame in many blocks and detecting the smoke block-by-block. Details of this aspect are provided in Section 4.
- Compared to R-CNN [30,31], and YOLO method [34,35], our block-based analysis makes the building of testing and training datasets easy because we mark only the blocks containing fire. We only need to label each block as fire or no-fire instead of marking the region of fire and smoke in a given frame using several bounding boxes. Making and updating the dataset in our method is much easier compared to R-CNN and YOLO method. In wildfire surveillance task, knowing the fire in which image block is sufficient for fire departments to take action. In addition, compared to frame-based methods, block-based analysis allows us to determine capture very small fire regions and smoke.
- Our system can detect very small smoke regions, while papers from related works [24–31] have not provided the test results on such small smoke regions. The details of this aspect are provided in Section 5.2.
- The input of our system is in 1080P and it can also be adjusted for higher resolution. Thus, our method matches common surveillance cameras, since down-sampling always causes information loss and may make small regions of smoke invisible. According to our experimental results, our method works well even if the smoke region is very small.
- After testing the performance on daytime surveillance and obtaining a very good result, we further tested our system system with night events, and it works on many video clips.

2. Dataset for Training

We use the dataset of Reference [28] which contains 4000 images gathered from the Internet (half with fire and half without fire) shown in Figure 1a,b, and FIRESENSE database [37] shown in Figure 1c,d which contains 11 fire videos and 16 no-fire videos. During practice, we notice that the system would false alarm at cloud region sometimes, so we also add 4 cloud video clips shown in Figure 1e,f into our training dataset to reduce the false-alarm rate. Figure 2 shows the distribution of our training dataset.

(a) (b) (c) (d) (e) (f)

Figure 1. Samples of our training dataset. (**a**,**b**) are gathered from the Internet; (**c**,**d**) are from FIRESENSE Database; (**e**,**f**) are cloud images.

(**a**) Fire images (**b**) No-Fire images

Figure 2. Image distribution of training dataset.

In order to avoid the over-fitting during training, we only keep the first frame in each 20–100 frames (depends on the length of the video) to maintain diversity because the Internet images are in variety of scenarios compare to the video clips. If we train too much on the video clips, the neural network may easily over-fit on the scenes in these video clips. All images and frames are resized into 224×224 as the input of the neural network, and 5% of them are picked randomly for validation. Data augmentation by shifting and translating wildfire images with filling by reflect translation as in Reference [28] is also adopted here. Let $\mathbf{Y} \in \mathbb{R}^{M \times N}$ denote the original image and index start from 0, if we want to shift the image by m length up and n length left (negative value means shifting in the opposite direction), then the augmented image $\tilde{\mathbf{Y}}$ can be represent as

$$\tilde{Y}_{i,j} = \begin{cases} Y_{\min(i+m,2M-i-m-2),\min(j+n,2N-j-n-2)} & m \geq 0, n \geq 0 \\ Y_{|i+m|,\min(j+n,2N-j-n)} & m < 0, n \geq 0 \\ Y_{\min(i+m,2M-i-m-2),|j+n|} & m \geq 0, n < 0 \\ Y_{|i+m|,|j+n|} & m < 0, n < 0, \end{cases} \tag{1}$$

where, $i = 0, 1, ..., M-1, j = 0, 1, ..., N-1$. Range of m and n depends on the location of smoke in each image and frame. The reason for using such an augmentation scheme is to ensure that the classifier will see examples in which wildfire starts from different locations during training.

3. Fourier Transform Based Pruning the Network

We use the MobileNet-V2 [36] for wildfire recognition after retraining. MobileNet-V2 is an efficient convolutional neural network for mobile vision applications. MobileNet's energy consumption is much smaller than other neural networks. Usually, wildfire monitoring cameras are placed in remote locations and/or UAVs. In such cases, energy consumption is very important. Therefore, MobileNet-V2 is suitable for wildfire detection. We replace the dense layers of MobileNet-V2 by a new dense layer with two neurons, and we apply softmax function as the activation function in the dense layer for decision making. The existence of fire is labeled as 1 and "no-fire" is labeled as 0. The probability P of fire can be represent as

$$P = \frac{\exp(P_1)}{\exp(P_1) + \exp(P_0)}, \tag{2}$$

where P_0 and P_1 represent the outputs of the two neurons before activating, respectively.

In this paper, we employ TensorFlow in Python 3 to train the network and then, convert the well-trained model to a TensorFlow lite model. We can install this model on a NVIDIA Jetson Nano board or a similar board.

In the following section we will describe how we can further trim the network.

3.1. Pruning Low-Energy Kernels

A well-trained CNN may contain some redundant neurons—no matter what the input is, the output of these neurons will always take small values. Such neurons have filter weights with too small values. As we all know, performing a convolution operation is relatively expensive, especially in a long-time surveillance task. Thus, efficiency can be significantly improved if we can avoid calculation with too small weights.

The MobileNet-V2 model we fine-tuned has 18 3×3 convolutional layers. When we extract weights from each 3×3 layer and calculate their DFT via MATLAB, we find that there are some with almost zero-magnitude response only in the first standard convolutional layer. It contains 32 3-channel 3×3 kernels, and some of their frequency responses are shown in Figure 3. We find that the magnitude responses of kernels with numbers 4, 8, 12–14, 20, 29 are close to 0 for all frequency values. Then, we calculate the energy of the first projection layer among the input direction as shown in Figure 4a, we notice that the energy with these indexes is also close to 0. After pruning these layers by removing those weights (the first standard convolutional layer is $3 \times 3 \times 3 \times 32$, the first depthwise separable convolutional layer is $3 \times 3 \times 32 \times 1$, and the projection convolutional layer is $1 \times 1 \times 32 \times 16$ before pruning, then each "32" reduces to "25"), we save 21.875% computation in these layers. Moreover, when we plot the energy distribution of the dense layer as shown in Figure 4b, we find that there are many weights having low energy. By removing these kernels in the dense layer with energy less than 0.0005 and corresponding kernels in the previous layers, we save $485/1280 = 37.89\%$ computation in these layers. Figure 5 shows the graph before and after pruning. By running the inference 100 times and calculating the average time consumption on the computer, we find that it takes 0.499921 s before pruning, and takes 0.464778 s after pruning. Thus, we totally save 7.04% time by pruning these layers.

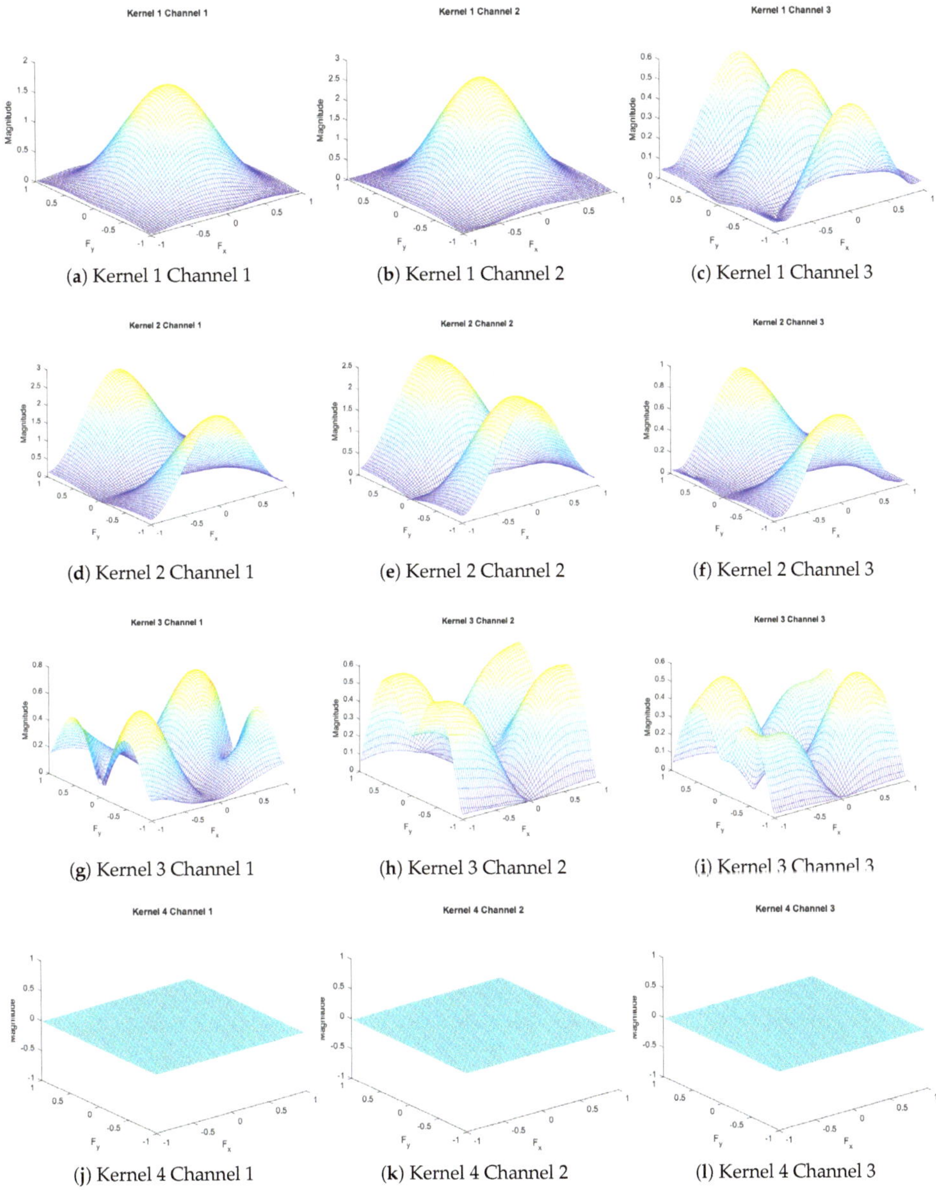

(**a**) Kernel 1 Channel 1 (**b**) Kernel 1 Channel 2 (**c**) Kernel 1 Channel 3

(**d**) Kernel 2 Channel 1 (**e**) Kernel 2 Channel 2 (**f**) Kernel 2 Channel 3

(**g**) Kernel 3 Channel 1 (**h**) Kernel 3 Channel 2 (**i**) Kernel 3 Channel 3

(**j**) Kernel 4 Channel 1 (**k**) Kernel 4 Channel 2 (**l**) Kernel 4 Channel 3

Figure 3. Frequency response of the first convolutional layer.

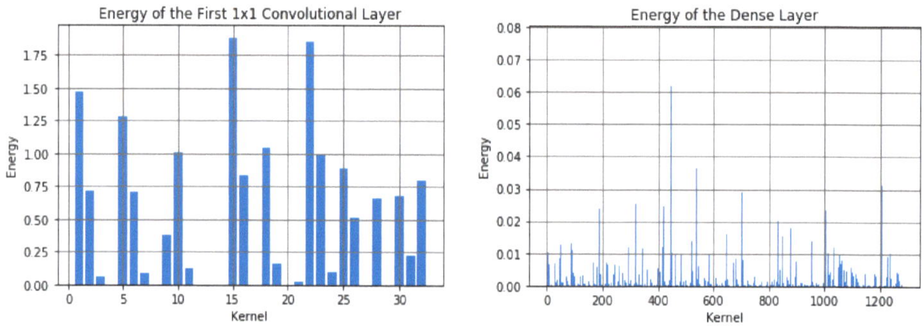

(**a**) First 1×1 conv. layer (**b**) Dense layer

Figure 4. Energy distribution.

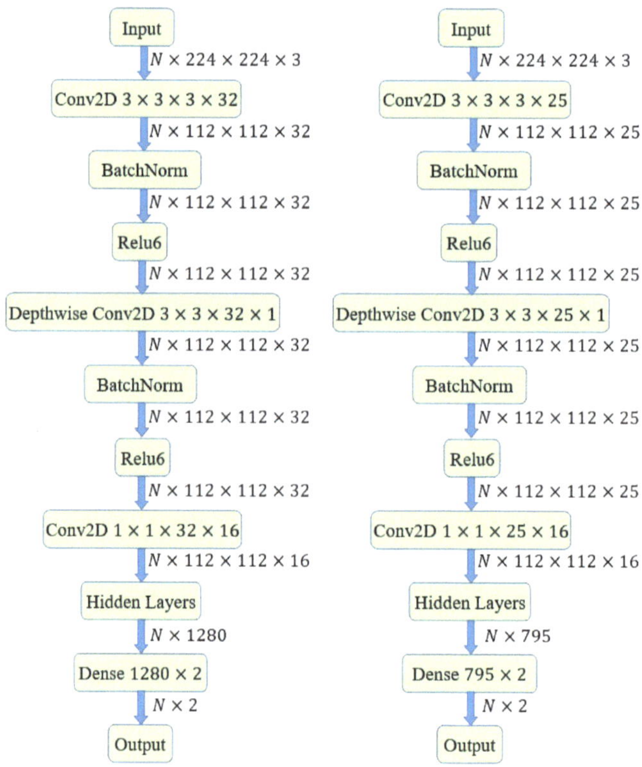

(**a**) Before Pruning (**b**) After Pruning

Figure 5. Pruning graph, "N" is batch size.

3.2. Slimming Similar Kernel Pairs

In other depthwise separable convolutional layers, on the other hand, when we plot the magnitude, we notice that some have similar shapes. For example, in the final convolutional layer, we notice that the magnitudes of kernel 259 and kernel 318, shown in Figure 6, are very similar in shape, though different in scale, which means we can slim the convolutional layer by saving only one of each similar kernel pairs and reconstruct the other one by scaling the output of the first one.

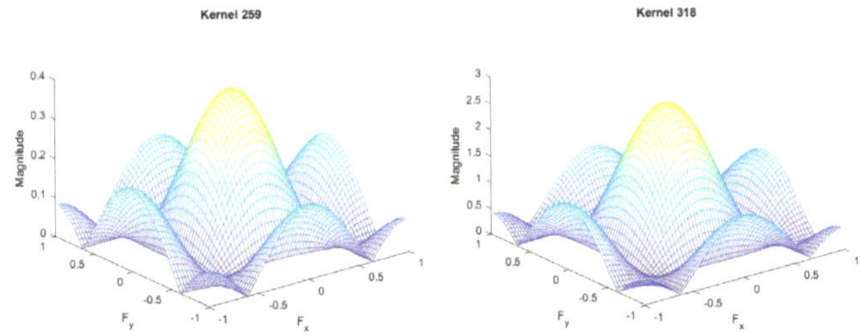

(**a**) Kernel 259 (**b**) Kernel 318

Figure 6. Frequency response of the final convolutional layer.

We compute the frequency response of the filters of the network and keep only one of the filters with similar magnitude frequency responses. We also discard filters with magnitude responses significantly smaller than the other filters. More specifically, we calculate cosine similarity of each kernel pair:

$$\text{Filter similarity} = \cos(\theta) = \frac{<\mathbf{X}, \mathbf{Y}>}{||\mathbf{X}|| \cdot ||\mathbf{Y}||}, \tag{3}$$

where **X** and **Y** are the Fourier transform magnitudes of the two filters in vector form, respectively. Since we use the magnitudes of the frequency responses, the filter similarity measure is in the range of $[0, 1]$. We treat the kernels with similarity larger than 0.99925 as a pair of similar kernels and store only one of them. We cannot compare the filters in spatial domain using their weights because there may be a phase shift between the two filters and the filters may look different but they can essentially be the same. Cosine similarity measure also handles the case of the two filters with

$$|Y[k_1, k_2]| \approx \alpha |X[k_1, k_2]|, \tag{4}$$

where X and Y are the DFT or the two kernels, respectively. Both of these filters will generate very similar outputs for a given input except that one of the outputs will be a shifted and scaled version of the other output.

Table 1 has the summary of our results of pruning and slimming. The first one is a standard layer named "conv" and others are depthwise separable named "expanded_conv", and between each depthwise separable convolution layer, there are two 1×1 standard convolutional layers to change the dimensions. In fact, this idea should be able to apply to all MobileNet-V2 based neural networks, because we find that the original MobileNet-V2 also contains both very low-energy kernels and similar kernels.

Figure 7 shows the output of a smoke frame before and after pruning and slimming. More statistics are provided in Sections 5.2 and 5.4. We can see that the result changes negligibly. In this way, although these layers cannot be accelerated even with similar kernels because the input of each kernel is still different, we save 22.59% storage of the weights, which is very important to edge devices.

Table 1. Pruning and slimming information of our model.

Layers Name	Kernels Num	Slimmed Num	Rate (%)
conv [a]	32	7	21.88
expanded_conv [a]	32	7	21.88
expanded_conv_1	96	2	2.08
expanded_conv_2 [b]	144	0	0.00
expanded_conv_3	144	74	51.39
expanded_conv_4	192	6	3.13
expanded_conv_5 [b]	192	0	0.00
expanded_conv_6	192	112	58.33
expanded_conv_7	384	9	2.34
expanded_conv_8	384	2	0.52
expanded_conv_9	384	3	0.78
expanded_conv_10	384	17	4.43
expanded_conv_11	576	2	0.35
expanded_conv_12	576	1	0.17
expanded_conv_13	576	490	85.07
expanded_conv_14	960	17	1.77
expanded_conv_15	960	45	4.69
expanded_conv_16	960	825	85.94
Slimming Overall [c]	7104	1605	22.59

Expanded_conv is the name of depthwise separable convolutional layers. [a] These layers are pruned by removing weights with too low energy. [b] These layers cannot be slimmed even with lower threshold (0.9990), while too low threshold may cause the loss of accuracy. [c] Layers with [a] are excluded because they are using pruning method.

(**a**) Before pruning and slimming (**b**) After pruning and slimming

Figure 7. Lyons Fire in San Diego, 1 September 2016.

4. Block-Based Analysis of Image Frames

Nowadays, with the development of technology, we can obtain forestry surveillance videos in 1080P or higher resolution. However, if we design the input of the neural network in these high resolutions, the network will be too huge to train. Most common method to use to solve this problem is to down-sample the images, which correspondingly will increase the difficulty to detect little smoke plumes and will cause information loss. To overcome this problem, we divide the image frames into small blocks as in [28]. Suppose that the width and height of the effective image region is M_i and N_i, and the width and height of each block is M_b and N_b, then the row number R and the column number C of each block can be represent as

$$R = \left\lfloor \frac{H_i}{H_b} \right\rfloor, C = \left\lfloor \frac{W_i}{W_b} \right\rfloor, \tag{5}$$

where $\lfloor\rfloor$ is the floor function. We divide the frames into many tiles as in Figure 7. The videos are in 1080 P (HPWREN videos are in 6 MP in fact, so we down-sample them into 1080 P for convenience.

In real-world applications, we can also use different division strategy to make good use of higher resolution), and then divide them into many tiles with 180 × 180 pixels, and every 2 by 2 tiles consist a block with 360 × 360 pixels. The score is the fire rate of each block, composed of the tile has score and its bottom, right and bottom-right tiles, rather than the forest fire rate of each tile. This is the reason that there is no score in the most right tiles and the most bottom tiles. In this way, a frame is divided into 5 × 9 = 45 blocks, and if the smoke exists at the edge of one block, it will also exist at the center of its neighbor block. Then, we resize each block into the input size and then feed the network with resized blocks. Therefore, the input of the neural network is 45 × 224 × 224 × 3.

5. Network Performance

In this section, we will test the speed and accuracy of our model with the HPWREN dataset [38] and some YouTube forest fire videos. Unlike other binary classification tasks (like the famous Kaggle's competition, cats VS dogs), forest fire detection requires a very low false-alarm rate, otherwise the system will produce false alarming in the long-term surveillance task. This may bother the security guards verifying the system outputs. If there are too many false alarms they may develop the bad habit of ignoring the wildfire monitoring system. To reduce the false-alarm rate, we set the alarm threshold value as 0.95, such that the system will alarm only if the score is large than this value. In practical the real-world applications, we may decrease the threshold value to 0.85, 0.9 or other values to increase the fire detection rate, if no dust, cloud or other confusing objects exist in the monitored area, or increase if they exist too much.

5.1. Speed Test

In this section, we test the speed of our model on the NVIDIA Jetson Nano board. We are only interested in the total time of each frame to finish frame pre-processing (division and resizing), and interpretation of the neural network, because time consumed by signal transmission depends on the surveillance cameras and the network and should be finished in an instant. Our system takes about 3 to 4 s to process a single frame. This delay is negligible compare to the time that fire departments take to take actions.

5.2. Daytime Fire Surveillance Test

Sample results of daytime fire surveillance test are shown in Figures 7–14, and Table 2 lists the frame number that fire starts in each video clips and the first frame number that our system manages to detect the fire. We also provide the result before pruning and slimming in Table 2. Only two videos a have one or two frame detection delay after pruning and slimming. It is worth pointing out that although smoke regions in Figures 9–11 and Figure 14 are very small, our system still works. According to Table 2, our system can detect smoke timely after fire occurs. Moreover, as shown in Table 3, our system has been tested with some YouTube daytime fire videos. All fire events in these videos have been detected successfully. However, fire occurs at the beginning of these videos and many are already large, so we only record the name and resolution of these videos.

Table 2. Daytime Fire Result of HPWERN Database.

Videos Name	Resolution	Fire Starts	First Detected [a]
Lyons Fire	1600 × 1200	156	164 (164)
Holy Fire East View	3072 × 2048	721	732 (732)
Holy Fire South View	3072 × 2048	715	725 (724)
Palisades Fire	3072 × 2048	636	639 (639)
Banner Fire	1440 × 1080	15	17 (17)
Palomar Mountain Fire	1440 × 1080	262	277 (275)
Highway Fire	1600 × 1200	4	6 (6)
Tomahawk Fire	1600 × 1200	32	37 (37)
DeLuz Fire	1440 × 1080	37	48 (48)

Fire appears some time after these videos start. Results in brackets in [a] are the result before pruning and slimming.

Table 3. Worked YouTube daytime fire video list.

Videos Name	Resolution
Barn Fire Overhaul in Marion County Oregon	2560 × 1440
Prairie Fire	1920 × 1080
Drone footage of DJI Mavic Pro Home Fire	1920 × 1080
Cwmcarn Forest Fire	3840 × 2160
Drone Footage of Kirindy Forest Fire	3840 × 2160
Drone Over Wild Fire	1920 × 1080
Fire in Bell Canyon	1920 × 1080
Forest Fire at the Grand Canyon	3840 × 2160
Forest Fire Puerto Montt by Drone	1920 × 1080
Forest Fire with Drone Support	1920 × 1080
Kirindy Forest Fire	3840 × 2160
Lynn Woods Reservation Fire	3840 × 2160
Prescribed Fire from Above	1920 × 1080
Semi Full of Hay on Fire I-70 Mile 242 KS Drone	1920 × 1080
Chimney Tops Fire	1920 × 1080

Fire occurs at the beginning of these videos and many are already large.

(**a**) East view, no-fire scene

(**b**) East view, fire scene

(**c**) South view, no-fire scene

(**d**) South view, fire scene

Figure 8. Holy Fire around Santiago Peak, 6 August 2018.

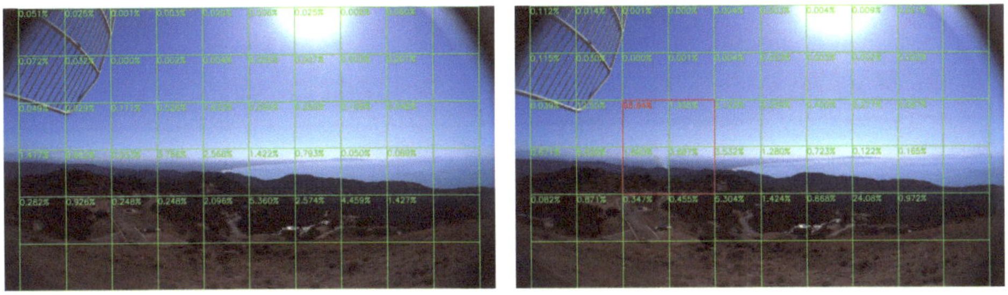

(a) No-fire scene (b) Fire scene

Figure 9. Palisades fire, 21 October 2019.

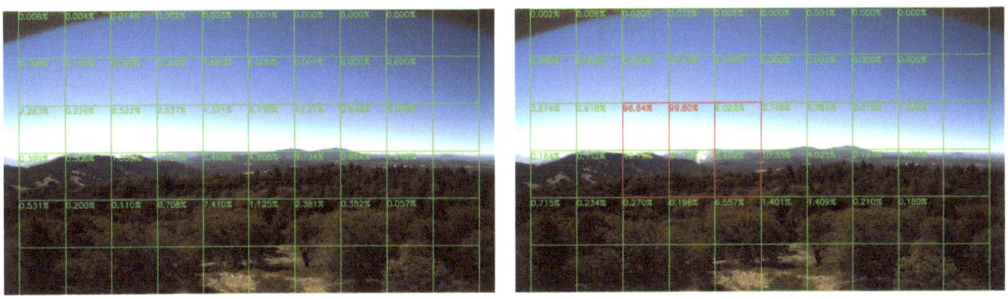

(a) No-fire scene (b) Fire scene

Figure 10. Banner fire, 3 June 2014.

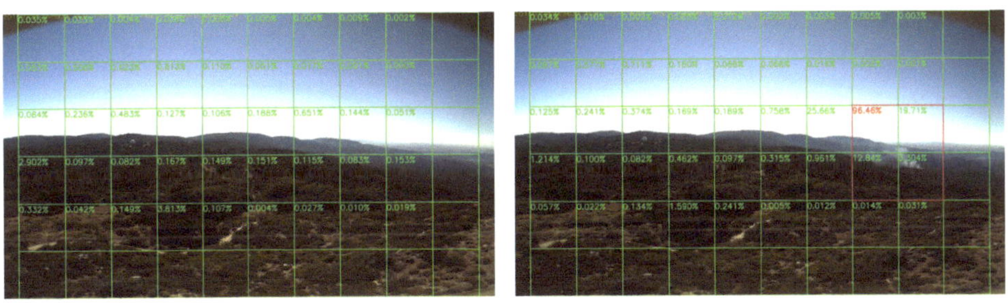

(a) No-fire scene (b) Fire scene

Figure 11. Palomar Mountain fire, 24 July 2015.

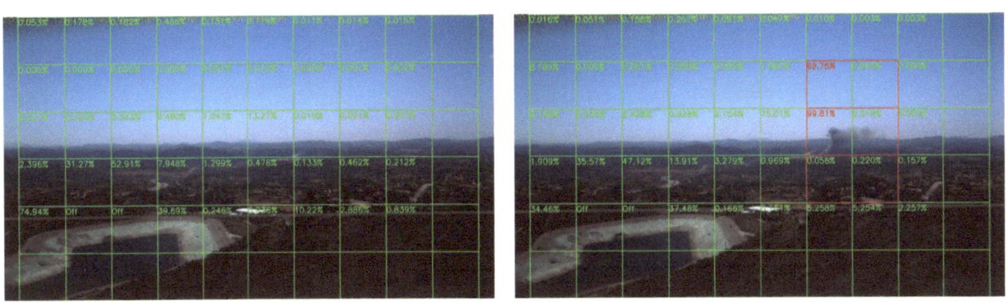

(a) No-fire scene (b) Fire scene

Figure 12. Highway fire, 14 May 2014.

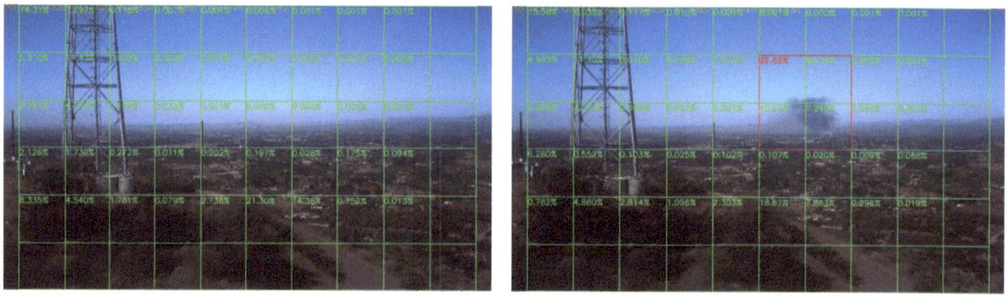

(a) No-fire scene (b) Fire scene

Figure 13. Tomahawk fire, 14 May 2014.

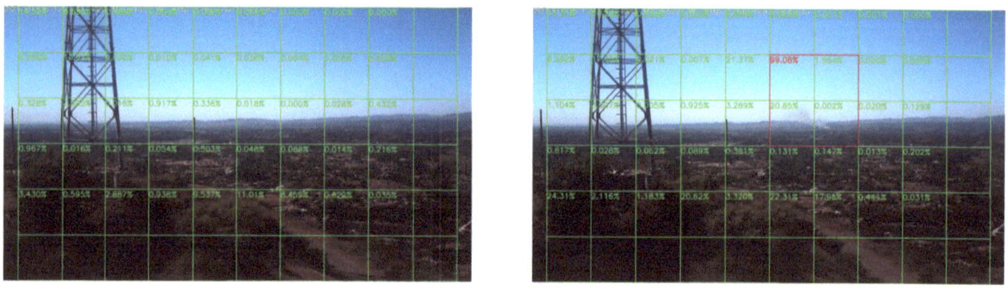

(a) No-fire scene (b) Fire scene

Figure 14. DeLuz fire, 5 October 2013.

5.3. Night Fire Surveillance Test

Although there are only a few images in night scenes in our training dataset, it is still worthy to test the network's performance at night. Unfortunately, compare to the daytime surveillance, smoke during the night is almost invisible, and the fire is very similar to the city lights like Figure 15, unless the fire is already very large like Figure 16. Thus, night fire detection under a color camera is very challenging, and performance will be better if infrared cameras are used instead of color cameras, but unluckily, there are few infrared video clips or images about night fire available on the Internet. Without sufficient data, it is difficult to train the network well.

Sample results of night fire detection are shown in Figures 15–18. Unlike the daytime surveillance, there are very few night fires, so we cannot conduct a detection table similar to Table 2 which lists the initial frame number when the fire starts in this section. Our system still detects all the night fire events in some videos.

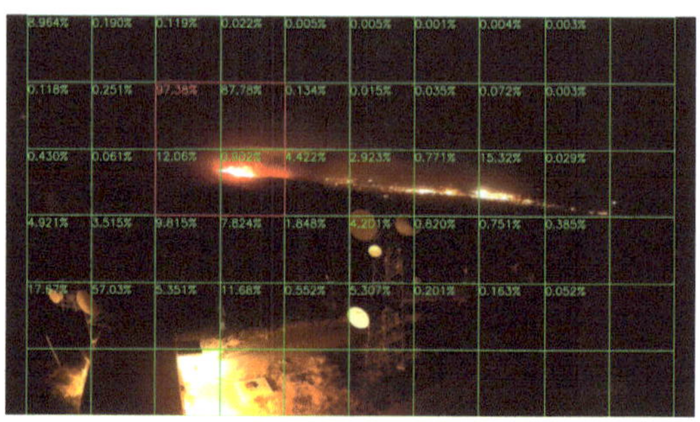

Figure 15. CaveFire, 25 November 2019.

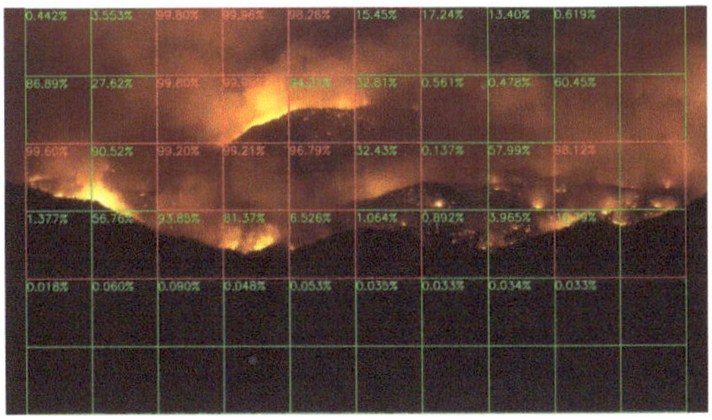

Figure 16. Boulder Colorado Fire, 6 September 2010.

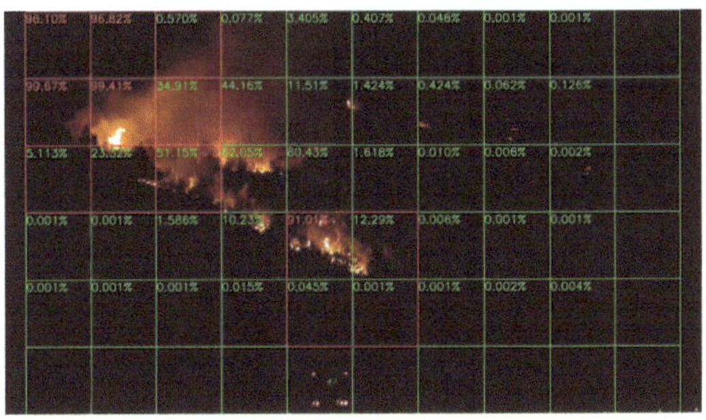

Figure 17. Night forest fire, 28 August 2011.

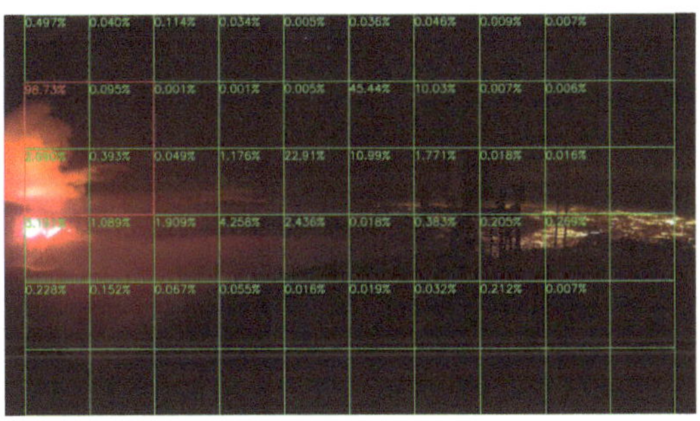

Figure 18. Holy Fire, 6 August 2018.

5.4. Performance on No-Fire Videos

In this section, we run our model on a group of long video clips to test our false-alarm rate, and the result is shown in Table 4. All videos record by whole days, which means they contain both daytime and night scenes. Figures 19 and 20 show a daytime no-fire scene and a night no-fire scene respectively. We can see that the false-alarm rates on all video clips are less than 0.22% except only one video clip with an unexpected light with a long duration at night, and more than half of them are less than 0.1%. Thus, our false-alarm rate is satisfying for a real-world forest fire surveillance task.

Table 4. No-Fire Video Result of HPWREN Database.

Videos Name	Frames Num	False-Alarm Num	False-Alarm Rate (%)
wilson-w-mobo-c	10,080	2	0.01984
wilson-s-mobo-c	10,074	2	0.01985
wilson-n-mobo-c	10,024	3	0.02993
wilson-e-mobo-c	10,028	43 [a]	0.4288
vo-w-mobo-c	10,009	5	0.04996
69bravo-e-mobo-c	1432	1	0.06983
69bravo-e-mobo-c	1432	0	0.0000
syp-e-mobo-c	1421	3	0.2111
sp-n-mobo-c	1252	2	0.1597
sp-w-mobo-c	1282	1	0.07800
sp-s-mobo-c	1272	2	0.1572
sp-e-mobo-c	1278	2	0.1565

[a] There is an unexpected long light shown in Figure 21b. We get same false-alarm result before and after pruning and slimming in threshold of 0.99925. With lower the slimming threshold (0.9990), the false-alarm rate will increase.

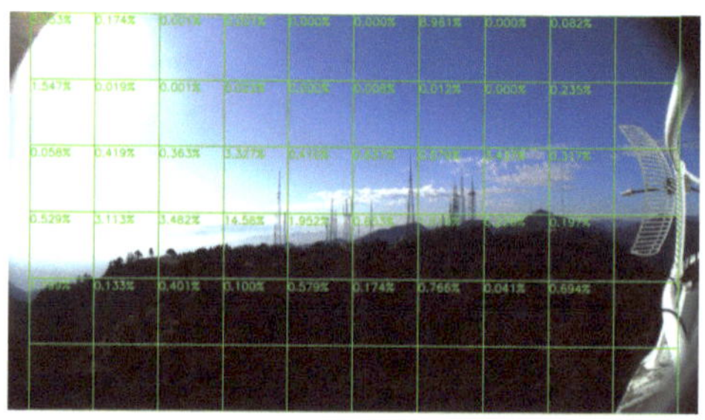

Figure 19. A HPWREN no-fire daytime frame, 3 November 2019.

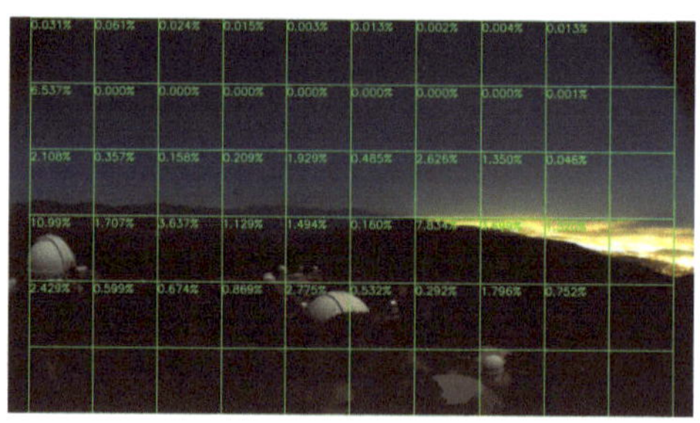

Figure 20. A HPWREN no-fire night frame, 11 November 2019.

Now we will explain some false-alarm scenarios. As shown in Figure 21a, a system releases false alarms because of dust. If it is omitted, we may also miss some small smoke. On the other hand, this kind of false alarm is in single frames and then disappears for a few more frames, so it will not cause too much trouble. Moreover, this kind of false alarm can be overcome by increasing the alarm threshold value as we mentioned in the first paragraph of Section 5.

Another scenario where our system releases false alarms is because of light like in Figure 21b. Unlike a large region of light like in Figure 15, light only existing in a very small region looks like a spark. If it is omitted, we may also miss the real spark. Moreover, in a fixed surveillance tower, we can manually set the system to ignore this block during the time the light is on.

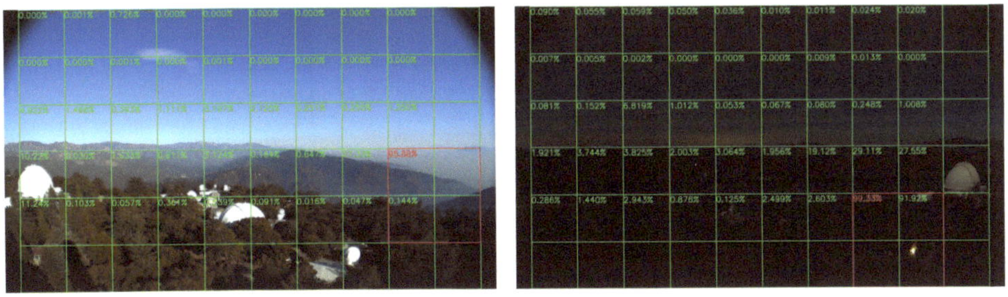

(**a**) (**b**)

Figure 21. False Alarms. (**a**) False alarm because of dust, 3 November 2019. (**b**) False alarm because of light, 10 November 2019.

5.5. Comparison with Other Methods

In this section, we will compare our method with some other related works. It is hard to compare the results from the HPWREN dataset with other methods because most of them are not open source, nor state their training setting. Here we use the BoWFire dataset [39] for comparison. It contains 119 fire images and 107 no-fire images, and it has been used as the test dataset by many related works [26–31]. The resolution of images in the BoWFire dataset are varied from 800×450 to 1365×1024, and many of them are portrait. So if we resize them into 1080 P, the images will be blurred and the portrait images will be deformed. So in this section, we resized them into 800×600 instead, or 600×800 if the image is portrait. Then we divide the frame into 5×7 or 7×5 blocks with block size is 200×200. Figure 22 shows two test result images, and Table 5 states our comparison with other methods. According to the table, our method and that of Reference [31] are tied for the highest accuracy, while our method works better with the lowest false-alarm rate.

(**a**) Fire image No.102 (**b**) No-fire image No.89

Figure 22. Test result on the BoWFire dataset.

Overall, in Section 5, we provided the test results on the daytime and night surveillance tasks. As shown in Figures 9–11 and Figure 14, our system canbstill capture those very small smoke regions. According to Table 2, our system can detect smoke soon after a fire occurs. We also provided the false-alarm information for many long videos in Section 5.4. It shows that our system is very reliable in long-term surveillance task. At the end, we also observed two false-alarm cases in Section 5.4 with the analyzed reason why they happened and solutions that to avoid them by changing the threshold from 0.95 to a higher value, or manually blocking the sky and other possible confusing regions, if the system is installed in a fixed surveillance tower. At the end, we compared our method with some related works about the performance on the BoWFire dataset. Our method has the advantage of the lowest false-alarm rate. As we have mentioned in the first paragraph of Section 5, a very low false-alarm rate is very important to a long-term surveillance system.

Table 5. Comparison with Other Methods.

Method	Detection Rate (%)	False-Alarm Rate (%)	Accuracy (%)
Muhammad et al. [26]	97.48	18.69	89.82
Muhammad et al. [33]	93.28	9.34	92.04
Chaoxia et al. [31]	92.44	5.61	93.36
Our Method	91.60	4.67 [a]	93.36

[a] There are two smoke images labeled as no-fire as shown in Figure 23. Our method managed to detect them, but we still count them as false-alarm cases here for comparison because they are not discussed in References [26–31]. If we count them as true-detected cases, then our three rates are 91.74%, 2.80% and 94.25%, respectively.

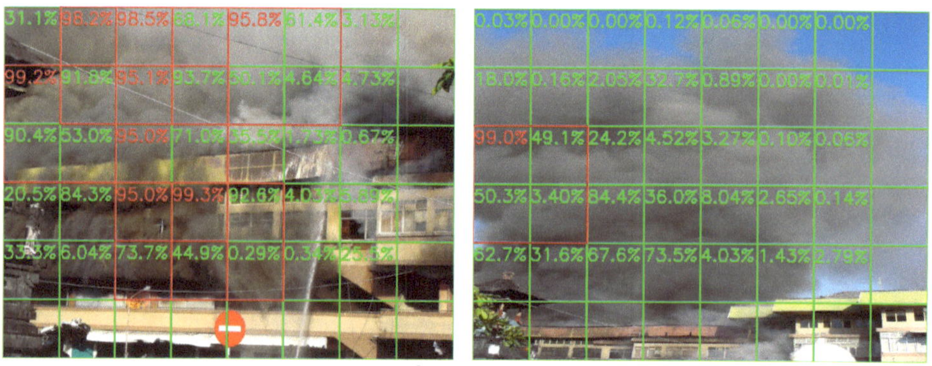

(**a**) No-Fire Image No.63 (**b**) No-Fire Image No.64

Figure 23. Two smoke images on the BoWFire no-fire test dataset.

6. Conclusions

In this paper, we proposed a block-based forest fire detection method via transfer learning. We introduced a novel pruning and slimming method using Fourier analysis to meet the lower time and storage requirements of edge devices. We use a block based strategy so it matches the high resolution of common surveillance cameras without performing any down-sampling. Down-sampling may cause information loss and small smoke plumes may become invisible. Therefore, our block based approach does not suffer from down-sampling and it can help locate the smoke locations as shown in Figure 7 We have tested various video clips in both daytime and night scenarios and obtained a satisfactory recognition performance. The system did not miss any smoke videos. The algorithm may not detect smoke in some frames but it detected all smoke events. This is because when the smoke is very large, its shape will be similar to the cloud, while we have added many cloud features in the training dataset to reduce false-alarms due to clouds. On the other hand, a well-developed forest fire surveillance system should be able to detect the fire immediately whenever it occurs within the viewing range of the camera. This can be achieved by fusing classical computer vision based wildfire detection methods and deep learning based methods.

In future work, we will add more images to our training dataset, which will improve the accuracy of our system. We will get more surveillance video clips from our installed wildfire monitoring cameras and new video clips from HPWREN. We will update our real-time forest fire surveillance system with the software described in this paper. Our NVIDIA Jatson Nano implementation was installed in one location and we will collect more data during the summer of 2020 and improve the proposed algorithm.

Author Contributions: Conceptualization, H.P. and A.E.C.; methodology, H.P. and A.E.C.; software, H.P. and D.B.; validation, H.P.; formal analysis, H.P., D.B. and A.E.C.; investigation, H.P. and D.B.; resources, H.P. and A.E.C.; data curation, H.P.; writing—original draft preparation, H.P.; writing—review and editing, D.B. and A.E.C.; visualization, H.P. and D.B.; supervision, A.E.C.; project administration, A.E.C.; funding acquisition, A.E.C. All authors have read and agreed to the published version of the manuscript.

Funding: This research is supported by National Science Foundation, (NO.1739396) and it is also supported in part by an award from the University of Illinois at Chicago Discovery Partners Institute Seed Funding Program.

Acknowledgments: We thank the developers of TensorFlow for publishing a so powerful machine learning library. We also appreciate FIRESENSE and HPWREN people for their contribution of making the database.

Conflicts of Interest: The authors declare no conflict of interest.

Abbreviations

The following abbreviations are used in this manuscript:

UAV	Unmanned Aerial Vehicle
CNN	Convolutional Neural Network
DFT	Discrete Fourier transform
MP	Million Pixels

References

1. Töreyin, B.U.; Dedeoğlu, Y.; Güdükbay, U.; Çetin, A.E. Computer vision based method for real-time fire and flame detection. *Pattern Recognit. Lett.* **2006**, *27*, 49–58. [CrossRef]
2. Töreyin, B.U.; Dedeoğlu, Y.; Cetin, A.E. Wavelet based real-time smoke detection in video. In Proceedings of the IEEE 2005 13th European Signal Processing Conference, Antalya, Turkey, 4–8 September 2005; pp. 1–4.
3. Habiboğlu, Y.H.; Günay, O.; Çetin, A.E. Covariance matrix-based fire and flame detection method in video. *Mach. Vis. Appl.* **2012**, *23*, 1103–1113. [CrossRef]
4. Habiboglu, Y.H.; Gunay, O.; Cetin, A.E. Real-time wildfire detection using correlation descriptors. In Proceedings of the IEEE 2011 19th European Signal Processing Conference, Barcelona, Spain, 29 August–2 September 2011; pp. 894–898.
5. Töreyin, B.U. Smoke detection in compressed video. In *Applications of Digital Image Processing XLI*; International Society for Optics and Photonics: Bellingham, WA, USA, 2018; Volume 10752, p. 1075232.
6. Aslan, S.; Güdükbay, U.; Töreyin, B.U.; Çetin, A.E. Early Wildfire Smoke Detection Based on Motion-based Geometric Image Transformation and Deep Convolutional Generative Adversarial Networks. In Proceedings of the ICASSP 2019–2019 IEEE International Conference on Acoustics, Speech and Signal Processing (ICASSP), Brighton, UK, 12–17 May 2019; pp. 8315–8319.
7. Borges, P.V.K.; Izquierdo, E. A probabilistic approach for vision-based fire detection in videos. *IEEE Trans. Circuits Syst. Video Technol.* **2010**, *20*, 721–731. [CrossRef]
8. Çelik, T.; Özkaramanlı, H.; Demirel, H. Fire and smoke detection without sensors: Image processing based approach. In Proceedings of the IEEE 2007 15th European Signal Processing Conference, Poznan, Poland, 3–7 September 2007; pp. 1794–1798.
9. Celik, T.; Demirel, H. Fire detection in video sequences using a generic color model. *Fire Saf. J.* **2009**, *44*, 147–158. [CrossRef]
10. Yuan, F. A fast accumulative motion orientation model based on integral image for video smoke detection. *Pattern Recognit. Lett.* **2008**, *29*, 925–932. [CrossRef]
11. Guillemant, P.; Vicente, J. Real-time identification of smoke images by clustering motions on a fractal curve with a temporal embedding method. *Opt. Eng.* **2001**, *40*, 554–563. [CrossRef]
12. Vicente, J.; Guillemant, P. An image processing technique for automatically detecting forest fire. *Int. J. Therm. Sci.* **2002**, *41*, 1113–1120. [CrossRef]

13. Gomez-Rodriguez, F.; Arrue, B.C.; Ollero, A. Smoke monitoring and measurement using image processing: Application to forest fires. In *Automatic Target Recognition XIII*; International Society for Optics and Photonics: Bellingham, WA, USA, 2003; Volume 5094, pp. 404–411.
14. Krstinić, D.; Stipaničev, D.; Jakovčević, T. Histogram-based smoke segmentation in forest fire detection system. *Inf. Technol. Control* **2009**, *38*, 237–244.
15. Luo, Q.; Han, N.; Kan, J.; Wang, Z. Effective dynamic object detecting for video-based forest fire smog recognition. In Proceedings of the IEEE 2009 2nd International Congress on Image and Signal Processing, Tianjin, China, 17–19 October 2009; pp. 1–5.
16. Toreyin, B.U.; Cetin, A.E. Computer vision based forest fire detection. In Proceedings of the 2008 IEEE 16th Signal Processing, Communication and Applications Conference, Aydin, Turkey, 20–22 April 2008; pp. 1–4.
17. Toreyin, B.U.; Cetin, A.E. Wildfire detection using LMS based active learning. In Proceedings of the 2009 IEEE International Conference on Acoustics, Speech and Signal Processing, Taipei, Taiwan, 19–24 April 2009; pp. 1461–1464.
18. Günay, O.; Taşdemir, K.; Töreyin, B.U.; Çetin, A.E. Fire detection in video using LMS based active learning. *Fire Technol.* **2010**, *46*, 551–577. [CrossRef]
19. Çetin, A.E.; Dimitropoulos, K.; Gouverneur, B.; Grammalidis, N.; Günay, O.; Habiboğlu, Y.H.; Töreyin, B.U.; Verstockt, S. Video fire detection—Review. *Digit. Signal Process.* **2013**, *23*, 1827–1843. [CrossRef]
20. Gunay, O.; Toreyin, B.U.; Kose, K.; Cetin, A.E. Entropy-functional-based online adaptive decision fusion framework with application to wildfire detection in video. *IEEE Trans. Image Process.* **2012**, *21*, 2853–2865. [CrossRef] [PubMed]
21. Huo, Y.; Lu, F.; Wu, F.; Dong, X. Multi-Beam Multi-Stream Communications for 5G and Beyond Mobile User Equipment and UAV Proof of Concept Designs. In Proceedings of the 2019 IEEE 90th Vehicular Technology Conference (VTC2019-Fall), Honolulu, HI, USA, 22–25 September 2019; pp. 1–5.
22. Valero, M.M.; Verstockt, S.; Mata, C.; Jimenez, D.; Queen, L.; Rios, O.; Pastor, E.; Planas, E. Image Similarity Metrics Suitable for Infrared Video Stabilization during Active Wildfire Monitoring: A Comparative Analysis. *Remote Sens.* **2020**, *12*, 540. [CrossRef]
23. Günay, O.; Çetin, A.E. Real-time dynamic texture recognition using random sampling and dimension reduction. In Proceedings of the 2015 IEEE International Conference on Image Processing (ICIP), Quebec City, QC, Canada, 27–30 September 2015; pp. 3087–3091.
24. Wu, X.; Lu, X.; Leung, H. An adaptive threshold deep learning method for fire and smoke detection. In Proceedings of the 2017 IEEE International Conference on Systems, Man, and Cybernetics (SMC), Banff, AB, Canada, 5–8 October 2017; pp. 1954–1959.
25. Zhao, Y.; Ma, J.; Li, X.; Zhang, J. Saliency detection and deep learning-based wildfire identification in UAV imagery. *Sensors* **2018**, *18*, 712. [CrossRef] [PubMed]
26. Muhammad, K.; Ahmad, J.; Mehmood, I.; Rho, S.; Baik, S.W. Convolutional neural networks based fire detection in surveillance videos. *IEEE Access* **2018**, *6*, 18174–18183. [CrossRef]
27. Yuan, F.; Zhang, L.; Xia, X.; Wan, B.; Huang, Q.; Li, X. Deep smoke segmentation. *Neurocomputing* **2019**, *357*, 248–260. [CrossRef]
28. Pan, H.; Badawi, D.; Zhang, X.; Cetin, A.E. Additive neural network for forest fire detection. *Signal Image Video Process.* **2019**, *14*, 675–682. [CrossRef]
29. Barmpoutis, P.; Dimitropoulos, K.; Kaza, K.; Grammalidis, N. Fire detection from images using faster R-CNN and multidimensional texture analysis. In Proceedings of the ICASSP 2019–2019 IEEE International Conference on Acoustics, Speech and Signal Processing (ICASSP), Brighton, UK, 12–17 May 2019; pp. 8301–8305.
30. Huang, J.; Chao-Xia, C.; Dong, X.; Gao, Y.; Zhu, J.; Yang, B.; Zhang, F.; Shang, W. Faster R-CNN based Color-Guided Flame Detection. *J. Comput. Appl.* **2020**, *8*, 58923–58932.
31. Chaoxia, C.; Shang, W.; Zhang, F. Information-Guided Flame Detection based on Faster R-CNN. *IEEE Access* **2020**, *8*, 58923–58932. [CrossRef]
32. Pratt, L.Y. Discriminability-based transfer between neural networks. In Proceedings of the Advances in Neural Information Processing Systems, Denver, CO, USA, 30 November–3 December 1992; pp. 204–211.
33. Muhammad, K.; Ahmad, J.; Lv, Z.; Bellavista, P.; Yang, P.; Baik, S.W. Efficient deep CNN-based fire detection and localization in video surveillance applications. *IEEE Trans. Syst. Man Cybern. Syst.* **2018**, *49*, 1419–1434. [CrossRef]

34. Park, M.; Ko, B.C. Two-step real-time night-time fire detection in an urban environment using Static ELASTIC-YOLOv3 and Temporal Fire-Tube. *Sensors* **2020**, *20*, 2202. [CrossRef] [PubMed]
35. Jiao, Z.; Zhang, Y.; Xin, J.; Mu, L.; Yi, Y.; Liu, H.; Liu, D. A Deep Learning Based Forest Fire Detection Approach Using UAV and YOLOv3. In Proceedings of the 2019 1st International Conference on Industrial Artificial Intelligence (IAI), Shenyang, China, 23–27 July 2019; pp. 1–5.
36. Sandler, M.; Howard, A.; Zhu, M.; Zhmoginov, A.; Chen, L.C. Mobilenetv2: Inverted residuals and linear bottlenecks. In Proceedings of the IEEE Conference on Computer Vision and Pattern Recognition, Salt Lake City, UT, USA, 18–23 June 2018; pp. 4510–4520.
37. Grammalidis, N.; Dimitropoulos, K.; Cetin, A.E. FIRESENSE Database of Videos for Flame and Smoke Detection. 2017. Available online: http://doi.org/10.5281/zenodo.836749 (accessed on 1 September 2018).
38. University of California San Diego, California, America: The High Performance Wireless Research and Education Network. 2019. Available online: http://hpwren.ucsd.edu/index.html (accessed on 27 November 2019).
39. Oliveira, W.D. BoWFire Dataset. Available online: https://bitbucket.org/gbdi/bowfire-dataset/downloads/ (accessed on 3 May 2020).

© 2020 by the authors. Licensee MDPI, Basel, Switzerland. This article is an open access article distributed under the terms and conditions of the Creative Commons Attribution (CC BY) license (http://creativecommons.org/licenses/by/4.0/).

MDPI
St. Alban-Anlage 66
4052 Basel
Switzerland
Tel. +41 61 683 77 34
Fax +41 61 302 89 18
www.mdpi.com

Sensors Editorial Office
E-mail: sensors@mdpi.com
www.mdpi.com/journal/sensors

www.ingramcontent.com/pod-product-compliance
Lightning Source LLC
Chambersburg PA
CBHW041407210926
51850CB00013B/3207
9783036519852